To my nephew
Adam Muller

CONTENTS

PREFACE

Local area networks (LANs) were developed in the 1980s, starting with Ethernet and quickly followed by Token Ring. They enable members of an organization to share databases, applications, files, messages, and resources such as servers, printers, and Internet connections. The promised benefits of LANs are often too compelling to ignore: improved productivity, increased flexibility, and cost savings. These benefits sparked the initial move from mainframe-based data centers to a more distributed model of computing that continues today. The impetus for this "downsizing" can come from several directions, including:

- Senior managers, who are continuously looking for ways to streamline operations to improve financial performance.
- End users, who are becoming more technically proficient, resent the gatekeeper function of data center staff, and want immediate access to data that they perceive as belonging to them. In the process, they benefit from being more productive and from the ability to make better and faster decisions, which comes from increased job autonomy.
- Information technology (IT) managers, who are responding to budget cutbacks or scarce resources and are looking for ways to do more using less powerful computers.

From the perspectives of these people, LANs represent the most feasible solution. With PCs now well entrenched in corporate offices, individuals, work groups, and departments have become acutely aware of the benefits of controlling information resources and of the need for data coordination. In becoming informationally self-sufficient and being able to share resources via LANs, users can better control their own destinies within an organization. For instance, they can increase the quality and timeliness of their decision making, execute transactions faster, and become more responsive to internal and external constituencies—all without the need to confront a gatekeeper in the data center.

In many cases, this arrangement has the potential of moving accountability to the lowest common point in the organization, where many end users think it properly belongs. This scenario also has the potential of peeling back layers of bureaucracy that traditionally have stood between users and centralized resources. IT professionals eventually discovered that it was in their best interest to gain control over LANs, enabling them to justify their existence within the organization by using their technical expertise to keep LANs secure and operating at peak performance. Further, there was the need to assist users who were not technically savvy. Rendering assistance helped companies get the most out of their technology investments.

The latest trend is the growth of LANs within homes. The number of multiple-PC homes is growing faster than the number of single-PC homes. According to some industry estimates, the number of homes in the United States with at least two computers stands at between 35 and 40 million. Setting up a network in the home can save money by eliminating the need for duplicate peripherals, software, phone lines, and Internet accounts that normally would be required for each computer. With cost savings on all these items, there is plenty of money left over for purchasing a broadband Internet connection such as Digital Subscriber Line (DSL), which offers Internet access at up to multimegabit-per-second speeds. This is good news for businesses because it means more employees can now telecommute without contending with family members for use of the peripherals or a slow Internet connection. Employees with high-capacity connections can be as productive at home as they are at the office.

LANs have become so popular worldwide and sufficiently sophisticated and complex as to merit dozens of books on the topic that are published every year. This encyclopedia is a quick reference that clearly explains the essential concepts of these networks, including services, applications, protocols, access methods, development tools, administration and management, and standards. It is designed as a companion to other books about LANs you may want to read, providing clarification of concepts that may not be covered fully elsewhere.

The information contained in this book, especially as it relates to specific vendors and products, is believed to be accurate at the time it was written and is, of course, subject to change with continued advancements in technology and shifts in market forces. Mention of specific products and services is for illustration purposes only and does not constitute an endorsement of any kind by either the author or the publisher.

Nathan J. Muller

A

ACCESS POINTS

An access point (AP) provides the connection between one or more wireless client devices and a wired local area network (LAN). The AP is usually connected to the LAN via a Category-5 cable connection to a hub or switch. Client devices communicate with the AP over the wireless link, giving them access to all other devices through the hub or switch, including a router on the other side of the hub, which provides Internet access (Figure A-1).

An AP that adheres to the IEEE 802.11b Standard for operation over the unlicensed 2.4-GHz band supports a wireless link with a data transfer speed of up to 11 Mbps, while an AP that adheres to the IEEE 802.11a Standard for operation over the unlicensed 5-GHz band supports a wireless link with a data transfer speed of up to 54 Mbps. Access points include a number of functions and features, including

- Radio power control for flexibility and ease of networking setup.
- Dynamic rate scaling, mobile Internet Protocol (IP) functionality, and advanced transmit/receive technology to enable multiple APs to serve users on the move.

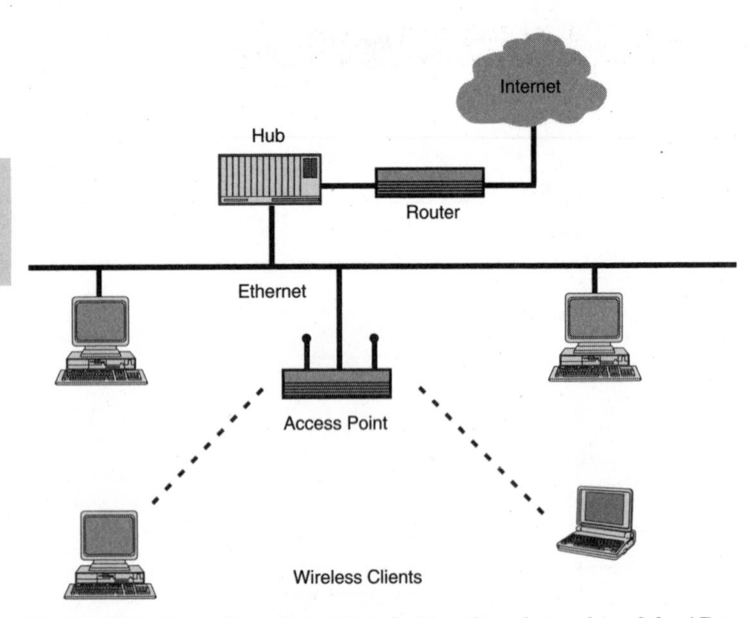

Figure A-1 A simple configuration showing the relationship of the AP to the wired and wireless segments of the network.

- Built-in bridging and repeating features to connect buildings miles apart. The use of specialty antennas increases range. The AP can support simultaneous bridging and client connections.
- Wired Equivalent Privacy (WEP), which helps protect data in transit over the wireless link between the client device and the AP via 64-, 128-, or 256-bit encryption.
- Access control list (ACL) and virtual private network (VPN) compatibility, which helps guard the network from intruders.
- Statistics on the quality of the wireless link (Figure A-2).
- Configurable using the embedded Web browser.

Consumer-level APs stress ease of setup and use (Figure A-3). Many products are configured with default settings

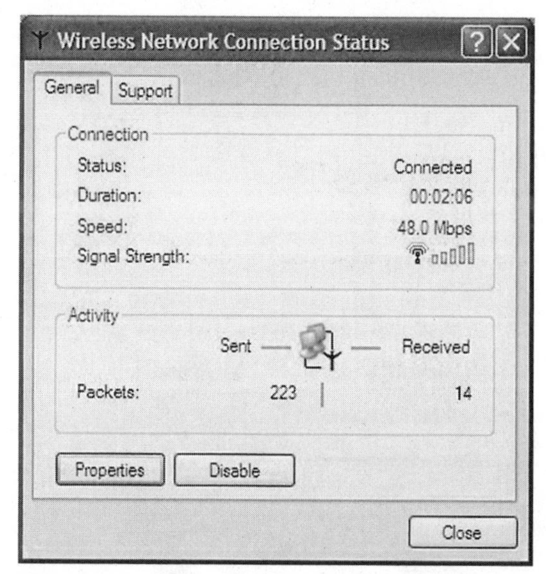

Figure A-2 The 5-GHz DWL-5000 AP from D-Link Systems, Inc., keeps the client device notified of the status of the wireless link. In this case, the signal is at maximum strength and is capable of a data transfer rate of 48 Mbps.

that allow the user to plug in the device and use the wireless connection immediately. Later, the user can play with the configuration settings to improve performance and set up security.

Although APs adhere to the IEEE 802.11 Standards, manufacturers can include some proprietary features that improve the data transfer speed of the wireless link. For example, one vendor advertises a "turbo mode" that optionally increases the maximum speed of IEEE 802.11b wireless links from 11 to 22 Mbps. When this turbo feature is applied to IEEE 802.11a wireless links, the maximum speed is increased from 54 to 72 Mbps.

Enterprise-level APs provide more management features, allowing LAN administrators to remotely set up and configure

Figure A-3 An example of a consumer AP is this 5-GHz Wireless AP (WAP54A) from Linksys, which features antenna with a range of up to 328 feet indoors.

multiple APs and clients from a central location. For monitoring and managing an entire wireless LAN infrastructure consisting of hundreds or even thousands of APs, however, a dedicated management system is usually required. Such systems automatically discover every AP on the network and provide real-time monitoring of an entire wireless network spread out over multiple facilities and subnets. These management systems support the Simple Network Management Protocol (SNMP) and can be tied into higher-level management platforms such as Hewlett-Packard's OpenView.

Among the capabilities of these wireless managers is support of remote reboot, group configuration, or group software uploads for all the wireless infrastructure devices on the network. In addition, the LAN administrator can see how many client devices are connected to each AP, monitor those connections to measure link quality, and monitor all the APs for performance.

Some enterprise APs provide dual-band wireless connections to support both IEEE 802.11a and IEEE 802.11b client users at the same time. This is accomplished by equipping the AP with two plug-in radio cards—one that supports the 2.4-GHz frequency specified by the IEEE 802.11b Standard and one that supports the 5-GHz frequency specified by the IEEE 802.11a Standard.

The choice of a dual-band AP provides organizations with a migration path to the higher data transfer speeds available with IEEE 802.11a while continuing to support their existing investment in IEEE 802.11b infrastructure. Depending on manufacturer, these dual-band APs are modular so that they can be upgraded to support future IEEE 802.11 technologies as they become available, which further protects an organization's investment in wireless infrastructure.

Summary

APs points are the devices that connect wireless client devices to the wired network. They are available in consumer and commercial versions, with the latter generally costing more because of more extensive management capabilities and troubleshooting features. They may have more security features as well and support both the 2.4- and 5-GHz frequency bands with separate radio modules that plug into the same unit.

See also

Bluetooth
Wireless LANs

ADVANCED PEER-TO-PEER NETWORKING

Advanced Peer-to-Peer Network (APPN) is IBM's enhanced Systems Network Architecture (SNA) technology for linking devices without requiring the use of a mainframe. Specifically, it is IBM's proprietary SNA routing scheme for client-server computing in multiprotocol environments. As such, it is part of IBM's LU 6.2 architecture, also known as Advanced Program-to-Program Communications (APPC), which facilitates communications between programs running on different platforms.

SNA Routing

Generally, APPN is used when SNA traffic must be prioritized by class of service (CoS) in order to get it to its destination with minimal delay or when SNA traffic must be routed peer to peer without going through a mainframe.

Included in the APPN architecture are Automatic Network Routing (ANR) and Rapid Transport Protocol (RTP) features. These features route data around network failures and provide performance advantages, closing the gap with the Transmission Control Protocol/Internet Protocol (TCP/IP). ANR provides end-to-end routing over APPN networks, eliminating the intermediate routing functions of early APPN implementations, while RTP provides flow control and error recovery. A more advanced feature called Adaptive Rate Based (ARB) is available with IBM's High Performance Routing (HPR) for congestion prevention.

HPR can be added to streamline SNA traffic so that routers can move the data around link failures or outages. HPR is used when traffic must be sent through the distributed network without disruptions. HPR provides link-utilization features that are important when moving SNA traffic over the wide area network (WANs) and provides congestion control for optimizing bandwidth. HPR's performance gain comes from its end-to-end flow controls, which are an improvement over APPN's hop-by-hop flow controls.

The ARB feature available with HPR uses three inputs to determine the sending rate for data. As data are sent into the network, the rate at which they are sent is monitored. At the destination node, that rate is also monitored and reported back to the originating node. The third input is the allowed sending rate. Together these inputs determine the optimal throughput rate, which minimizes the potential for packet discards to alleviate congestion.

By enabling peer-to-peer communications among all network devices, APPN helps SNA users connect to LANs and more effectively create and use client-server applications. APPN supports multiple protocols, including TCP/IP, and allows applications to be independent of the transport protocols that deliver them.

APPN's other benefits include allowing information routing without a host, tracking network topology, and simplifying network configuration and changes. For users still supporting 3270 applications, APPN can address dependent Logical Unit (LU) protocols as well as the newer LU 6.2 sessions, which protects existing investment in applications relying on older LU protocols.

Summary

There are other SNA routing techniques available. Data Link Switching (DLSw), for example, is used in environments consisting of a large installed base of mainframes and TCP/IP backbones. DLSw assumes the characteristics of APPN and HPR routing and combines them with TCP/IP and other LAN protocols. DLSw encapsulates TCP/IP and supports synchronous data link control (SDLC) and high-level data link control (HDLC) applications (Figure A-4). It prevents session timeouts and protects SNA traffic from becoming susceptible to link failures during periods of heavy congestion.

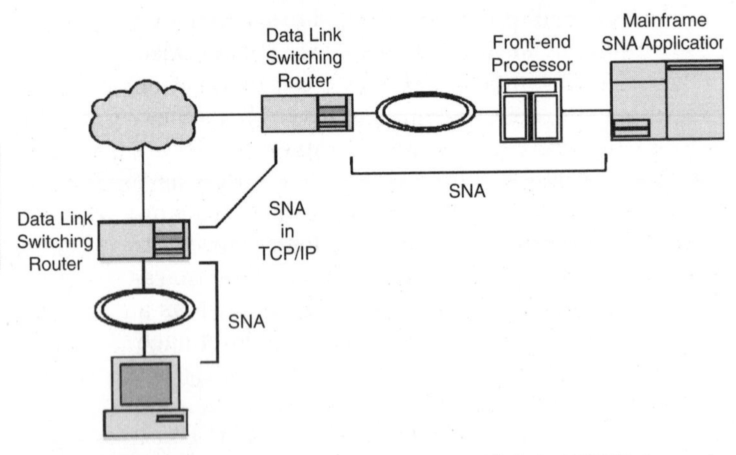

Figure A-4 A local DLSw router encapsulates SNA in TCP/IP. A remote DLSw router deencapsulates TCP/IP so that SNA traffic can be accepted by the target FEP or mainframe.

SNA traffic also can be routed over Frame Relay. Like DLSw, SNA and APPN protocols are encapsulated—in this case, within Frame Relay frames. Frame Relay provides SNA traffic with guaranteed bandwidth through permanent virtual circuits (PVC) and, compared with DLSw, uses very little overhead in the process.

See also

Advanced Program-to-Program Communications

Frame Relay

Peer-to-Peer Networks

ADVANCED PROGRAM-TO-PROGRAM COMMUNICATIONS

Advanced Program-to-Program Communication (APPC) is an IBM protocol that allows IBM network nodes to communicate as peers instead of the hierarchical arrangement dictated by IBM's Systems Network Architecture (SNA). Also known as LU 6.2, APPC enables high-speed communications between programs on different computers, from portables and workstations to midrange and host computers. APPC software is available for many different systems, either as part of the operating system or as a separate software package.

In the SNA scheme of things, a logical unit (LU) provides a software-defined AP through which users interact over the SNA network rather than a physical AP such as a port. LUs allow communication between users without each user having to know detailed information about the other's device type and characteristics.

The SNA network uses physical units (PUs) to indicate categories of devices and the resources they present to the network. The resources associated with a particular device category include the communications links. A combination of hardware and software in the device implements the PU, of which there are four types. Type 1 (PU 1) devices are "dumb" terminals, whereas type 2 (PU 2) devices are user-programmable and have processing capabilities. Type 4 (PU 4) refers to the host node, and type 5 (PU 5) refers to a communications controller. There is no type 3 physical unit (PU 3).

The SNA network also includes System Service Control Points (SSCPs), which provide the services required to manage the network as well as establish and control the interconnections that allow users to communicate with each other. The SSCP provides broader functionality than an LU, which represents a single user, or a PU, which represents a device and its associated resources. The relationship of all three SNA components is illustrated in Figure A-5.

APPC introduced a new LU and a new PU. LU 6.2 is a type of LU that supports program-to-program communication, including communication between programs in peripheral nodes for file transfers between workstation database programs, for example. PU 2.1 is a type of PU that supports communication between peripheral nodes, such as a PC emulating an IBM 3174 cluster controller. PU 2.1 supports

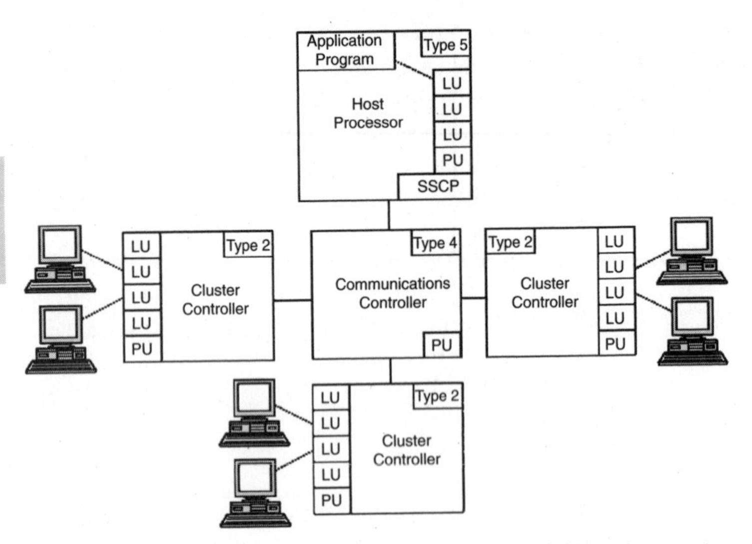

Figure A-5 Terminal-to-host communications in the SNA environment.

connections to other PU 2.1 nodes, as well as conventional hierarchical connections to the mainframe. PU 2.1 also supports simultaneous multiple links and parallel sessions over a given link. PU 2.1 is used in conjunction with LU 6.2 in implementing APPC.

Summary

Originally developed by IBM as a remote transaction-processing tool, APPC is now used to provide distributed services within a heterogeneous computing environment. For example, APPC overcomes the inefficiencies that result when a PC is forced to emulate a 3270 terminal to access data on the mainframe. In the terminal emulation mode, the PC and mainframe must devote processing resources to servicing screen-by-screen data transfers. The PC can be appropriately equipped with an emulation board that uses its own processor to handle the increased load. But the mainframe can get bogged down when it is forced to handle requests from many PCs in the emulation mode. One solution to this problem is to use departmental or workgroup systems to service local PC users, thus offloading the mainframe, but adding APPC capabilities in the distributed environment makes network computing even more efficient and economical.

See also

Advanced Peer-to-Peer Networking

AMERICAN STANDARD CODE FOR INFORMATION INTERCHANGE

American Standard Code for Information Interchange (ASCII) refers to the 7-bit binary encoding scheme that is used to assign a number to the most frequently used characters in American English. ASCII, which also includes the encoding for common keyboard functions, is understood by almost all applications used by PCs, including e-mail and communications software.

Although ASCII uses a 7-bit binary encoding scheme, which makes for 128 possible characters and functions, there is provision for an eighth bit, the left-most bit, which is reserved for parity. In ASCII, the capital letter *C,* for example, is assigned the decimal code 67 and is assigned 01000011 in binary code. The 0 bit in the eighth position to the left is reserved for the parity bit, which is used for error checking when data are sent via modem over phone lines. Table A-1 compares the decimal, octal, hex, and binary encoding schemes.

International ASCII

Although ASCII's 7 bits are enough to encode the common characters used in American English, this is not enough to include the symbols frequently used in other countries, such as

TABLE A-1 Comparison of Common Encoding Schemes

Decimal	Octal	Hex	Binary	Value	Definition
000	000	000	00000000	NUL	Null character
001	001	001	00000001	SOH	Start of header
002	002	002	00000010	STX	Start of text
003	003	003	00000011	ETX	End of text
004	004	004	00000100	EOT	End of transmission
005	005	005	00000101	ENQ	Enquiry
006	006	006	00000110	ACK	Acknowledgment
007	007	007	00000111	BEL	Bell
008	010	008	00001000	BS	Backspace
009	011	009	00001001	HT	Horizontal tab
010	012	00A	00001010	LF	Line feed
011	011	00B	00001011	VT	Vertical tab
012	012	00C	00001100	FF	Form feed
013	013	00D	00001101	CR	Carriage return
014	014	00E	00001110	SO	Serial in/shift out
015	015	00F	00001111	SI	Serial out/shift out
016	016	010	00010000	DLE	Data link escape
017	017	011	00010001	DC1/XON	Device control 1
018	018	012	00010010	DC2	Device control 2
019	019	013	00010011	DC3/XOFF	Device control 3
020	020	014	00010100	DC4	Device control 4
021	021	015	00010101	NAK	Negative acknowledgement
022	022	016	00010110	SYN	Synchronous idle
023	023	017	00010111	ETB	End of transmission block

024	024	018	00011000	CAN	Cancel
025	025	019	00011001	EM	End of medium
026	026	01A	00011010	SUB	Substitute
027	027	01B	00011011	ESC	Escape
028	028	01C	00011100	FS	File separator
029	029	01D	00011101	GS	Group separator
030	030	01E	00011110	RS	Request to send/record separator
031	031	01F	00011111	US	Unit separator
032	032	020	00100000	SP	Space
033	033	021	00100001		!
034	034	022	00100010		“””””
035	035	023	00100011		#
036	036	024	00100100		$
037	037	025	00100101		%
038	038	026	00100110		&
039	039	027	00100111		ë
040	040	028	00101000		
041	041	029	00101001		
042	042	02A	00101010		*
043	043	02B	00101011		+
044	044	02C	00101100		“,”
045	045	02D	00101101		-
046	046	02E	00101110		.
047	047	02F	00101111		/
048	048	030	00110000		0
049	049	031	00110001		1
050	050	032	00110010		2
051	051	033	00110011		3

TABLE A-1 *(Continued)*

Decimal	Octal	Hex	Binary	Value	Definition
052	052	034	00110100	4	
053	053	035	00110101	5	
054	054	036	00110110	6	
055	055	037	00110111	7	
056	056	038	00111000	8	
057	057	039	00111001	9	
058	058	03A	00111010	:	
059	059	03B	00111011	;	
060	060	03C	00111100		
061	061	03D	00111101	=	
062	062	03E	00111110	>	
063	063	03F	00111111	?	
064	064	040	01000000	@	
065	065	041	01000001	A	
066	066	042	01000010	B	
067	067	043	01000011	C	
068	068	044	01000100	D	
069	069	045	01000101	E	
070	070	046	01000110	F	
071	071	047	01000111	G	
072	072	048	01001000	H	
073	073	049	01001001	I	
074	074	04A	01001010	J	
075	075	04B	01001011	K	

076	076	04C	01001100	L
077	077	04D	01001101	M
078	078	04E	01001110	N
079	079	04F	01001111	O
080	080	050	01010000	P
081	081	051	01010001	Q
082	082	052	01010010	R
083	083	053	01010011	S
084	084	054	01010100	T
085	085	055	01010101	U
086	086	056	01010110	V
087	087	057	01011111	W
088	088	058	01011000	X
089	089	059	01011001	Y
090	090	05A	01011010	Z
091	091	05B	01011011	[
092	092	05C	01011100	\
093	093	05D	01011101	]
094	094	05E	01011110	^
095	095	05F	01011111	_
096	096	060	01100000	`
097	097	061	01100001	a
098	098	062	01100010	b
099	099	063	01100011	c
100	100	064	01100100	d
101	101	065	01100101	e
102	102	066	01100110	f
103	103	067	01100111	g

TABLE A-1 *(Continued)*

Decimal	Octal	Hex	Binary	Value	Definition
104	104	068	01101000	h	
105	105	069	01101001	i	
106	106	06A	01101010	j	
107	107	06B	01101011	k	
108	108	06C	01101100	l	
109	109	06D	01101101	m	
110	110	06E	01101110	n	
111	111	06F	01101111	o	
112	112	070	01110000	p	
113	113	071	01110001	q	
114	114	072	01110010	r	
115	115	073	01110011	s	
116	116	074	01110100	t	
117	117	075	01110101	u	
118	118	076	01110110	v	
119	119	077	01110111	w	
120	120	078	01111000	x	
121	121	079	01111001	y	
122	122	07A	01111010	z	
123	123	07B	01111011	{	
124	124	07C	01111100	|	
125	125	07D	01111101	}	
126	126	07E	01111110	~	
127	127	07F	01111111	DEL	

the British pound symbol or the German umlaut. There is a version of ASCII standardized by the International Organization for Standardization (ISO) that includes the original 128 characters along with an additional 128 characters, such as the British pound symbol and the American cent symbol. This increase to 256 characters is achieved by dispensing with the need for a parity bit so that the full 8 bits can be used for encoding characters. There are several variations of this ISO-8859 Standard, which can be applied to different language families:

- Latin-1 (western European languages)
- Latin-2 (non-Cyrillic central and eastern European languages)
- Latin-3 (southern European languages and Esperanto)
- Latin-5 (Turkish)
- Latin-6 (northern European and Baltic languages)
- 8859-5 (Cyrillic)
- 8859-6 (Arabic)
- 8859-7 (Greek)
- 8859-8 (Hebrew)

EBCDIC

Extended Binary-Coded Decimal Interchange Code (EBCDIC) is IBM's 8-bit scheme for representing 256 possible characters as numbers. The operating system of the AS/400 midrange computer, for example, processes data internally in EBCDIC format. Although EBCDIC is used widely on large IBM computers, most other computers, including PCs and Macintoshes, use ASCII codes or Unicode.

Unicode

While ISO ASCII uses 8 bits to encode characters, Unicode uses 16 bits, which means that it can represent more than

65,000 characters, the first 256 characters of which are identical to Latin-1. Thus Unicode provides a unique number for every character, no matter what the platform, program, or language. While this is hardly necessary for English and western European languages, it is necessary for Chinese, Japanese, Korean, and other languages with a large number of ideographic characters.

An extension mechanism allows for the encoding of as many as 1 million additional characters. This capacity is sufficient for all known character encoding requirements, including full coverage of all the world's historic scripts. There is even a proposal to accommodate within Unicode Sumero-Akkadian cuneiform, the ancient Near Eastern writing system used for a number of languages from the end of the fourth millennium B.C. until the first century B.C. The purpose of encoding cuneiform in Unicode is to broaden the study and translation of ancient texts via computers and networks.

Work on Unicode began in 1988, but the project was not given wide exposure until the Unicode Consortium was formed in 1991. Also known as ISO 10646, Unicode has been adopted for use in the products of such industry leaders as Apple, Hewlett-Packard, IBM, Microsoft, Oracle, PeopleSoft, Sun, Sybase, Unisys, and many others. With business becoming more global and national economies becoming more interdependent, many operating systems, databases, and programming languages now use Unicode as the character set instead of ASCII.

Summary

ASCII is an encoding scheme that allows computers to read, process, store, and move information between computers. However, the inability of ASCII to correctly represent the characters of a variety of languages has resulted in a proliferation of encoding schemes worldwide. Unicode is the ultimate encoding standard. It can potentially encode the characters of every language, past and present. The current edition of the Unicode Standard, version 3.0, contains over

49,000 characters, covering the principal written languages of the Americas, Europe, the Middle East, Africa, Asia, and the Pacific Rim. The Unicode Standard is fully synchronized with the ISO 10646, providing a formal, internationally recognized basis for its character encoding.

See also

Asynchronous Communication

ARCNET

The Attached Resource Computer Network (ARCnet), introduced by Datapoint Corp. in 1977, was the first LAN technology. Over the years, ARCnet has been overshadowed by higher-speed LAN technologies, notably Ethernet and Token Ring. Today, ARCnet products are still available and embedded in many companies' products, but they may not be advertised as ARCnet. Unlike other network technologies, there is no upward migration path for ARCnet to higher speeds that approach the Ethernet standard at 10 Gbps.

Currently, ARCnet products for the LAN are available that operate at 2.5 Mbps. Higher-speed ARCnet products are available at 5 and 10 Mbps, but these are used as general-purpose communications controllers for networking microcontrollers and intelligent peripherals in industrial, automotive, and embedded control environments using an ARCnet protocol engine. An ARCnet protocol engine is the ideal solution for embedded control applications because it provides a deterministic token-passing protocol, a highly reliable and proven networking scheme.

Operation

For LANs, most ARCnet products operate at 2.5 Mbps. When an ARCnet node receives the token, it is permitted to send packets of data to other stations. While the Token

Ring protocol passes its token around a physical cable ring, ARCnet passes its token from node to node in order of each node's address.

Nodes could be located up to 2000 feet from an ARCnet hub through the use of RG-62 coaxial cabling. The total end-to-end length of the network could be 20,000 feet—nearly 4 miles. With twisted-pair wiring, each node could be located up to 400 feet from an ARCnet hub. As many as 254 connections may be supported on a single ARCnet LAN via interconnected active hubs.

ARCnet was designed originally to operate in a distributed star configuration, which entailed each node being directly connected to a hub (Figure A-6), with several hubs connected to each other. This design suited organizations that terminated cables in centrally located wiring closets. Later, an Ethernet-like bus topology was introduced for ARCnet. This allowed nodes to be interconnected via a sin-

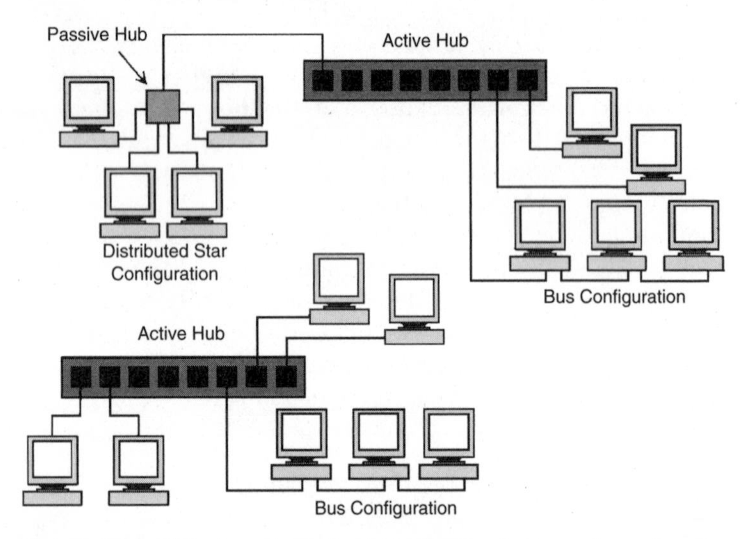

Figure A-6 ARCnet configuration incorporating both passive and active hubs.

gle run of cable. The two topologies could even be combined for maximum configuration flexibility. For example, instead of connecting a single ARCnet node to a port on an ARCnet hub, a bus cable with a maximum of eight nodes attached could be connected to the hub.

Passive and Active Hubs

ARCnet makes use of two types of hubs: passsive and active. Passive hubs are small four-port devices (nonpowered) that support workstations at distances of up to 100 feet using coaxial cabling. Active hubs are eight-port units that support workstations at distances of up to 200 feet using coaxial cabling and up to 400 feet with twisted-pair wiring. By attaching passive hubs to each of an active hub's eight ports, a single active hub can support 24 workstations.

The primary advantages of this distributed-star arrangement include cost savings on cable installation and hub ports. A central active hub that uses twisted-pair wiring offers the best protection against network failure, since each station has its own dedicated connection to the active hub. Furthermore, the central hub approach makes all the wiring accessible at one point, which simplifies troubleshooting, fault isolation, and network expansion.

From the beginning, ARCnet used internal transceivers in its hubs. Under Datapoint's concept of "conjoint networks," there was no need for bridges and routers to interconnect multiple LANs. Selected workstations and/or file servers could be configured to participate directly in up to six LANs at the same time. Access to each LAN or group of LANs was effectively controlled through hardware.

Summary

Although ARCnet is rarely considered today by companies seeking LAN solutions, the technology is still employed for such niche applications as data acquisition, plant monitoring and control, closed-circuit cameras, commercial heating and air conditioning, waterway controls, automotive navigation systems, building automation, and motor-drive communications. The vendors of these applications use ARCnet as the network for linking various system components.

See also

Ethernet
StarLAN
Token Ring

ASSET MANAGEMENT

Asset management entails the proper accounting of various hardware and software assets within an organization. Without a thorough understanding of what assets an organization has, it will not be possible to accurately plan departmental budgets or allocate costs. The failure to account for and manage such assets as desktop hardware and software, in-house cabling, and network lines has other ramifications as well. It can lead to cost overruns on projects, leave the door open to employee theft ("asset shrinkage"), and lead to misuse or abuse of the network. The lack of controls can expose the company to financial penalties for copyright infringement, such as when employees copy software or the organization allows more concurrent usage than the vendor's license permits.

Types of Assets

There are several types of assets that organizations must track. These fall into the general categories of hardware, software, network, and cable assets. With the right asset management tool, these assets can be tracked through their life cycles.

Hardware Assets Hardware inventory starts with identification of the major kinds of systems that are in use in the distributed computing environment—from the servers all the way down to the desktop, laptop, and handheld computers, as well as their various components, including the CPU, memory, boards, and disk drives. The asset management utilities that come with servers generally scan connected computers for this kind of information.

Most asset management products provide the following basic hardware configuration information:

- *CPU* Model and vendor
- *Memory* Type (extended or expanded) and amount (in kilobytes or megabytes)
- *Hard disk* Amount and percentage of disk space used and available, volume number, and directories
- *Ports* In use and available

Most hardware identification is based on the premise that if a driver is loaded, then the associated hardware must be present. However, many of these drivers go unused and are not removed, resulting in inaccurate inventory. This situation is remedied by industry standards, such as the Desktop Management Interface (DMI) and Plug and Play (PnP).

Hardware inventories can be updated automatically on a scheduled basis—daily, weekly, or monthly. Typically included as part of the hardware inventory is the physical location of the unit, owner (workgroup or department), and name of the user. Other information may include vendor contact information and the unit's maintenance history. All this information is entered and updated manually. When all resources are scanned, an inventory report can be printed.

In addition to providing inventory and maintenance management, some products provide procurement management as well. They maintain a catalog of authorized products from preferred suppliers, as well as list and discount prices. They track all purchase requests, purchase orders, and deliveries. With some products, even the receipt of new equipment can be automated, with the system collecting information from scans of asset tags and bar codes. Warranty information also can be added.

Still other asset packages accommodate additional information for financial reporting, such as

- *Cost* Purchase price of the unit and add-in components
- *Payment schedule* Principal and interest.
- *Depreciation* One-time expense or multiyear schedule.
- *Taxes* Local, state, and federal (as applicable).
- *Lease* Terms and conditions.
- *Chargeback* Cost charged against the budgets of departments, workgroups, users, or projects.

This kind of information is entered and updated manually in the asset management database. Depending on the product, this information can be exported to spreadsheets and other financial applications and used for budget monitoring, expense planning, and tax preparation.

Software Assets Another technology asset that must be tracked is client software. Not only can software tracking (also called "applications metering") reduce support costs, it also can protect the company from litigation resulting from claims of copyright infringement, such as when users copy and distribute software on the network in violation of the vendor's license agreement.

Asset management products that support software tracking automatically discover what software is being used on each system on the network by scanning local hard drives and file servers for all installed software. They do this by looking for the names of all executable files and arranging them in alphabetical order. They determine how many copies of the executable files are installed and look into them to provide the product name and the publisher. Files that cannot be identified absolutely are listed as found but flagged as unidentified. Once the file is eventually identified, the administrator can fill in the missing information.

The accumulated asset information can be used to build a software distribution list. The administrator can then automatically install future upgrades on each workstation appearing on the distribution list.

The administrator also can monitor the usage status of all software on the network to enforce license compliance. If several copies of an application are not being used, they can be made available to other users. If all copies of an application are in use, a queue is started, and the application is made available on a first-come, first-served basis. This kind of control may qualify the organization for discounts on software. A growing number of software vendors will not sell network licenses unless the customer has a metering system in place. If a metering system is not in place, the organization may be forced to pay a higher price for unlimited software usage.

Network Assets The kinds of network assets that must be monitored, controlled, and accounted for in inventory include repeaters, bridges, routers, gateways, hubs, and switches—any device that is used to implement the network. An enterprise management system may offer asset management as a native function or permit integration of a third-party application that offers this function.

The entire chassis of a hub or switch, for example, can be viewed via the network management system, which shows the types of cards that are inserted into each slot. With a zoom feature, any card can be isolated, and a representation of the ports, LEDs, and configuration switches can be displayed at the management console. A list of devices attached to any given port also can be displayed and/or printed. Some of the other views available from the management system include

- *Configuration view* Organizes the configuration values for a device and its model, including device location, model name, firmware version, IP address, and security string.
- *Resource view* A special-purpose view for end point devices showing where the end point device accesses its application resources, such as primary and secondary print servers, electronic mail server and file server.
- *Cable walk view* Illustrates the connections that exist along a segment of cable (e.g., Ethernet, Token Ring, Fiber Distributed Data Interface) and the devices connected to each segment.
- *Diagnostic view* Organizes diagnostic and troubleshooting information for a device, including errors, collisions, events, and alarms.
- *Performance view* Displays performance statistics, including load, hard and soft errors, and frame traffic.
- *Port performance views* For each individual port or board, summarizes port-specific or board-specific performance statistics.
- *Application view* Organizes device application information, including device IP and Internet Control Message Protocol (ICMP) statistics.
- *Assigns view* Allows a specific technician to be assigned to devices owned by network users.

The management system also may provide a method for documenting the equipment and cable plant inventory through a third-party cable asset management system. This and other third-party applications typically are integrated with application programming interfaces (APIs).

Cable Assets Among the assets that also must be managed is the cabling that connects all the devices on the network, including coaxial cable (thick and thin), twisted-pair (shielded and unshielded), and optical fiber (single-mode and multimode). There are a number of specialized applications available that keep track of the wiring associated with connectors, patch panels, cross-connects, and wiring hubs. They use a graphical library of system components to display a network. Clicking on any system component brings up the entire data path with all its connection points. These cable management products offer color maps and floor plans that are used to illustrate the cabling infrastructure in one or more buildings. A zoom feature can isolate backbone cables within a building, on a floor, or within an office.

Some cable asset management products can generate work orders for moving equipment or rewiring. Managers can create both logical and physical views of their facilities and even view a complete data path simply by clicking on a connection. Some cable asset management products automatically validate the cabling architecture by checking the continuity of the data paths and the type of network for every wire. In addition, a complete picture of the connections can be generated and printed out. With this information, network administrators and technicians know where new equipment should go, what needs to be disconnected, and what should be reconnected. Some cable asset management products can even calculate network load statistics to facilitate proactive management and troubleshooting.

Like other types of asset management applications, cable management applications can be run as stand-alone systems or may be integrated with help desk products, hub management systems, and major enterprise management platforms such as Hewlett-Packard's OpenView.

Web-Based Asset Management

For organizations that understand the value of asset management and just do not have the time to do it, there is a new option available—a hosted PC inventory service that provides instant, accurate inventory over the Web. Tally Systems, for example, offers WebCensus as a fast, centralized, low-maintenance way to audit enterprise PCs without the requirement to install or manage anything. The tool securely and efficiently audits PCs and returns hardware and software inventory results in minutes (Figure A-7).

Information technology (IT) managers collect inventory data by sending users an e-mail with an embedded link to the service provider's Web site. When users click on the link, an agent is installed onto the PC, performs an inventory, and sends it to a database on the service provider's site before uninstalling itself. All data are encrypted during transport. IT administrators then log onto the site and run reports against the database. The reports can be used to determine such things as how many copies of Microsoft Office are deployed and which versions. The reports can be saved as an Excel spreadsheet. Aside from tracking inventory, the data can be used to plan for operating system and application upgrades. However, the necessity for end users to initiate the inventory collection is a weak link in the process. If the link in the e-mail is ignored, the inventory process is not initiated.

If everyone cooperates, the advantage of a subscription-based inventory service is that it allows IT managers to offload management costs and pay only for the services they use rather than invest in a software product and incur its associated installation and maintenance costs. As a hosted service, it completely eliminates the need for installing, configuring, and updating complex software. This reduces stress on IT staff, protects information system (IS) infrastructure, and eliminates the need for in-house upgrades. Web-based inventory services are available in 1-, 3-, and 12-month subscriptions and are priced from $3 to $15 per PC on the basis of length of subscription and type of service.

Summary

There are several approaches to asset management. Organizations can buy one or more software packages, use an integrated approach available with some help desk or network management systems, or outsource the asset management task to a systems integrator or computer vendor. In addition

Details for Workstation BBBouchard

User	Login	Group	
Bruce Benjamin Bouchard	BBBouchard	Finance	
MAC Address	**IP Address**	**Serial Number**	
00C04F8B677	192.169.4.654	SN16173	
Total Disk Space (MB)	**Free Disk Space (MB)**	**Total Memory (MB)**	**Last Inventory**
7683	3438	64	12/14/1999 12:43:29 PM

System

Dell OptiPlex G1 266MTbr+

Operating System

Microsoft Windows 95 4.00

Hardware Components

Category	SubCategory	Manufacturer	Product
			System Board
		Novell	NetWare Shell Driver 3.26-0
			Memory Module
			Memory Module
			Memory Module
BIOS	BIOS	Phoenix	ROM BIOS
CD/DVD	CD-ROM		CD-ROM Drive
Diskette	Diskette		Diskette Drive
Hard Drive	Hard Drive	Quantum	Fireball ST2.1A
Keyboard	101/102 key		101/102 keyboard
LAN Adapter	LAN Adapter	Intel	EtherExpress PRO/100B
Logical Drive	Logical Drive		FAT-16 Partition – Big DOS
Logical Drive	Logical Drive		FAT-16 Partition – Big DOS
Logical Drive	Logical Drive		FAT-16 Partition – Big DOS
Monitor	VGA	Dell	D825TM
Monitor	VGA	Gateway	CrystalScan 1572DG
Mouse	Serial Mouse	IBM compatible	PS/2 Mouse
Parallel Port	Parallel Port		Parallel Ports
Processor	Pentium II	Intel	Pentium II
Serial Port	Serial Port		Serial Ports
Video Adapter	VGA	ATI	3D RAGE IIC Controller

Software Components

Category	SubCategory	Manufacturer	Product
Asset Mgmt	PC Inventory	Tally Systems	NETCensus Win32 Collector Unknown
Comm Software	Integrated	Microsoft	Exchange Client for Win32 5.0
Comm Software	Internet Tools	Microsoft	NetMeeting for Win32 2.1
Comm Software	Internet Tools	Netscape Communications	Netscape Communicator for Win32 4.5
Comm Software	Fax	Cracchiolo and Feder	RightFAX Client for Windows 5.0
Database	DB Managers	Sybase	Sybase SQL Anywhere for Win32 5.0
Data Base	DB Managers	Microsoft	Access 97 8.0 SR-2
Games	Misc	Microsoft	Minesweeper 95
Games	Misc	Microsoft	Hearts Network for Win32 95
Games	Misc	Microsoft	FreeCell 95
Games	Misc	Microsoft	Solitaire 95
Graphics	Presentation	Microsoft	Powerpoint 97 8.0 SR-2
Integrated	Integrated	Microsoft	Office 97 Taskbar 8.0
Multimedia	Integrated	Macromedia	Shockwave 7 7.0
Spreadsheet	Spreadsheets	Microsoft	Excel 97 8.0 SR-2
Utility	PIM/Cont. Mgr.	Microsoft	Schedule+ for Win95 7.5
Utility	PIM/Cont. Mgr.	Microsoft	Outlook 98 8.5
Utility	Data Compress.	Nico Mak Computing	WinZip for Win32 7.0 SR-1
Word Processor	Word Processors	Microsoft	Word 97 8.0 SR-2

Open in Excel

Figure A-7 Workstation Details is one of about 20 reports offered by Tally Systems' WebCensus, an online inventory service.

to containing the cost of technology acquisitions and reining in hidden costs, asset management can improve help desk operations, enhance network management, assist with technology migrations, minimize asset shrinkage, and provide essential information for planning a corporate reengineering strategy.

See also

Electronic Software Distribution

Network Design Tools

Network Management Systems

ASYNCHRONOUS COMMUNICATION

Asynchronous communication is used for the transmission of non-time-sensitive data. The term *asynchronous* means that the bits in the serial data stream are not locked to a specific clock that guides the timing of bits between the sending and receiving devices, as in "synchronous" communication. Instead, asynchronous communication relies on start and stop bits that bracket the characters. These bits are used to assist the receiving device in determining where each character in the data stream of 1s and 0s begins and ends. This makes the asynchronous method of serial data transmission ideal for terminal-to-host connections, where characters are generated at irregular intervals from a keyboard.

Asynchronous communication is a simple, economical way to connect to a wide variety of systems and services, especially when the end device may be different with every connection attempt. This form of communication is commonly used for applications that are not real time in nature, such as e-mail or requesting Web pages. Any communication that is not time sensitive is considered asynchronous. Synchronous communication, on the other hand, is used for real-time applications. It relies on precise timing from a clock source rather than start and stop bits to make sense of the data stream.

Parity Bit

In the PC environment, 7- or 8-bit characters are often used to read, process, store, and transmit information. Seven bits are enough to encode all upper- and lowercase characters, symbols, and function keys, which number 128, in conformance with the American Standard Code for Information Interchange (ASCII). An optional eighth bit, called the "parity bit," is used to check data integrity. When used, it is inserted between the last bit of a character and the first "stop" bit (Figure A-8).

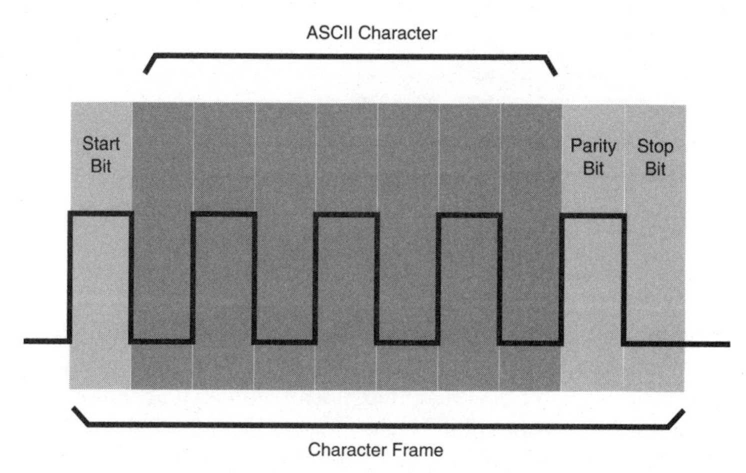

Figure A-8 This 10-bit character frame consists of 2 bits for start-stop, a parity bit, and 7 bits of user data.

The parity bit is included as a simple means of error checking. There is even and odd parity. The devices at each end of the connection must have the same parity setting. The idea is that parity is agreed on before the start of transmission. The actual configuration is done from within an operating environment such as Windows when setting up the connection preferences of the modem (Figure A-9).

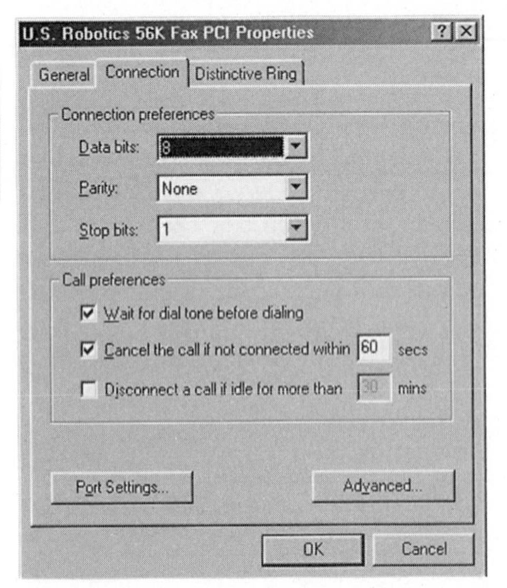

Figure A-9 The connection preferences for a U.S. Robotics 56-kbps fax PCI modem as configured from within Windows 98.

Suppose that the parity chosen is odd. The transmitter will then set the parity bit in such a way as to make an odd number of 1s among the data bits and the parity bit. For example, if there are five 1s among the data bits, already an odd number, the parity bit will be set to 0. If errors are detected at the receiving device, a notification is sent in the header of the return packet so that only corrupt bytes need to be retransmitted.

While asynchronous communication is a relatively simple and, therefore, inexpensive method of serial data transmission, it is very inefficient. This is so because asynchronous transmissions include high overhead in that each byte carries at least 2 extra bits for the start-stop functions, which results in a 20 percent loss of useful bandwidth (2/10 = 0.20, or 20 percent). For large amounts of data, this adds up quickly. For example, to transmit 1000 characters, or 8000 bits, 2000 extra bits must be transmitted for the start and stop functions, bringing the total number of bits sent to 10,000. The 2000 extra bits is equivalent to sending 250 more characters over the link.

Summary

Asynchronous communication overcomes the problem of how to synchronize the sending and receiving devices so that the receiver can detect the beginning of each new character in the bit stream being presented to it. Without synchronization, the receiver will not be able to interpret the incoming bit stream correctly. This problem is overcome through the use of a start and stop bit that brackets each character. A start bit is sent by the sending device to inform the receiving device that a character is being sent. The character is then sent, followed by a stop bit, which indicates that the transfer of that character is complete. This process continues for the duration of the session.

See also

American Standard Code for Information Interchange
Synchronous Communication

ASYNCHRONOUS TRANSFER MODE

Asynchronous Transfer Mode (ATM) is a cell switching technology that offers low-latency transmission with quality-of-service (QoS) guarantees in support of data, voice, and video traffic at multimegabit-per-second speeds. ATM is also highly scalable, making it equally suited for interconnecting legacy systems and LANs and for building WANs over today's high-performance optical fiber infrastructures. ATM-based networks may be accessed through a variety of standard interfaces, including Frame Relay.

Applications

There are many applications that are particularly well suited for ATM networks, including

- *LAN internetworking* ATM can be used to interconnect LANs over the WAN. Special protocols make the connection-oriented ATM network appear as a connectionless Ethernet or Token Ring LAN segment.
- *Videoconferencing or broadcasting* ATM can be provisioned for interactive videoconferencing between two or more locations or to support point-to-multipoint video broadcasts.
- *Telemedicine* With ATM, large amounts of bandwidth can be provisioned to support the rapid exchange of high-resolution diagnostic images and multimedia patient records while permitting interactive consultations among medical specialists at different locations.
- *Private-line connectivity* An ATM virtual circuit can be used to provide a more economical way to provision leased lines on the WAN. ATM protocols can emulate $N \times 64$ kbps DS0 transport.
- *PBX voice trunking* An ATM virtual circuit can be used to interconnect Private Branch Exchanges (PBXs) and maintain full PBX feature support, call routing, and switching. Voice trunking combines multiple calls onto a single virtual circuit for further bandwidth optimization, reduced delay, and lower cost. PBX voice trunking requires an integrated access device at the customer premises between the PBX and ATM switch that performs the protocol conversions necessary to extend feature signaling across the ATM network.

ATM also offers a consolidation solution for any company that maintains separate networks for voice, video, and data. The reason for separate networks is to provide appropriate bandwidth and preserve performance standards for the different applications. But ATM can eliminate the need for separate networks, providing a unified platform for multiservice networking that meets the bandwidth and quality-of-service (QoS) needs of all applications. Although the startup cost for ATM is high, the economics of network consolidation mean that companies do not have to wait very long to realize return on their investment.

Quality of Service

ATM serves a broad range of applications very efficiently by allowing an appropriate QoS to be specified for each application. Various categories have been developed to help characterize network traffic, each of which has its own QoS requirements. These categories and QoS requirements are summarized in Table A-2.

CBR is intended for applications where the PVC requires special network timing requirements (i.e., strict PVC cell loss, cell delay, and cell delay variation performance). For example, CBR would be used for applications requiring circuit emulation (i.e., a continuously operating logical channel) at transmission speeds comparable to DS1 and DS3. Such applications would include private line–like service or voice-type service where delays in transmission cannot be tolerated.

Variable Bit Rate–real time (VBR-rt) is intended for applications where the PVC requires low cell delay variation. For example, VBR-rt would be used for applications such as variable-bit-rate video compression and packet voice and video, which are somewhat tolerant of delay. Variable Bit Rate–near real time (VBR-nrt) is intended for applications where the PVC can tolerate larger cell delay variations than VBR-rt. For example, VBR-nrt would be used for applications such as data file transfers.

ABR is intended for routine applications and when the customer seeks a low-cost method of transporting bursty data for noncritical applications that can tolerate delay variations. The traffic goes out into the network when bandwidth becomes available; otherwise, it is held back until other applications with higher priority are finished using

TABLE A-2 ATM Quality of Service (QoS) Categories

		Quality of Service Requirements			
Category	Bandwidth Application	Delay Variation Guarantee	Throughput Guarantee	Congestion Guarantee	Feedback
Constant Bit Rate (CBR)	Provides a fixed virtual circuit for applications that require a steady supply of bandwidth, such as voice, video, and multimedia traffic.	Yes	Yes	Yes	No
Variable Bit Rate (VBR)	Provides enough bandwidth for bursty traffic such as transaction processing and LAN interconnection, as long as rates do not exceed a specified average.	Yes	Yes	Yes	No
Available Bit Rate (ABR)	Makes use of available bandwidth and minimizes data loss through congestion notification. Applications include e-mail and file transfers.	Yes	No	Yes	Yes
Unspecified Bit Rate (UBR)	Makes use of any available bandwidth for routine communications between computers but does not guarantee when or if data will arrive at their destination.	No	No	No	No

the bandwidth. If congestion builds up in the network, ABR traffic is held back to help relieve the congestion condition.

UBR is intended for routine applications and when the customer seeks a low-cost method of transporting bursty data for noncritical applications that can tolerate delay variations. Although the carrier will attempt to deliver all ATM cells received over the PVC, if there is any network congestion, this may result in loss of ATM cells to relieve congestion in the network.

Operation

QoS enables ATM to admit a CBR voice connection while protecting a VBR connection for a transaction processing application and allowing an ABR or UBR data transfer to proceed over the same network. Each virtual circuit will have its own QoS contract, which is established at the time of connection setup at the user-to-network interface (UNI). The network will not allow any new QoS contracts to be established if they will adversely affect its ability to meet existing contracts. In such cases, the application will not be able to get on the network until the network is fully capable of meeting the new contract.

When the QoS is negotiated with the network, there are performance guarantees that go along with it: maximum cell rate, available cell rate, cell transfer delay, and cell loss ratio. The network reserves the resources needed to meet the performance guarantees, and the user is required to honor the contract by not exceeding the negotiated parameters. Several methods are available to enforce the contract. Among them are traffic policing and traffic shaping.

Traffic policing is a management function performed by switches or routers on the ATM network. To police traffic, the switches or routers use a buffering technique referred to as a "leaky bucket." This technique entails traffic flowing (leaking) out of the buffer (bucket) at a constant rate (the negotiated rate) regardless of how fast it flows into the buffer. If the traffic flows into the buffer too fast, the cells will be allowed onto the network only if enough capacity is available. If there is not enough capacity, the cells are discarded and must be retransmitted by the sending device.

Traffic shaping is a management function performed at the UNI of the ATM network. It ensures that traffic matches the contract negotiated between the user and network during connection setup. Traffic shaping helps guard against cell loss in the network. If too many cells are sent at once, cell discards can result, which will disrupt time-sensitive applications. Because traffic shaping regulates the data transfer rate by evenly spacing the cells, discards are prevented.

Cell Structure

Voice, video, and data traffic is usually composed of bytes, packets, or frames. These larger payloads are chopped up into smaller fixed-length cells by the customer's router or the carrier's ATM switch. ATM cells are fixed at 53 octets[1] and consist of a 5-octet header and 48-octet payload (Figure A-10).

The cell header contains the information needed to route the information field through the ATM network. The header has several fields, which add up to 40 bits (5 bytes), as follows:

- *Generic Flow Control (GFC)* This 4-bit field has only local significance; it enables customer premises equipment at the UNI to regulate the flow of traffic for different grades of service.
- *Addressing* An 8-bit Virtual Path Identifier (VPI) is used in conjunction with the VCI to identify the next destination of a cell as it passes through a series of switches on the ATM network. The Virtual Channel Identifier (VCI) is

---

[1] This odd cell size was the result of a compromise among international standards bodies. The United States wanted the cell's data payload size to be 64 bytes, and Europe wanted a data payload of 32 bytes. The compromise was simply to average the two, which equals 48 bytes. The cell's header required 5 bytes, providing an overall cell size of 53 bytes.

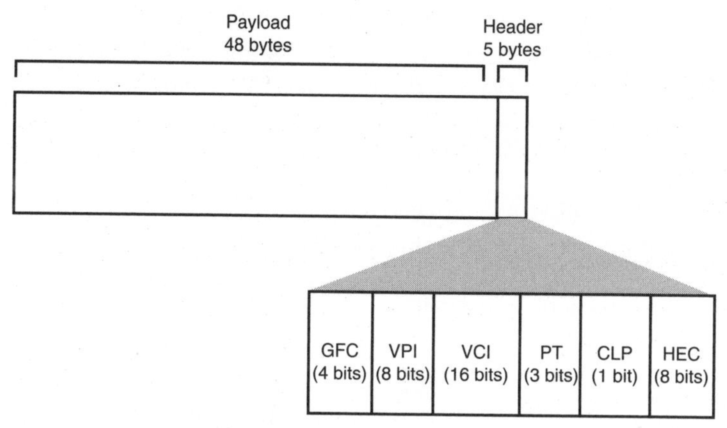

Figure A-10 ATM cell structure.

a 16-bit field used to identify the virtual channel on a particular virtual path.

- *Payload Type (PT)* This 3-bit field is used to indicate whether the cell contains user information or connection management information. This field also provides for network congestion notification.
- *Cell Loss Priority (CLP)* This 1-bit field, when set to a 1, indicates that the cell may be discarded in the event of congestion. When set to 0, it indicates that the cell is of higher priority and should not be discarded.
- *Header Error Check (HEC)* This 8-bit field is used by the physical layer for detection and correction of bit errors in the cell header. The header carries its own error check to validate the VPIs and VCIs and prevent delivery of cells to the wrong UNI at the remote end. Cells received with header errors are discarded. Higher-layer protocols are responsible for initiating lost cell recovery procedures.

Initially, there was concern about the high overhead of cell relay, with its ratio of 5 header octets to 48 data octets. However, with innovations in Wave Division Multiplexing

(WDM) to increase fiber's already high capacity, ATM's overhead is no longer a serious issue. Instead, the focus is on ATM's unique ability to provide a QoS in support of all applications on the network.

Virtual Circuits

ATM virtual circuits (VCs) can be bidirectional or unidirectional, meaning that each VC can be configured for one-way or two-way operation. The VCs can be configured as point to point (i.e., permanent virtual circuit), switched, or multipoint. They also can be symmetric or asymmetric in nature. In other words, each bidirectional VC can be configured for symmetric operation (same speed in both directions) or asymmetric operation (different speeds in each direction).

A VC has two components: a virtual path and a virtual channel. In this simplified view of an ATM network (Figure A-11), the customer has two locations connected together by a virtual path, which contains a bundle of virtual channels. Each of the three virtual channels is assigned to a particular

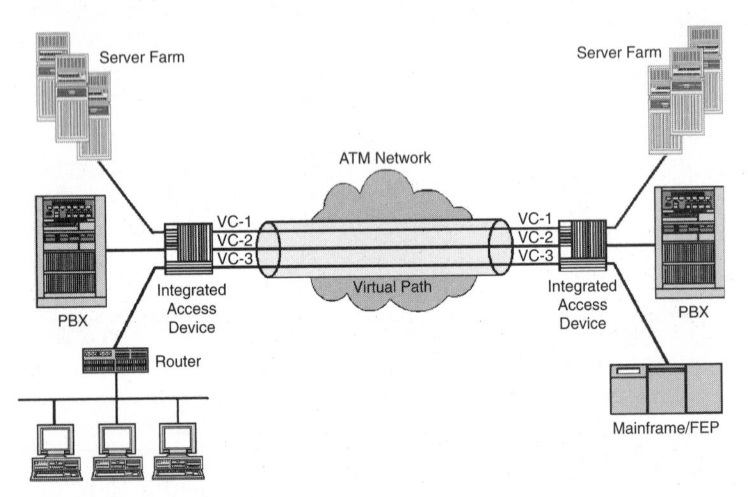

Figure A-11 A simplified view of VCs through an ATM network.

end system, such as a PBX, server, or router. The individual connections between the end devices at each location are identified by ATM addresses, which consist of a virtual channel identifier and a virtual path identifier (VCI/VPI).

In this example, an integrated access device (IAD) is used to consolidate the VCs and deliver them to the ATM switch via a dedicated line. VC-1 provides LAN users with access to a mainframe. VC-2 provides trunking between the PBXs. VC-3 provides LAN users with access to a remote server.

In a large network, there may be hundreds of virtual paths. ATM standards allow up to 65,000 virtual channels to share the same virtual path. This scheme simplifies network management and network recovery. When a virtual path must be reconfigured to bypass a failed port on an ATM switch, for example, all its associated virtual connections go with it, eliminating the need to reconfigure each VC individually.

ATM Layers

Like other technologies, ATM uses a layered protocol model. ATM exists at Layer 2 of the Open Systems Interconnection (OSI) reference model and has only four layers (Figure A-12), which typically operate above Synchronous Optical Network (SONET) at Layer 1.

The ATM Adaptation Layer (AAL) provides the necessary services to support the higher-layer protocols. The functions of this layer are summarized in Table A-3 and may exist in the end stations, servers, or network switches. Among other things, this layer is responsible for segmenting the information into 53-byte cells and, at the receiving end, reassembling it back into its native format (SAR).

Inverse Multiplexing over ATM

Today even midsize companies with multiple traffic types and three or more distributed locations can benefit from ATM's

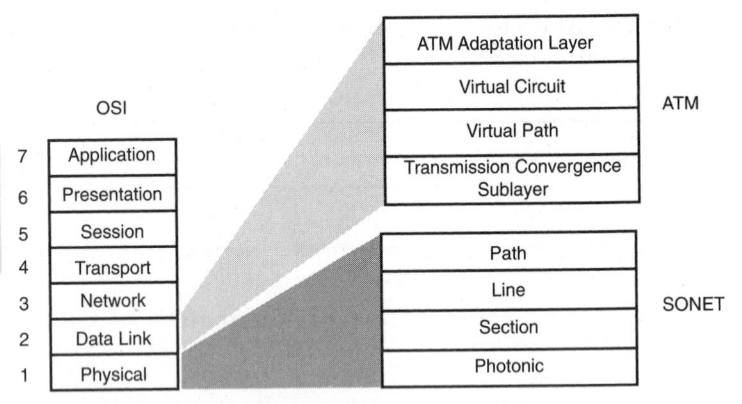

Figure A-12 ATM protocol model in relation to the OSI reference model and SONET physical layer protocols.

sustained throughput, low latency, and adept traffic handling via appropriate QoS mechanisms. The availability of ATM-based inverse multiplexers and $N \times$ T1 access makes ATM suitable for mainstream use, particularly for companies that appreciate the benefits of ATM but have been locked out of the service because of its high cost of implementation.

In the past, T3 links were the minimum bandwidth required to access ATM networks, making the cost prohibitive for the vast majority of companies. Inverse Multiplexing over ATM (IMA) solves the bandwidth gap problem. With IMA, companies can aggregate multiple DS1 circuits to achieve just the right amount of bandwidth they need for their applications and pay for only that amount on an $N \times$ T1 basis. The advantage of IMA is that such companies can scale up to the bandwidth they need, starting with a single T1, and then add links as more bandwidth is justified.

For example, when the bandwidth of four T1s is bonded by the IMA device, the virtual connection through the service provider's network is provisioned at 6 Mbps. When the bandwidth of eight T1s is bonded by the IMA device, the virtual connection through the service provider's network is provisioned at 12 Mbps. Regardless of the number of T1 access

TABLE A-3 ATM Adaptation Layer (AAL) Functions

AAL	Application Examples	Notes
1	Isochronous CBR services, such as audio and video streaming.	Supports connection-oriented services that require CBRs and have specific timing and delay requirements.
2	Isochronous VBR services, such as interactive videoconferencing.	Supports connection-oriented services that do not require CBRs, in other words, VBR applications.
3/4	Near-real-time VBR data, such as SMDS (connectionless) or X.25 (connection oriented).	Originally intended as two layers: one for connection-oriented services and the other for connectionless services. Consolidation into one layer supports VBR services over both types of connections.
5	Near-real-time VBR data, such as datagrams and signaling messages.	Supports connection-oriented VBR data services. It is a leaner AAL compared with AAL3/4, at the expense of error recovery and built-in retransmission. It uses less bandwidth overhead, has simpler processing requirements, and reduces implementation complexity.

links in place, the IMA device bonds them together, combining the bandwidth into a fatter logical pipe that can support mixed-media applications running over interconnected LANs (Figure A-13).

Summary

A solid base of standards now exists to allow equipment vendors, service providers, and end users to implement a wide

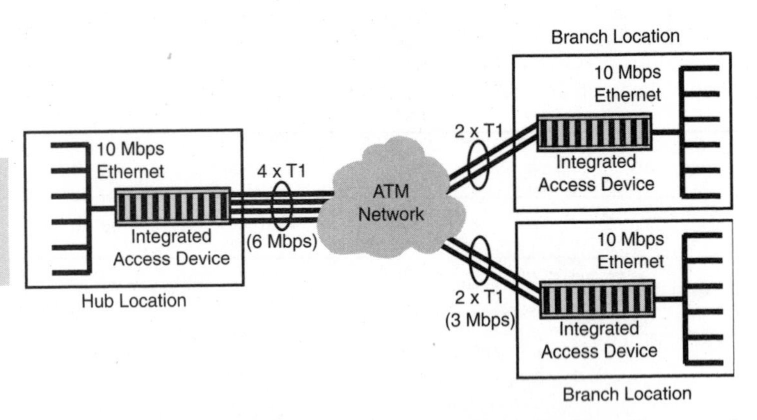

Figure A-13 Inverse Multiplexing over ATM (IMA) allows the use of bonded multiple T1 access lines into and out of the service provider's ATM network instead of forcing companies to use more expensive T3 access lines at each location. This makes it cost-effective for midsize companies to take advantage of ATM services to support a variety of applications.

range of applications via ATM. The standards will continue to evolve as new applications emerge. The rapid growth of the Internet is one area where ATM can have a significant impact. With the Internet forced to handle a growing number of multimedia applications—telephony, videoconferencing, faxes, and collaborative computing, to name a few—congestion and delays are becoming ever more frequent and prolonged. ATM backbones will play a key role in alleviating these conditions, enabling next-generation networks to be used to their full potential.

See also

Frame Relay
Integrated Access Devices
Inverse Multiplexers

B

BLUETOOTH

Bluetooth is an omnidirectional wireless technology that provides limited-range voice and data transmission over the unlicensed 2.4-GHz frequency band, allowing connections with a wide variety of fixed and portable devices that normally would have to be cabled together. Up to eight devices—one master and seven slaves—can communicate with one another in a so-called piconet at distances of up to 30 feet. Table B-1 summarizes the performance characteristics of Bluetooth products that operate at 1 Mbps in the 2.4-GHz range.

Origins of Bluetooth

Since its development in 1994 by the Swedish telecommunications firm Ericsson, more than 1800 companies worldwide have signed on as members of the Bluetooth Special Interest Group (SIG) to build products to the wireless specification and promote the new technology in the marketplace.

The engineers at Ericsson code-named the new wireless technology Bluetooth to honor a tenth-century Viking king, Harald Bluetooth, who reigned from 940 to 985 and is cred-

TABLE B-1 Performance Characteristics of Bluetooth Products

Feature/Function	Performance
Connection type	Spread spectrum (frequency hopping)
Spectrum	2.4-GHz ISM (industrial, scientific, and medical) band
Transmission power	1 milliwatt (mW)
Aggregate data rate	1 Mbps using frequency hopping
Range	Up to 30 feet (9 meters)
Supported stations	Up to eight devices per piconet
Voice channels	Up to three synchronous channels
Data security	For authentication, a 128-bit key; for encryption, the key size is configurable between 8 and 128 bits
Addressing	Each device has a 48-bit Media Access Control (MAC) address that is used to establish a connection with another device

ited with uniting Denmark and bringing order to that country. Harald's name was actually Blåtand, which roughly translates into English as "Bluetooth." This has nothing to do with the color of his teeth—Blåtand actually referred to Harald's very dark hair, which was unusual for Vikings.

Applications

Among the many things users can do with Bluetooth is swap data and synchronize files merely by having the devices come within range of one another. Images captured with a digital camera, for example, can be dropped off at a PC for editing or a color printer for output on photo-quality paper—all without having to connect cables, load files, open applications, or click buttons.

The technology is a combination of circuit switching and packet switching, making it suitable for voice as well as data. Instead of fumbling with a cell phone while driving, for example, the user can wear a lightweight headset to answer a call and engage in a conversation even if the phone is tucked away in a briefcase or purse.

While useful in minimizing the need for cables, wireless local area networks (LANs) are not intended for interconnecting the range of mobile devices people carry around everyday between home and office. For this, Bluetooth is needed. And in the office, a Bluetooth portable device can be in motion while connected to the LAN access point as long as the user stays within the 30-foot range.

Bluetooth can be combined with other technologies to offer wholly new capabilities, such as automatically lowering the ring volume of cell phones or shutting them off as users enter quiet zones like churches, restaurants, theaters, and classrooms. On leaving the quiet zone, the cell phones are returned to their original settings.

Topology

The devices within a piconet play one of two roles: that of master or slave. The master is the device in a piconet whose clock and hopping sequence are used to synchronize all other devices (i.e., slaves) in the piconet. The unit that carries out the paging procedure and establishes a connection is by default the master of the connection. The slaves are the units within a piconet that are synchronized to the master via its clock and hopping sequence.

The Bluetooth topology is best described as a multiple-piconet structure. Since Bluetooth supports both point-to-point and point-to-multipoint connections, several piconets can be established and linked together in a topology called a "scatternet" whenever the need arises (Figure B-1).

Piconets are uncoordinated, with frequency hopping occurring independently. Several piconets can be established and linked together ad hoc, where each piconet is identified by a different frequency-hopping sequence. All users participating on the same piconet are synchronized to this hopping

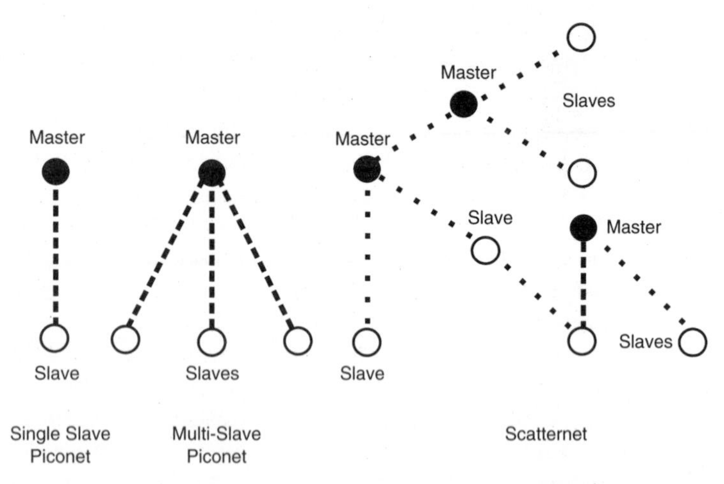

Figure B-1 Possible topologies of networked Bluetooth devices, where each is either a master or slave.

sequence. Although synchronization of different piconets is not permitted in the unlicensed ISM band, Bluetooth units may participate in different piconets through Time Division Multiplexing (TDM). This enables a unit to sequentially participate in different piconets by being active in only one piconet at a time.

With its service discovery protocol, Bluetooth enables a much broader vision of networking, including the creation of personal area networks, where all the devices in a person's life can communicate and work together. Technical safeguards ensure that a cluster of Bluetooth devices in public places, such as an airport lounge or train terminal, would not suddenly start talking to one another.

Technology

Two types of links have been defined for Bluetooth in support of voice and data applications: an asynchronous connectionless (ACL) link and a synchronous connection-oriented

(SCO) link. ACL links support data traffic on a best-effort basis. The information carried can be user data or control data. SCO links support real-time voice and multimedia traffic using reserved bandwidth. Both data and voice are carried in the form of packets, and Bluetooth devices can support active ACL and SCO links at the same time.

ACL links support symmetric or asymmetric, packet-switched, point-to-multipoint connections, which are typically used for data. For symmetric connections, the maximum data rate is 433.9 kbps in both directions, send and receive. For asymmetric connections, the maximum data rate is 723.2 kbps in one direction and 57.6 kbps in the reverse direction. If errors are detected at the receiving device, a notification is sent in the header of the return packet so that only lost or corrupt packets need to be retransmitted.

SCO links provide symmetric, circuit-switched, point-to-point connections, which are typically used for voice. Three synchronous channels of 64 kbps each are available for voice. The channels are derived through the use of either Pulse Code Modulation (PCM) or Continuous Variable Slope Delta (CVSD) modulation. PCM is the standard for encoding speech in analog form into the digital format of 1s and 0s. CVSD is another standard for analog-to-digital encoding but offers more immunity to interference and therefore is better suited than PCM for voice communication over a wireless link. Bluetooth supports both PCM and CVSD; the appropriate voice-coding scheme is selected after negotiations between the link managers of each Bluetooth device before the call takes place.

Voice and data are sent as packets. Communication is handled with Time Division Duplexing (TDD), which divides the channel into time slots, each 625 microseconds (μs) in length. The time slots are numbered according to the clock of the piconet master. In the time slots, master and slave can transmit packets. In the TDD scheme, master and slave alternatively transmit (Figure B-2). The master starts its

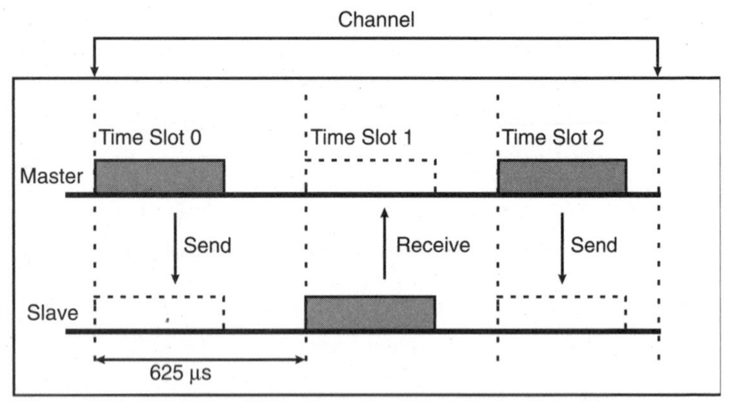

Figure B-2 With the TDD scheme used in Bluetooth, packets are sent over time slots of 625 microseconds (μs) in length between the master and slave units within a piconet.

transmission in even-numbered time slots only, and the slave starts its transmission in odd-numbered time slots only. The start of the packet is aligned with the slot start. Packets transmitted by the master or the slave may extend over as many as five time slots.

With TDD, bandwidth can be allocated on an as-needed basis, changing the makeup of the traffic flow as demand warrants. For example, if the user wants to download a large data file, as much bandwidth as is needed will be allocated to the transfer. Then, at the next moment, if a file is being uploaded, that same amount of bandwidth can be allocated to that transfer.

No matter what the application—voice or data—making connections between Bluetooth devices is as easy as powering them up. In fact, one advantage of Bluetooth is that it does not need to be set up—it is always on, running in the background, and looking for other devices that it can communicate with.

When Bluetooth devices come within range of one another, they engage in a service discovery procedure, which entails the exchange of messages to become aware of each another's service and feature capabilities. Having located available services within the vicinity, the user may select from any of them. After that, a connection between two or more Bluetooth devices can be established.

The radio link itself is very robust, using frequency-hopping spread-spectrum technology to overcome interference and fading. Spread spectrum is a digital coding technique in which the signal is taken apart or "spread" so that it sounds more like noise as it is sent through the air. With the addition of frequency hopping—having the signals skip from one frequency to another—wireless transmissions are made even more secure. Bluetooth specifies a rate of 1600 hops per second among 79 frequencies. Since only the sender and receiver know the hopping sequence for coding and decoding the signal, eavesdropping is virtually impossible. For enhanced security, Bluetooth also supports device authentication and encryption.

Other frequency-hopping transmitters in the vicinity will be using different hopping patterns and much slower hop rates than Bluetooth devices. Although the chance of Bluetooth devices interfering with non-Bluetooth devices that share the same 2.4-GHz band is minimal, should non-Bluetooth transmitters and Bluetooth transmitters coincidentally attempt to use the same frequency at the same moment, the data packet transmitted by one or both devices will become garbled in the collision, and a retransmission of the affected data packets will be required. A new data packet will be sent again on the next hopping cycle of each transmitter. Voice packets, because of their sensitivity to delay, are never retransmitted.

Points of Convergence

In some ways Bluetooth competes with infrared, and in other ways the two technologies are complementary. With both infrared and Bluetooth, data exchange is considered to be a fundamental function. Data exchange can be as simple as transferring business card information from a mobile phone to a palmtop or as sophisticated as synchronizing personal information between a palmtop and desktop PC. In fact, both technologies can support many of the same applications, raising the question: why would users need both technologies?

The answer lies in the fact that each technology has its advantages and disadvantages. The very scenarios that leave infrared falling short are the ones where Bluetooth excels and vice versa. Take the electronic exchange of business card information between two devices. This application usually will take place in a conference room or exhibit floor where a number of other devices may be attempting to do the same thing. This is the situation where infrared excels. The short-range, narrow angle of infrared—30 degrees or less—allows each user to aim his or her device at the intended recipient with point-and-shoot ease. Close proximity to another person is natural in a business card exchange situation, as is pointing one device at another. The limited range and angle of infrared allows other users to perform a similar activity with ample security and no interference.

In the same situation, a Bluetooth device would not perform as well as an infrared device. With its omnidirectional capability, the Bluetooth device must first discover the intended recipient. The user cannot simply point at the intended recipient—a Bluetooth device must perform a discovery operation that probably will reveal several other Bluetooth devices within range, so close proximity offers no advantage here. The user will be forced to select from a list of discovered devices and apply a security mechanism to prevent unauthorized access. All this makes the use of Bluetooth for business card exchange an awkward and needlessly time-consuming process.

However, in other data-exchange situations Bluetooth might be the preferred choice. Bluetooth's ability to penetrate solid objects and its ability to communicate with other devices in a piconet allows for data-exchange opportunities that are very difficult or impossible with infrared.

For example, Bluetooth allows a user to synchronize a mobile phone with a notebook computer without taking the phone out of a jacket pocket or purse. This would allow the user to type a new address at the computer and move it to the mobile phone's directory without unpacking the phone and setting up a cable connection between the two devices. The omnidirectional capability of Bluetooth allows synchronization to occur instantly, assuming that the phone and computer are within 30 feet of each other.

Using Bluetooth for synchronization does not require that the phone remain in a fixed location. If a phone is carried about in a briefcase, the synchronization can occur while the user moves around. This is not possible with infrared because the signal is not able to penetrate solid objects and the devices must be within a few feet of each other. Furthermore, the use of infrared requires that both devices remain stationary while the synchronization occurs.

When it comes to data transfers, infrared does offer a big speed advantage over Bluetooth. While Bluetooth moves data between devices at an aggregate rate of 1 Mbps, infrared offers 4 Mbps of data throughput. A higher-speed version of infrared is now available that can transmit data between devices at up to 16 Mbps—a four times improvement over the previous version. The higher speed is achieved with the Very Fast Infrared (VFIR) protocol, which is designed to address the new demands of transferring large image files between digital cameras, scanners, and PCs. Even when Bluetooth is enhanced for higher data rates in the future, infrared is likely to maintain its speed advantage for many years to come.

Bluetooth complements infrared's point-and-shoot ease of use with omnidirectional signaling, longer-distance communications, and capacity to penetrate walls. For some users, having both Bluetooth and infrared will provide the optimal short-range wireless solution. For others, the choice of adding Bluetooth or infrared will be based on the applications and intended usage.

Summary

Communicator platforms of the future will combine a number of technologies and features in one device, including mobile Internet browsing, messaging, imaging, location-based applications and services, mobile telephony, personal information management, and enterprise applications. Bluetooth will be a key component of these platforms. Since Bluetooth radio transceivers operate in the globally available ISM (industrial, scientific, and medical) radio band of 2.4 GHz, products do not require an operator license from a regulatory agency, such as the Federal Communications Commission (FCC) in the United States. The use of a generally available frequency band means that Bluetooth-enabled devices can be used virtually anywhere in the world and link up with one another for ad-hoc networking when they come within range.

See also

Infrared Networking

BRIDGES

Bridges are used to extend or interconnect LAN segments. At one level, they are used to create an extended network that greatly expands the number of devices and services available to each user. At another level, bridges can be used for segmenting LANs into smaller subnets to improve performance, control access, and facilitate fault isolation and testing without affecting the overall user population.

The bridge does this by monitoring all traffic on the subnets that it links. It reads both the source and destination addresses of all the packets sent through it. If the bridge encounters a source address that is not already contained in its address table, it assumes that a new device has been added to the local network. The bridge then adds the new address to its table.

In examining all packets for their source and destination addresses, bridges build a table containing all local addresses. The table is updated as new packets are encountered and as addresses that have not been used for a specified period of time are deleted. This self-learning capability permits bridges to keep up with changes on the network without requiring that their tables be updated manually.

The bridge isolates traffic by examining the destination address of each packet. If the destination address matches any of the source addresses in its table, the packet is not allowed to pass over the bridge because the traffic is local. If the destination address does not match any of the source addresses in the table, the packet is discarded onto an adjacent network. This filtering process is repeated at each bridge on the internetwork until the packet eventually reaches its destination. Not only does this process prevent unnecessary traffic from leaking onto the internetwork, it also acts as a simple security mechanism that can screen unauthorized packets from accessing various corporate resources.

Bridges also can be used to interconnect LANs that use different media, such as twisted-pair, coaxial, and fiberoptic cabling, and various types of wireless links. In office environments that use wireless communications technologies such as spread spectrum and infrared, bridges can function as an access point to wired LANs. On the wide area network (WAN), bridges even switch traffic to a secondary port if the primary port fails. For example, a full-time wireless bridging system can establish a modem connection on the public network if the primary wire or wireless link is lost because of environmental interference.

In reference to the Open Systems Interconnection (OSI) model, a bridge operates at Layer 2; specifically, it operates at the Media Access Control (MAC) sublayer of the Data Link Layer. It routes by means of the Logical Link Control (LLC), the upper sublayer of the Data Link Layer (Figure B-3).

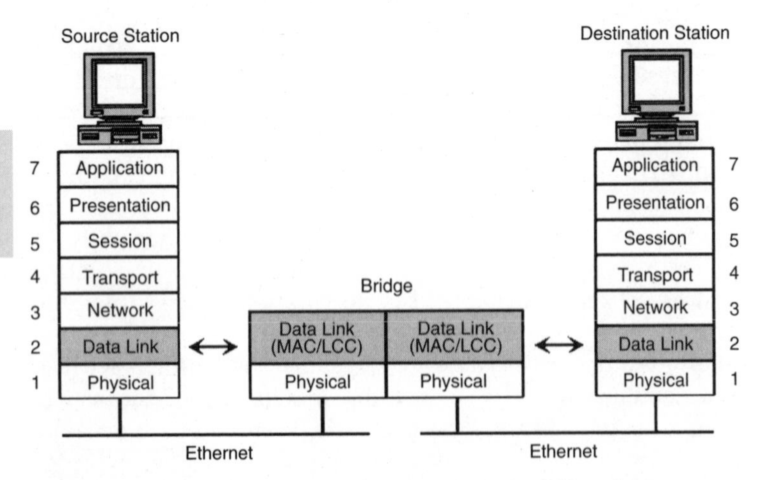

Figure B-3 Bridge functionality in reference to the OSI model.

Because the bridge connects LANs at a relatively low level, throughput often exceeds 30,000 packets per second (pps). Multiprotocol routers and gateways, which provide LAN interconnection over the WAN, operate at higher levels of the OSI model and provide more functionality. In performing more protocol conversions and delivering more functionality, routers and gateways are generally more processing-intensive and, consequently, slower than bridges.

See also

Gateways

Open Systems Interconnection (OSI)

Repeaters

Routers

BUILDING LOCAL EXCHANGE CARRIERS

A Building Local Exchange Carrier (BLEC) designs, constructs, deploys, and manages high-speed broadband networks inside commercial office buildings to meet the communications needs of tenants, which typically are small and midsize businesses. Typically, the BLEC links a tenant's LAN to the riser cabling to distribute its services to each office on the various floors. Among the services the BLEC may offer are high-speed Internet access, enhanced conference calling services, Web hosting, managed network security, remote access, and information technology services. The BLEC has the capability to provide services using wireless, optical, and copper-based technologies. The BLEC may seek to enhance existing customer relationships by also offering similar broadband services to customers' branch offices and other businesses located in buildings in which it does not have an installed network.

The in-building broadband data network transmits data to and from each customer at a variety of speeds. Utilizing information on the demand characteristics of customers in a particular building, the BLEC installs broadband data equipment by connecting each building network to a central facility in each metropolitan area, usually over lines leased from other carriers. At this metropolitan hub, traffic is aggregated and distributed to the appropriate locations.

The BLEC partners with large-scale real estate owners and secures the right to install its broadband data network inside office buildings that meet its criteria. It generally targets buildings with more than 100,000 rentable square feet and 10 or more tenants. The BLEC pays real estate owners either a modest portion of the gross revenue generated from tenants in their buildings or a fixed rental fee. In having its services available in the building, BLECs believe they assist the real estate owner with their tenant leasing and retention efforts. A typical lease or license agreement with a real estate owner is for a term of 10 or more years.

Once an agreement is reached with the building owner, the BLEC's sales and marketing efforts focus primarily on tenants. The BLEC develops building-specific marketing and promotional techniques, such as lobby events and advertising in landlord newsletters. In many cases they work with building managers to demonstrate their services to tenants and prospective tenants.

The BLEC provides affordable services and products with a range of choices for tenants. In addition to pricing based on the number of desktops connected to the network, the BLEC may offer tenants data communication service bundles that combine a broad range of high-speed connectivity, business communication applications, and professional network management services, including Internet connections, multiple customer-branded e-mail accounts, Web hosting, remote access, a desktop business portal with a customer-specific design, and professional information technology services. This approach permits customers to obtain the services they need for a low monthly cost and add on as the BLEC demonstrates the ability of its services to enhance productivity.

The BLEC supports its services through a national customer care center and a national operations control center, which are staffed 24 × 7 and continuously monitor the network to detect disruptions in service, remotely resolve problems, configure networks, and compile data on customer service levels. In addition, field operations personnel augment the customer care center by providing on-site support.

Network Architecture

Inside the building, the BLEC installs a broadband data infrastructure that typically runs from the basement of the building to the top floor inside the building's vertical utility shaft. This broadband data infrastructure is designed to carry data and voice traffic for all the building's tenants for the foreseeable future. Tenants receive services through a data cable from their LAN to the BLEC's infrastructure in the vertical utility shaft.

Inside the building, usually in the basement, the BLEC establishes a point of presence (POP). In each building POP, the BLEC connects the broadband data cable to routers and other equipment that enables the transmission of data and video traffic and the aggregation and dissemination of traffic to and from those cables. The BLEC typically negotiates with the building owner for the right to use a small amount of space in the basement to establish the POP.

Within each market the BLEC serves, it has a metropolitan POP where traffic is aggregated and distributed to and from all the building networks via broadband data circuits. These broadband data lines typically are leased from carriers that have previously installed local transport capacity in that market. The POP contains all the equipment necessary to provide services in the metropolitan area and may include network computer servers and traffic routers. Through a collocation arrangement, each metropolitan POP is connected to multiple major service providers that provide Internet connectivity.

Building Access

The telecommunications industry has come to see property owners as a key bottleneck in the emergence of competition. With exclusive agreements, property owners have limited access to their buildings by alternative service providers. Facilities-based competition in multiunit buildings is viewed as crucial to promoting consumer choice and advancing all the economic benefits of competition. Although the FCC has considered rules that would force property owners to allow new entrants into buildings to wire customers, real estate interests have argued that the FCC lacks jurisdiction and that the market can be relied on to deliver competition.

In considering this matter, the FCC has recognized that the real estate industry has taken positive steps to facilitate tenant choice of telecommunications providers by working toward the development of best practices and model agreements. In particular, a coalition of 11 trade associations representing over 1 million owners and operators has committed to a best practices implementation plan regarding these issues. The FCC will closely monitor these industry efforts.

Meanwhile, the FCC in October 2000 ruled that telecom carriers are forbidden from signing exclusive contracts with multitenant building owners. The new rule also outlaws carriers from signing the functional equivalent of exclusive contracts—deals in which building owners effectively restrict access to necessary facilities such as wiring closets and risers to all but that carrier. The new FCC rule directs the ban at carriers, not building owners, which alleviates the jurisdictional concerns of landlords who charged that the rule equates to confiscation of private property.

The FCC was prepared to take additional action, but some members of Congress argued that the FCC lacks any express statutory authority to take up this matter itself. Under the FCC's original plan, building owners would have been required to let in not only more than one carrier but also every carrier who thought it might want to serve tenants in a building. To do this, the FCC had proposed to regulate the building owners themselves, ordering them to sell wiring closet and rooftop space to potentially dozens of carriers at equivalent prices and contract terms.

As the FCC sees it, unless landlords open their doors to new telephone entrants, the market will remain dominated by the entrenched Bell companies, which still control the vast majority of wires reaching homes and businesses. The BLECs have argued that many landlords effectively delay or impede service with demands for expensive fees and engineering studies before they can gain access to buildings. Sometimes the landlords deny access altogether. Property owners counter that they are simply seeking to safeguard their buildings and services for their tenants by ensuring that there is no damage from installing antennas on rooftops or new wiring.

Summary

Ever since passage of the Telecommunications Act of 1996, which was aimed at deregulating the industry, the FCC has been playing referee. The FCC has mostly sided with the competitors, forcing the Bell companies to lease key components of their networks to their rivals. FCC officials see the battle over building access in the same context. When FCC staff took up the issue in late 2000, it considered imposing rules that would force landlords to provide access to the rival companies in exchange for fair compensation. But some commissioners voiced concerns that the FCC would be on shaky legal ground if it sought to regulate landlords. Members of Congress, under intense lobbying from building owners, have since echoed this concern. Although FCC staff remain convinced that they have the authority to regulate on this issue as part of their mandate to foster telecommunications competition, a negotiated resolution between private parties has become the more palatable approach.

See also

Competitive Local Exchange Carriers

Incumbent Local Exchange Carriers

C

CABLE TELEVISION NETWORKS

Cable television networks are broadband local area networks (LANs) that distribute television programs and other services over a shared facility within a neighborhood. Cable systems were designed originally to deliver broadcast television signals efficiently to subscribers' homes. To ensure that consumers could obtain cable service with the same TV sets they use to receive over-the-air broadcast TV signals, cable operators recreate a portion of the over-the-air radiofrequency (RF) spectrum within the coaxial cable.

Traditional coaxial cable systems typically operate with 330 or 450 MHz of capacity, whereas modern hybrid fiber/coax (HFC) systems operate to 750 MHz or more. Each standard television channel occupies 6 MHz of RF spectrum. A traditional cable system with 400 MHz of downstream bandwidth can carry the equivalent of 60 analog TV channels, and a modern HFC system with 750 MHz of downstream bandwidth has the capacity for over 200 channels when digital compression is applied.

Many cable companies are migrating their networks from analog to digital. This provides customers with greater programming diversity, better picture quality, improved reliability, and enhanced service. Advanced compression techniques can be applied to digital signals, allowing up to 12 digital channels to be inserted into the space of only one traditional 6-MHz analog channel and enabling cable companies to greatly increase the capacity of their networks. The larger cable companies now offer 250 channels, including enhanced pay-per-view service, digital music channels, new networks grouped by genre, and an interactive program guide.

In the Beginning . . .

Cable television began around 1949 in Lansford, Pennsylvania. A local TV shop owner noticed a decrease in television sales and wanted to find out the reason. After talking to town residents, he discovered that low sales were due in large part to the poor reception in the area. The closest station was in Philadelphia, about 65 miles away, and there was a mountain overlooking Lansford that blocked reception.

After helping some residents in the outlying areas set up antennas to help with their reception, the shop owner came up with an idea to help with the town's reception problem. He built an antenna on top of the mountain and, with an amplifier, boosted the signal back to full strength. Next he ran coaxial cable down the mountain and into the town, charging people a fee to connect to the cable. This was the first community antenna television (CATV) system, consisting of three channels and a few hundred subscribers.

At around the same time, other towns claimed to have put the first CATV system in operation. Regardless of which town was actually first, the intent was the same—to provide television service to remote areas where over-the-air signal reception was difficult or impossible. Almost immediately, CATV began moving into metropolitan areas, such as New York City, where reception was difficult because the tall buildings caused multipath signal interference or blocked the signal altogether.

By 1952, there were 70 CATV systems in operation nationwide with 14,000 subscribers. However, by the 1960s, growth in the CATV market had all but stopped. Cable service had been installed in most of the major market areas. The existing technology also limited cable's growth. Most cable systems only had enough capacity for 12 channels until the mid-1970s.

Major growth in the cable market began to take off after 1975. The availability of satellite receivers allowed cable operators to take specific signals and insert them into their channel lineup. This led to cable-only programming. Cable system operators began adding programs such as movie (HBO), sports (ESPN), and shopping (HSN) channels, as well as superstations (TBS). The technology also allowed cable companies to give subscribers pay-per-view programming. With this new service, a subscriber pays a one-time fee to view a special event, such as a concert or sporting event, or watch a first-run movie.

Program Delivery

The actual video signals delivered to the cable system can be generated from three basic sources:

- *Satellite or microwave receivers* Program sources include national networks such as CNN, HBO, and ESPN and local sources such as commercial and public television. Usually, these program sources run 24 hours a day but may be interrupted by inserting locally originated programming or commercials.
- *Videotape* Prerecorded material such as commercials, infomercials, public-service programs, and movies may be delivered via videotape. The use of videotape is undesirable because of the labor involved in getting the tapes made, moved to the broadcast site, and played. Instead, multimedia servers are being used increasingly, which automate program delivery.

- *Multimedia servers* Servers store and play multimedia programming that includes graphics, animation, sound, text, and digital Moving Pictures Experts Group (MPEG) video. These computers may accept real-time data from weather services, Internet information sources, computer databases, and satellite data networks for automated delivery on a scheduled or demand basis.

Today CATV is the primary method of program distribution in the United States, where more than 75 million subscribers access programming from a cable TV network. There are 11,000 networks nationwide and over 1 million miles of cable plant. These networks pass 95 percent of all households, making information, entertainment, and education available to almost everyone who chooses to subscribe.

Subscribers pay a monthly fee for a set of basic services and may select optional packages of premium services, including Internet access and telephone service, for an additional monthly fee. In addition, subscribers may choose pay-per-view programs by calling the cable operator to request a specific program from a menu of choices that changes daily. Usually, there is a nominal extra charge for each additional television set that is set up to receive cable programming. All services are itemized on the monthly bill from the cable TV operator.

The CATV market generates about $25 billion in revenue per year from subscribers. The funds are generally split two ways: (1) financing the operating costs of existing networks and constructing new systems and (2) providing payment for programming like HBO, MTV, and the Disney Channel.

Operating Environment

Cable companies operate in an industry that is undergoing rapid change due to consolidation and technological innovation. Complicating matters is the fact that many cable companies are getting into technologies with which they do not have much previous experience, such as business-class telephony and broadband data. Businesses today expect the most from their vendors. Businesses want superior service and products delivered at an excellent value. If cable companies expect to succeed in these areas, they must acquire the expertise and support infrastructure necessary to ensure that the needs of businesses are addressed in a timely manner.

The industry is moving from the phase of consolidation to the phase of swapping. As the list of big cable properties that are for sale continues to shrink, property swapping is likely to become more common as companies seek to build regional holdings. The rush to swap is being driven by the need to assemble "clusters" of cable property. Clusters allow cable companies to more economically provision new broadband technologies and reap faster return on capital investments. By swapping assets, cable companies can concentrate on one particular region of the country, much like local phone companies do.

While cable TV makes up 99 percent of cable operators' revenues, some operators have advanced hybrid fiber coaxial (HFC) networks that can be leveraged for new revenue streams. In addition to interactive television and video on demand (VOD), cable operators offer business-class broadband Internet access and telecommunications services that compete with Internet service providers (ISPs) and telephone companies. One cable company, Cox, offers a diversified portfolio of services, illustrating the new service direction of the cable industry, as well as new choices for corporate network managers.

In early 2002, Cox completed the development of its own nationwide Internet Protocol (IP) network to deliver high-speed data and Internet services to its subscribers. This network utilizes Cox's HFC infrastructure to connect subscribers to the Internet and other online services and includes standard ISP functionality, such as Web page hosting for subscribers, access to Internet news groups, multiple e-mail accounts, and remote access.

Cox delivers telecommunications services to businesses through its Competitive Local Exchange Carrier (CLEC) operation, Cox Business Services. Through both its dedicated fiberoptic networks and its HFC cable networks, Cox Business Services provides business customers with video, telephony, and high-speed Internet access services.

The company's Network Operations Center (NOC) provides 24 × 7 active monitoring, diagnostics, and remote control. The center continually tests all customer lines and troubleshoots any problems before they become severe enough to disrupt service. When a service error occurs, technicians immediately reroute network traffic and dispatch local crews to ensure minimal or no network downtime. The NOC supports all the metro fiber and HFC networks, including the management of building connections

Cable Modems

One of the services most in demand from cable companies is broadband Internet access. To deliver data services over a cable network, one television channel in the 50- to 750-MHz range is typically allocated for downstream traffic to homes and businesses, while another channel in the 5- to 42-MHz band is used to carry upstream signals. A head-end cable modem termination system (CMTS) communicates through these channels with cable modems to create a virtual local area network (VLAN) connection. Most cable modems are external devices that connect to a PC through a standard 10Base-T Ethernet card or Universal Serial Bus (USB) connection, although internal peripheral component interconnect (PCI) modem cards are also available.

A single downstream 6-MHz television channel supports up to 27 Mbps of downstream data throughput from the cable head-end using 64 QAM (Quadrature Amplitude Modulation) transmission technology. This downstream bandwidth is shared by all subscribers on the cable segment, giving users between 1 and 3 Mbps of bandwidth at any given time, assuming that not all will access the bandwidth at the same time. The speed can be boosted to 36 Mbps using 256 QAM. Upstream channels may deliver 500 kbps to 10 Mbps depending on the amount of spectrum allocated for the service. This upstream and downstream bandwidth is shared by the active data subscribers connected to a given cable network segment, which typically range from 250 to 500 on a modern HFC network.

Most cable systems are not yet equipped for two-way capability. To get broadband Internet access, subscribers must still use a dial-up modem and telephone line for the upstream data path. Simple requests for Web pages are issued over the low-speed modem connection, and the rich content is returned over the high-speed cable connection (Figure C-1).

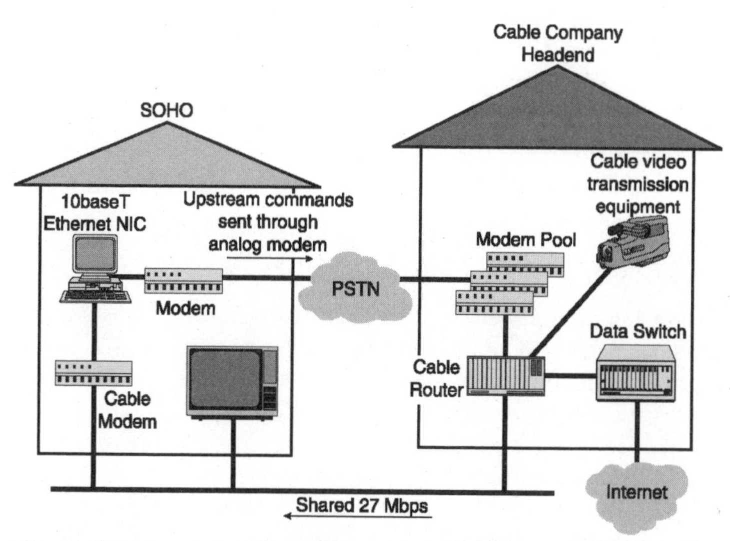

Figure C-1 One-way cable systems require Internet subscribers to use a separate phone line and modem for the upstream path. Sometimes the cable modem contains an integral dialup modem.

Performance suffers when users want to send large files upstream, in which case the data are sent at the speed of the modem—no more than 56 kbps—over the dial-up line.

The key standard for cable modems is the Data Over Cable Service Interface Specification (DOCSIS) developed by CableLabs and approved by the International Telecommunication Union (ITU) in 1998. DOCSIS specifies downstream traffic transfer rates between 27 and 36 Mbps over an RF path in the 50- to 750+-MHz range and upstream traffic transfer rates between 320 kbps and 10 Mbps over an RF path between 5 and 42 MHz. It also defines interface standards for cable modems and supporting equipment. With certification from CableLabs, manufacturers can produce cable modems for retail sale, so consumers no longer have to depend on leased cable modems from their cable providers.

Cable operators are typically charging between $20 and $60 per month for the broadband Internet service, which includes cable modem rental and unlimited Internet access. To qualify for the lower price, the subscriber usually must choose a premium package of entertainment services. For the amount of bandwidth, even the higher price is cheaper than the Digital Subscriber Line (DSL) services offered by telephone companies.

Cable versus DSL

When facing competition from cable operators, DSL service providers like to point out that cable is a shared service. Actual performance of the Internet connection deteriorates as new customers are brought online to share the bandwidth. This point is exaggerated because cable provides much greater bandwidth than DSL, and most cable operators limit the number of subscribers on a segment. In any case, the same argument can be applied to DSL service, since the fiber on the network side of the DSL Access

Multiplexer (DSLAM) is routinely overprovisioned and becomes a bottleneck during peak demand.

Cable operators point out that DSL has distance limitations from customer premises to the DSLAM—the farther away the customer is from the DSLAM, the less bandwidth is available. Since online prequalification is only 70 percent accurate at best, customers really do not know what speed their DSL service will operate at until after installation. Furthermore, since DSL is provisioned over an analog line, it will not work if load coils are attached to the line, if there is a bridge tap along the line, or if the Incumbent Local Exchange Carrier (ILEC) has attached the line to pair-gain equipment.

Summary

To survive in the new competitive climate ushered in by the Telecommunications Act of 1996, cable companies are investing billions of dollars to upgrade their networks for full-duplex operation to support advanced services, such as local and long-distance voice services in competition with the telephone companies, as well as broadband Internet access, VOD, and interactive television. Among the technology choices for upgrading CATV networks for advanced services are HFC and Fiber in the-Loop (FITL) systems and Synchronous Optical Network (SONET) rings.

See also

Digital Subscriber Line Technologies

CLIENT-SERVER NETWORKS

For much of the 1990s, the client-server architecture has dominated corporate efforts to downsize, restructure, and otherwise reengineer for survival in an increasingly global economy. Frustrated with the restrictive access policies of traditional Management Information Services (MIS) managers and the slow pace of centralized, mainframe-centric applications development, client-server grew out of the need to bring computing power and decision making down to the user so that businesses could respond faster to customer needs, competitive pressures, and market dynamics.

Architectural Model

The client-server architecture is not new. A more familiar manifestation of the architecture is the decades-old corporate telephone system, with the Private Branch Exchange (PBX) acting as a server and the telephones acting as the clients. All the telephones derive their features and user access privileges from the PBX, which also processes incoming and outgoing calls. What is relatively new is the application of this model to the LAN environment, which is data-oriented. Here, an application program is broken out into two parts—client and server—that exchange information over the network (Figure C-2).

Client The client portion of the program, or front end, is run by individual users at their desktops and performs such tasks as querying a database, producing a printed report,

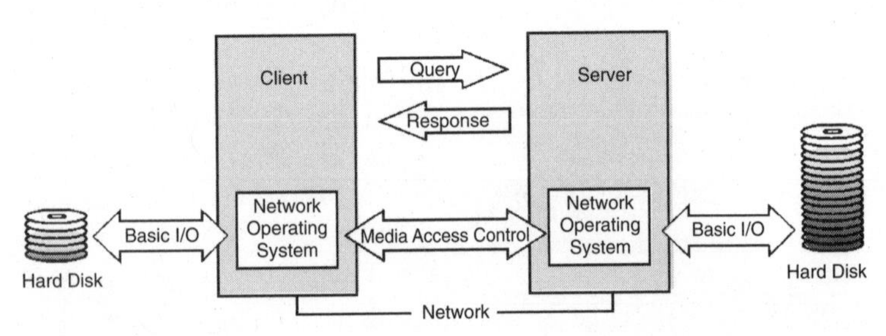

Figure C-2 Simplified model of the client-server architecture.

and entering a new record. These functions are carried out through a database specification and access language, better known as Structured Query Language (SQL), that operates in conjunction with existing applications. The front-end part of the program executes on the user's workstation, drawing on its random access memory (RAM) and central processing unit (CPU).

Server The server portion of the program, or back end, is resident on a computer that is configured to support multiple clients, offering them shared access to numerous application programs as well as to printers, file storage, database management, communications, and other resources. The server not only handles simultaneous requests from multiple clients but also performs such administrative tasks as transaction management, security, logging, database creation and updating, concurrency management, and maintaining the data dictionary. The data dictionary standardizes terminology so that database records can be maintained across a broad base of users.

Network The network consists of the transmission facility—usually a LAN. Among the commonly used media for LANs are coaxial cable (thick and thin), twisted-pair wiring (shielded and unshielded), and optical fiber (single- and multimode). In some cases, wireless links such as infrared and spread spectrum are used to link clients and servers.

A medium access control protocol is used to regulate access to the transmission facility. Ethernet and Token Ring are the two most popular medium access control protocols. When linking client-server computing environments over the wide area network (WAN), other communications facilities come into play, such as private T1 links, which provide a transmission rate of up to 1.544 Mbps. The Internet Protocol (IP) remains the most common transport protocol used over the WAN, while Frame Relay and other fast packet technologies are also growing in popularity.

Client-Server Telephony

One of the newest applications that leverage the client-server architecture is IP telephony. Vendors offer scalable IP-based PBXs that run on the Windows NT platform and can transport voice in packet format over an Ethernet LAN, a managed IP network, or the Public Switched Telephone Network (PSTN) via an IP/PSTN gateway. Full-featured digital phones link directly to the Ethernet LAN via a 10BaseT interface without requiring direct connection to a desktop computer. Phone features can be configured using a Web browser (Figure C-3).

The digital phones have access to the calling features offered through the IP PBX management software running on the LAN server. The call management software supports each IP phone with functions such as call hold, call transfer, call forward, call park, and caller ID. Even such advanced PBX functions as multiple lines per phone or multiple phones per line are determined in software and may be reconfigured from any location through a Web browser.

Summary

To date, the promises behind client-server are somewhat mixed. With client-server networks, administration and management complexity increases while costs become more difficult to track. According to some industry estimates, the total cost of owning a client-server system is about 3 to 6 times greater than it is for a centralized mainframe system, while the software tools for managing and administering client-server cost 2 times more than comparable mainframe tools. Regardless of these factors, many organizations have implemented client-server networks successfully and have achieved significant efficiency and productivity gains.

See also

Advanced Peer-to-Peer Networking
Ethernet
Token Ring

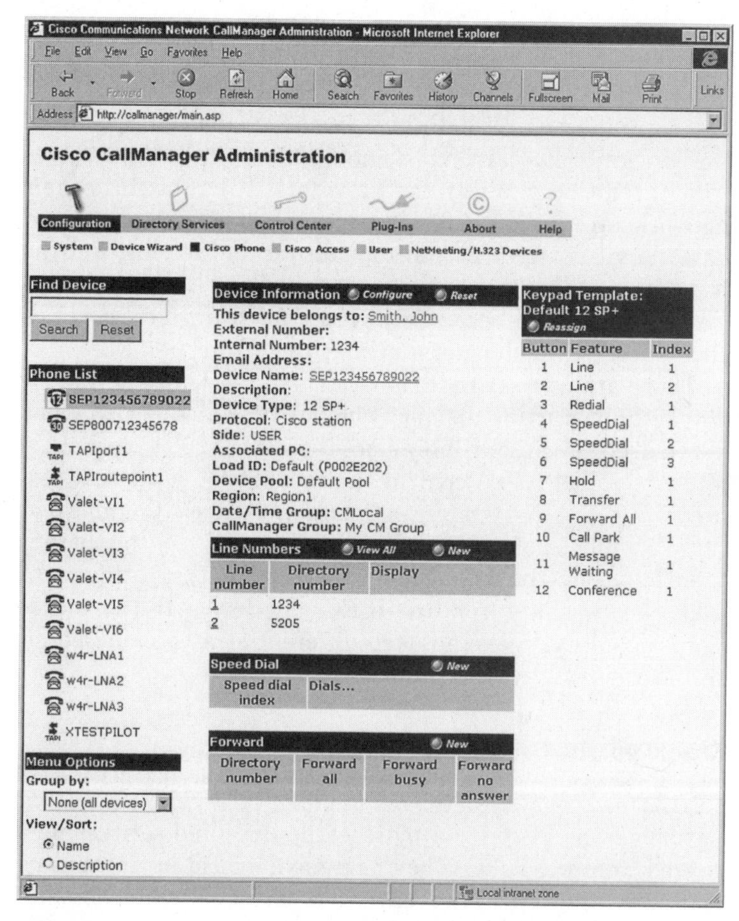

Figure C-3 A view of the administrative interface of Cisco Systems' CallManager, which allows phone features delivered through a server-based IP PBX to be set up from within a Web browser.

COMPETITIVE LOCAL EXCHANGE CARRIERS

Competitive Local Exchange Carriers (CLECs) offer voice, data services, and value-added services at significantly lower prices than the Incumbent Local Exchange Carrier (ILEC), enabling residential and business users to save money on such things as local calls, call handling features, lines, and Internet access. Typically, CLECs offer service in major cities, where traffic volumes are greatest and, consequently, users are hardest hit with high local exchange charges from the incumbent carrier. Some CLECs call themselves Integrated Communications Providers (ICPs) because their networks are designed from the outset to support voice and data services, such as metropolitan Ethernet, as well as Internet access. Others call themselves Data Local Exchange Carriers (DLECs) because they specialize in data services such as DSL, which is used primarily for Internet access.

There are more than 2000 CLECs in the United States—400 of them starting up operations in 1999 alone, with a few dozen declaring bankruptcy since then. As of 2001, the ILECs still controlled 97 percent of the market for local services, according to the Federal Communications Commission (FCC), which means that the CLECs are trying to sustain themselves on the remaining 3 percent as they attempt to take market share from the ILECs. To deal with this situation, the CLECs have adopted different strategies based on resale and facilities ownership.

Resale versus Ownership

CLECs may compete in the market for local services by setting up their own networks or by reselling lines and services purchased from the ILEC. They may have hybrid arrangements for a time, which are part resale and part facilities ownership. Most CLECs prefer to have their own networks because the profit margins are higher than for resale. However, many CLECs start out in new markets as resellers. This enables them to establish a local presence, build brand awareness, and begin building a customer base while they assemble their own facilities-based network.

Although this strategy is used by many CLECs, many fail to carry it out properly. They get into financial trouble by using their capital to expand resale arrangements to capture even more market share instead of using that capital to quickly build their own networks and migrate customers to the high margin facilities. Depending on the service, it could take a carrier 3 to 4 years to break even on a pure resale customer versus only 6 to 9 months on a pure facilities-based customer. With capital markets drying up for telecom companies and customers deferring product and services purchases, prolonged dependence on resale could set the stage for bankruptcy.

CLECs employ different technologies for competing in the local services market. Some set up their own Class 5 central office switches, enabling them to offer "dial tone" and the usual voice services, including Integrated Services Digital Network (ISDN), as well as features such as caller ID and voice messaging. The larger CLECs build their own fiber rings to serve their metropolitan customers with high-speed data services. Some CLECs have chosen to specialize in broadband data services by leveraging existing copper-based local loops, offering DSL services for Internet access. Others bypass the local loop entirely through the use of broadband wireless technologies, such as Local Multipoint Distribution Service (LMDS), enabling them to feed customer traffic to their nationwide fiber backbone networks without the incumbent carrier's involvement.

Despite the risks, some CLECs view resale as a viable long-term strategy. It not only allows them to enter into new markets more quickly than if they had initially deployed their own network, but it also reduces initial capital requirements in each market, allowing them to focus capital resources initially on the critical areas of sales, marketing, and operations support systems (OSS). In addition, the strategy allows them to avoid deployment of conventional circuit switches and maintain design flexibility for the next generation of telecommunications technology.

Unfortunately, the resale strategy also results in lower margins for services than for facilities-based services. This means the CLEC must pass much of its customer revenues back to the ILEC to pay the monthly fees for access lines. When investors stopped stressing market growth over profits in 2000, these CLECs found that capital was hard to get. By then, many had no money to invest in their own facilities, where margins are greater. Most financial analysts doubt that CLECs can rely strictly on resale and survive. Although the ILECs have a vested interest in survival of some resale CLECs in order to receive regulatory approval to provide in-region long distance, once that approval is gained, some analysts believe that the ILECs may have no further interest in cooperating with the CLECs.

Summary

With the Telecommunications Act of 1996, CLECs and other types of carriers are allowed to compete in the offering of local exchange services and must be able to obtain the same service and feature connections as the ILECs have for themselves—and on an unbundled basis. If the ILEC does not meet the requirements of a 14-point checklist to open up its network in this and other ways, it cannot get permission from the FCC to compete in the market for long-distance services. In the 5 years (1996–2001) since the Telecom Act was passed by Congress and signed into law by President Clinton, only two ILECs have qualified for limited entry into the long-distance market—Verizon and SBC Communications.

See also

Building Local Exchange Carriers
Interexchange Carriers

COMPUTER-TELEPHONY INTEGRATION

Computer-Telephony Integration (CTI) combines the use of the advanced call-processing capabilities of digital telephone systems and the open application environment of PCs and LANs via intelligent computer-to-telephone system interfaces to increase productivity and customer response. The resulting benefits are especially suited to telephone-intensive environments such as customer service, technical support, and telemarketing. Today's CTI systems are quite sophisticated and can handle a variety of incoming and outgoing communications, including phone calls, faxes, Internet messages, and Web content.

In rudimentary form, CTI has been around for more than a decade. The original CTI approach was to link the PBX with a host, such as a mainframe or minicomputer. With today's LAN capabilities, a server with a database management system can act as the host, bringing integrated voice-data applications to every agent desktop. It is not just database records that can be matched with a customer call for delivery to the desktop, but e-mail, text chat, and voice calls from Web sessions as well.

Many of the attributes of CTI trace their lineage from the call center environment; specifically

- *Automatic Call Distributor (ACD)* Manages incoming calls in a variety of ways, including holding them in queue and parceling them out to the next available agent.
- *Automatic Number Identification (ANI)* Provides the system with the telephone number of the incoming call to identify the caller.
- *Database Matching* Provides the means to look up customer data, based on ANI, for delivery to the agent by the time the call is answered.
- *Call Accounting* Entails the collection of call-related information for cost containment, internal billing, trend analysis, and agent performance.

Today's concept of CTI takes this integration further, treating voice as simply another data type that can be manipulated by the user. In this integrated environment, voice is a messaging format on a par with electronic mail, facsimile, and even paper. Once a link to a voice-processing system or PBX is in place, it becomes possible to work with voice-mail and other telephone functions as simply desktop applications, tying them into e-mail messaging systems and creating entirely new categories of applications that are telephone-enabled.

Applications

The CTI architecture allows one or more computer applications to communicate with the telephone switch, either an ACD or PBX. Among the possible applications of CTI are

- *Inbound call information* Information passed from the telephone network to the telephone switch, such as the caller's telephone number and the number dialed, is passed to computer applications. Applications can then identify the caller (by the calling number) and the purpose of the call (from the number dialed). This allows the application to automatically deliver caller information and data specific to the purpose of the call to a workstation as the telephone rings.
- *Computerized call processing* Commands passed from computer applications instruct the telephone system to perform call-processing functions such as make a call, answer a call, or other call functions. This allows for application controlled call routing based on inbound call information and numbers in a computer database.
- *Outbound calling* CTI increases productivity in outbound calling environments. With predictive dialing systems acting off of call lists in a database, agents proceed from one active call to the next. No time is wasted manually dialing numbers or listening to busy signals, unanswered ringing, or operator messages.

In these and other cases, computer applications use a call-processing server and application programming interfaces (APIs) to originate, answer, and manipulate calls. The call-processing server interfaces to the telephone switch and invokes the required function, as requested by the client applications. The server keeps track of call status information on the telephone switch side and session status on the application side, making the logical association between the two. The most advanced systems even track the pace at which call center agents complete transactions over the phone so that automatic dialing activity can be speeded up or slowed down.

The Role of APIs

The challenge for CTI vendors has been to come up with a standardized way of allowing developers to build and implement integrated voice-data solutions that work across vendor domains. Their strategy has been to establish standardized APIs that work across vendor boundaries.

Intel and Microsoft developed the Telephony Application Programming Interface (TAPI), which is intended to create a single specification for Windows application developers to use in connecting their products to the telephone network. At the LAN server level, AT&T and Novell offer their jointly developed NetWare Telephony Services API, which is intended to make NetWare the platform for this kind of integration. Sun Microsystems offers a Java-based API for developing CTI applications.

Telephone Application Programming Interface TAPI allows custom applications to be built around inexpensive PCs; specifically, the Windows Telephony API provides a standard development interface between PCs and myriad telephone network APIs. TAPI is intended to insulate software developers from the underlying complexity of the telephone network. It allows developers to focus entirely on the application without having to take into account the type of telephone connection: PBX, ISDN, Centrex, cellular, or plain old telephone service (POTS). They can specify the features they want to use without worrying about how the hardware is ultimately linked.

Application Classes TAPI facilitates the development of three classes of Windows applications. The first class of applications offers telephone-enabled versions of existing applications. TAPI creates standard access to telephone functions such as call initiation, call answering, call hold, and call transfer for Windows applications. TAPI addresses only the control of the call, not its content. However, the specification can be applied to any type of call, whether voice, data, fax, or even video.

The second class of telephone-centric applications might embrace visual call control or telephone-based conferencing and collaborative computing. Although such applications have long been available, they relied on incompatible APIs. The third class of applications enables the telephone to act as an input-output device for audio data, including voice across data networks.

Application Components An actual TAPI product implementation comprises three distinct components:

- The TAPI-aware application.
- A TAPI dynamic link library (DLL).
- One or more Windows drivers to interface to the telephone hardware.

A TAPI application is any piece of software that makes use of the telephone system, such as a personal information manager (PIM) that could dial phone numbers automatically. An application becomes TAPI-compliant by writing to the applications programming interfaces defined in the TAPI specification.

The TAPI DLL is another major component. The application talks to the DLL using the standard APIs. The DLL translates those API calls and controls the telephone system using the device driver.

The final component is the Service Provider Interface (SPI), which is a driver that is unique to each TAPI hardware product. The TAPI specification supports more than one type of telephone adapter. In turn, the adapters can support more than one line.

TAPI-Enabled Features TAPI facilitates the development of applications that allow the user to control the telephone from a Windows PC. A number of possible control features are possible, including

- *Visual call control* Provides a Windows interface to such common PBX functions as call hold, call transfer, and call conferencing. Replacing difficult-to-remember dialing codes with Windows icons will make even the most complicated telephone system functions easy to implement.
- *Call filtering* In conjunction with ANI, this function allows the user to specify the telephone numbers allowed to get through. All others will be routed to an attendant, message center, or voice mailbox. Or the call can be automatically forwarded to another extension while the user is out of the office.
- *Custom menu systems* Allows users to build menu systems to help callers find the right information, agent, or department. Using the drag-and-drop technique, the menu system can be revised daily to suit changing business needs. The menu system can be interactive, allowing the caller to respond to voice prompts by dialing different numbers. A different voice message can be associated with each response. Voice messages can be created instantly via the PC's microphone.

NetWare Telephony Services API While TAPI defines the connection between a single phone and a PC, the NetWare Telephony Services API (TSAPI) defines the connection between a networked file server and a PBX. TSAPI is the result of a joint effort by Novell and AT&T to integrate computer and telephone functions at the desktop using a logical connection established over the LAN.

In connecting a NetWare server to the PBX, individual PCs are given control over telephone system functions.

TSAPI is implemented with NetWare Loadable Modules (NLMs) that run on Novell servers, along with another NLM containing a PBX driver. No special hardware is required at the desktop; the PBX supports its own physical connection and uses its own software. The physical link is an ISDN Basic Rate Interface (BRI) card in the server that allows for the connection between the NetWare server and the PBX.

NetWare Telephony Services consists of a Telephony Server NLM, a set of dynamic link libraries (DLLs) for the client, and a sample server application (a simple point-and-click telephone listing that is integrated with directory services). Novell also offers a driver for every major PBX. Alternatively, users can obtain a driver from their PBX vendor.

The NLM's features include drag-and-drop conference calling; the ability to put voice, facsimile, and electronic mail messages in one mailbox; third-party call control; and integration between telephones and computer databases. Noteworthy among these is third-party call control, which provides the ability to control a call without being a part of it. This feature would be used for setting up a conference call, for example.

Unlike Microsoft's TAPI, which allows only first-party call control, third-party constructs are an integral part of NetWare Telephony Services. The command "Make Call," for example, has two parameters: one for addressing the originating party and the other for addressing the destination party. An application using this command therefore would allow users to designate an address different from their own as the originating party and establish a connection without becoming a participant in the call. This third-party call control also lets users set up automatic routing schemes.

Java Telephony Application Programming Interface The platform independence of Java can be exploited for CTI applications; specifically, the Java Telephony Application Programming Interface (JTAPI) offers the means to build applications that will run on a variety of operating systems and hardware platforms over a variety of telephony networks.

JTAPI defines a reusable set of telephone call control objects, which enables application portability across computer platforms. The scalability of JTAPI enables it to be implemented on devices ranging from handheld phones to desktop computers to large servers. This allows enterprises to blend together Internet and telephony technology components within a single application environment as they design and deploy new business strategies for improving customer service levels, including launching their presence on the Web.

JTAPI is composed of a set of Java language packages. Each package provides a specific piece of functionality for a certain aspect of computer-telephony applications. Implementations of telephony servers choose the packages they support, depending on the capabilities of their underlying platform and hardware. Applications may query for the packages supported by the implementation they are currently using. Additionally, application developers may concern themselves with only the supported packages the application need to accomplish a task.

At the center of the Java Telephony API is the "core" package. The core package provides the basic framework to model telephone calls and rudimentary telephony features. These features include placing a telephone call, answering a telephone call, and disconnecting a telephone call. Simple telephony applications need to use the core to accomplish their tasks and do not need to concern themselves with the details of other packages. For example, the core package permits applet designers to easily add telephone capabilities to a Web page.

A number of standard packages extend the JTAPI core package. Among the extension packages are those for call control, call center, media, phone, private data, and capabilities:

- *Call control* Extends the core package by providing more advanced call-control features such as placing calls on hold, transferring telephone calls, and conferencing telephone calls.

- *Call center* Provides applications the ability to perform advanced features necessary for managing large call centers, such as routing, automated call distribution, predictive calling, and associating application data with telephony objects.
- *Media* Provides applications access to the media streams associated with a telephone call. They are able to read and write data from these media streams. DTMF (touch-tone) and non-DTMF tone detection and generation are also provided by this package.
- *Phone* Permits applications to control the physical features of telephone sets.
- *Private data* Enables applications to communicate data directly with the underlying hardware switch. These data may be used to instruct the hardware to perform a switch-specific action.
- *Capabilities* Allows applications to query whether certain actions may be performed. Capabilities take two forms: static capabilities indicate whether an implementation supports a feature, and dynamic capabilities indicate whether a certain action is allowed given the current state of the call.

JTAPI also defines call model objects that work together to describe telephone calls and the end points involved in a telephone call:

- *Provider* This object might manage a PBX connected to a server, a telephony/fax card in a desktop machine, or a computer networking technology such as IP. It hides the service-specific aspects of the telephony subsystem and enables Java applications and applets to interact with the telephony subsystem in a device-independent manner.
- *Call* This object represents a telephone call. In a two-party call, a telephone call has one Call object and two connections. A conference call is three or more connections associated with one Call object.

- *Address* This object represents a telephone number.
- *Connection* This object models the communication link, which is the relationship between a Call object and an Address object.
- *Terminal* This object represents a physical device such as a telephone and its associated properties. Each Terminal object may have one or more Address objects (telephone numbers) associated with it, as in the case of some office phones capable of managing multiple line appearances.
- *Terminal Connection* This object models the relationship between a Connection object and the physical end point of a call, which is represented by the Terminal object.

Applications built with JTAPI use the Java "sandbox" model for controlling access to sensitive operations. Callers of JTAPI methods are categorized as "trusted" or "untrusted" using criteria determined by the runtime system. Trusted callers are allowed full access to JTAPI functionality. Untrusted callers are limited to operations that cannot compromise the system's integrity. In addition, JTAPI may be used to access telephony servers or implementations that provide their own security mechanisms, such as user name and password.

Summary

CTI removes the barriers between telephony and other information and productivity tools, providing users with substantial gains in efficiency and information management in an easy-to-use environment. Under the CTI concept, the most appropriate pieces of technology are combined in practical applications for a more productive workplace. Vendors of CTI solutions typically provide a complete package, including networking platform, application software, and business support services.

See also

LAN Telephony

D

DATA COMPRESSION

Data compression, a standard feature of most bridges and routers, as well as modems, improves throughput by capitalizing on the redundancies found in the data to reduce frame size and thereby allow more data to be transmitted over a link. An algorithm detects repeating characters or strings of characters and represents them as a symbol or token. At the receiving end, the process works in reverse to restore the original data.

There are many different algorithms available to compress data, and they are designed for specific types of data sources and the redundancies found in them but do a poor job when applied to other sources of data. For example, the Moving Pictures Experts Group (MPEG) compression standards were designed to take advantage of the relatively small difference from one frame to another in a video stream and so do an excellent job of compressing motion pictures. On the other hand, MPEG would not be effective if applied to still images. For this data source, the Joint Photographic Experts Group (JPEG) compression standards would be applied.

JPEG is "lossy," meaning that the decompressed image is not quite the same as the original compressed image—there is some degradation. JPEG is designed to exploit known limitations of the human eye, notably that small color details are not perceived as well as small details of light and dark. JPEG eliminates the unnecessary details to greatly reduce the size of image files, allowing them to be transmitted faster and take up less space in a storage server.

On wide area network (WAN) links, the compression ratio tends to differ by application. The compression ratio can be as high as 6 to 1 when the traffic consists of heavy-duty file transfers. The compression ratio is less than 4 to 1 when the traffic is mostly database queries. When there are only "keep alive" signals or sporadic query traffic on a T1 line, the compression ratio can dip below 2 to 1.

Encrypted data exhibit little or no compression because the encryption process expands the data and uses more bandwidth. However, if data expansion is detected and compression is withheld until the encrypted data are completely transmitted, the need for more bandwidth can be avoided.

The use of data compression is particularly advantageous in the following situations:

- When data traffic is increasing due to the addition or expansion of local area networks (LANs) and associated data-intensive, bursty traffic.
- When LAN and legacy traffic are contending for the same limited bandwidth.
- When reducing or limiting the number of 56/64-kbps lines is desirable to reduce operational costs.
- When lowering the Committed Information Rate (CIR) for Frame Relay services or sending fewer packets over an X.25 network can result in substantial cost savings.

The greatest cost savings from data compression occurs most often at remote sites, where bandwidth is typically in short supply. Data compression can extend the life of 56/64-kbps leased lines, thus avoiding the need for more expensive fractional T1 lines or $N \times 64$ services. Depending on the application, a 56/64-kbps leased line can deliver 112 to 256 kbps or higher throughput when data compression is applied.

Types of Data Compression

There are several different data compression methods in use today over WANs—among them are Transmission Control Protocol/Internet Protocol (TCP/IP) header compression, link compression, and multichannel payload compression. Depending on the method used, there can be a significant tradeoff between lower bandwidth consumption and increased packet delay.

TCP/IP Header Compression With TCP/IP header compression, the packet headers are compressed, but the data payload remains unchanged. Since the TCP/IP header must be replaced at each node for IP routing to be possible, this compression method requires hop-by-hop compression and decompression processing. This adds delay to each compressed/decompressed packet and puts an added burden on the router's CPU at each network node.

TCP/IP header compression was designed for use on slow serial links of 32 kbps or less, and to produce a significant performance impact, it needs highly interactive traffic with small packet sizes. In such traffic, the ratio of Layer 3 and 4 headers to payload is relatively high, so just shrinking the headers can result in a substantial performance improvement.

Payload Compression Payload compression entails compression of the payload of a Layer 2 WAN protocol, such as Point-to-Point Protocol (PPP), Frame Relay, High-Level Data Link Control (HDLC), X.25, and Link Access Procedure, Balanced (LAPB). The Layer 2 packet header is not compressed, but the entire contents of the payload, including higher-layer-protocol headers (i.e., TCP/IP), are compressed. They are compressed using the industry standard Lemple-Ziv algorithm or some variation of that algorithm.

Layer 2 payload compression applies the compression algorithm to the entire frame payload, including the TCP/IP headers. This method of compression is used on links operating at speeds from 56 to 1.544 Mbps and is useful on all traffic types, as long as the traffic has not been compressed previously by a higher-layer application. TCP/IP header compression and Layer 2 payload compression, however, should not be applied at the same time because it is redundant and wasteful and could result in the link not coming up to not passing IP traffic.

Link Compression With link compression, the entire frame—both protocol header and payload—is compressed. This form of compression is typically used in LAN-only or legacy-only environments. However, this method requires error-correction and packet-sequencing software, which adds to the processing overhead already introduced by link compression and results in increased packet delays. Also, like TCP/IP header compression, link compression requires hop-by-hop compression and decompression, so processor loading and packet delays occur at each router node the data traverses.

With link compression, a single data compression vocabulary dictionary or history buffer is maintained for all virtual circuits compressed over the WAN link. This buffer holds a running history about what data have been transmitted to help make future transmissions more efficient. To obtain optimal compression ratios, the history buffer must be large, requiring a significant amount of memory. The vocabulary dictionary resets at the end of each frame. This technique offers lower compression ratios than multichannel, multihistory buffer (vocabularies) data compression methods. This is particularly true when transmitting mixed LAN and serial protocol traffic over the WAN link and frame sizes are 2 kilobytes or less. This translates into higher costs, but if more memory is added to get better ratios, this increases the upfront cost of the solution.

Mixed-Channel Payload Data Compression By using separate history buffers or vocabularies for each virtual circuit, multichannel payload data compression can yield higher compression ratios that require much less memory than other data compression methods. This is particularly true in cases where mixed LAN and serial protocol traffic traverses the network. Higher compression ratios translate into lower WAN bandwidth requirements and greater cost savings.

But performance varies because vendors define payload data compression differently. Some consider it to be compression of everything that follows the IP header. However, the IP header can be a significant number of bytes. For overall compression to be effective, header compression must be applied. This adds to the processing burden of the CPU and increases packet delays.

External Data Compression Solutions Bridges and routers can perform data compression with optional software or add-on hardware modules. While software-based compression capabilities can support fractional T1/E1 rates, hardware-based compression offloads the bridge/router's main processor to deliver even higher levels of throughput. With a compression ratio of up to 4 to 1, a data compression module can support up to 16 Mbps of compressed data throughput without imposing additional traffic latency. This is enough to keep four T1/E1 circuits full of compressed data in both directions simultaneously.

The use of a separate digital signal processor (DSP) for data compression, instead of the software-only approach, enables the router to perform all its core functions without any performance penalty. This parallel-processing approach minimizes the packet delay that can occur when the router's CPU is forced to handle all these tasks by itself.

If there is no vacant slot in the bridge/router for the addition of a data compression module, there are two alternatives: the software-only approach or an external compression device. The software-only approach could bog down the overall performance of the router, since its processor would be used to implement compression in addition to core functions. Although an external data compression device would not bog down the router's core functions, it means that one more device must be provisioned and manages at each remote site.

Summary

Data compression will become increasingly important to most organizations as the volume of data traffic at branch locations begins to exceed the capacity of the wide area links. Multichannel payload solutions provide the highest compression ratios and reduce the number of packets transmitted across the network. Reducing packet latency can be achieved effectively via a dedicated processor like a DSP and by employing end-to-end compression techniques rather than node-to-node compression/decompression. All these factors contribute to reducing WAN circuit and equipment costs as well as improving the network response time and availability for user applications.

See also

LAN Telephony

DATA SWITCHES

Data switches have been in mainframe computer environments since their introduction in 1972 and were the forerunners of today's hubs and LAN switches. Through port selection or port contention, data switches permit a larger number of users to share a limited number of host ports. Data switches also perform the necessary protocol conversions that allow PCs to communicate with the mainframe as terminals. They have evolved to become a very economical means of controlling access to the mainframe via partitioning (Figure D-1) and other security features. Typically, these switches support data transmission speeds of up to 19.2 kbps for asynchronous data and 64 kbps for synchronous data.

Connectivity

As the architecture of data switches became more modular, they permitted incremental growth of up to thousands of

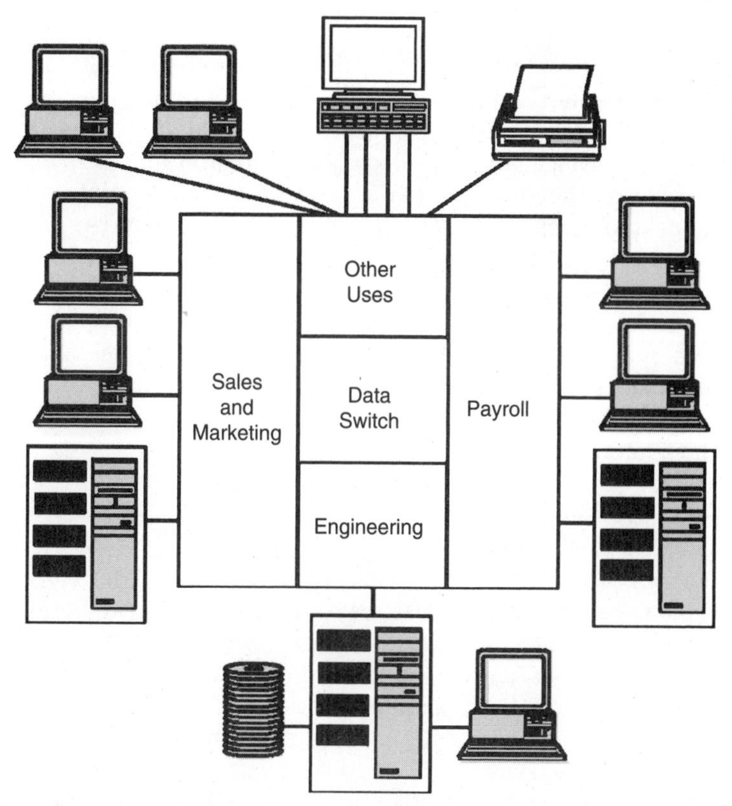

Figure D-1 PC-to-mainframe connectivity, with partitioning, implemented by a data switch.

computer ports worldwide. The data switch also can serve as a LAN server, or gateway, to packet and T1 networks for remote host access.

PCs and other devices may be connected to the data switch in a variety of ways, including direct connection with EIA-232C cabling for distances of up to 50 feet. For longer distances of 1 to 2 miles, between buildings, for example, PCs and peripherals may be connected to the data switch via line drivers or local multiplexers. Some data switch vendors have integrated these functions into their data switches in the form of optional plug-in cards. Connections between PCs and the data switch are usually accomplished via the extra twisted pairs of telephone wire already in place in most office environments. Remote data switches may be accessed via dedicated or dial-up phone lines.

When a user enters the connect command and personal password, the data switch will attempt to complete the requested connection. If the requested port is busy, the user is put into queue and notified of changes in position with a screen message. Some users may be assigned a higher priority in the queue than others. When a high-priority user attempts to access a busy port, other waiting users are bumped, and the priority user assumes first place in the queue. All other contenders for that port are then notified of their new status.

Features

Not only do some data switches permit different configurations to be loaded, they also can implement one or the other automatically on a scheduled basis. A late-night crew of CAD/CAM people can plod along with cumbersome design work on high-performance workstations, for example, and be restricted to accessing the mainframe from 6 P.M. to 8 A.M. The class-of-service (CoS) feature keeps these users from being able to access sensitive financial information that also may be stored in the mainframe. When the workday begins at 8 A.M., the time-of-day clock changes the data switch back to the primary configuration for general access—but without interrupting existing connections. This enhances overall network performance in that available resources may be reallocated during the day for optimal usage based on the varying needs of different classes of users. Various alternative configurations may be stored on disk and implemented by an authorized PC with only a few keystrokes.

A session toggling feature enhances operator productivity by permitting two connections—one designated primary and the other designated secondary. The operator can toggle back and forth between two host sessions on different links to perform multiple tasks simultaneously. For example, a batch file transfer can be in progress over the secondary link while a real-time database search is being performed over the primary link. When the batch file transfer is completed, the primary link can be put on hold while another file transfer is initiated over the secondary link.

The data switch automatically adapts to different transmission rates. The network administrator does not have to match terminals with computer ports; each computer port may be set at its highest rate. The data switch uses a buffer to perform the rate conversion for any device that communicates with another device at a faster or slower rate. This means that users do not have to be concerned about speed and network administrators do not have to waste time changing the transmission speeds of computer ports to accommodate lower-speed devices. Thus a computer port set at 19.2 kbps may send data to a much slower device. But for reliable data rate conversion, the connecting devices must be capable of flow control; if not, there is a risk of losing data.

When XON/XOFF is used for flow control, the switch buffer will be prevented from overflowing. When the buffer is in danger of overflowing, an XOFF signal is sent to the computer, telling it to suspend transmission. When the buffer clears, an XON signal is sent to the computer, telling it to resume transmission. These settings are also used for reformatting character structures, enabling devices of different manufacturers to communicate with each other through the data switch.

Administration

Instead of being confined to one terminal, many data switches allow the network administrator to log onto the computer from any terminal. Once connected, there are a variety of functions that can be invoked to enhance operating efficiency.

A broadcast feature allows the network administrator to transmit messages to individual users, with delivery controlled by the time-of-day clock. The same message may be sent every hour, or a different message may be sent to different users simultaneously.

A special link feature lets the network administrator make permanent connections between a terminal and a port. Sometimes called "nail-up," this feature allows the continuous access to certain devices like printers.

With the force disconnect feature, the network administrator can disconnect any port, any time, for any reason. Open files are even closed automatically with the proper disconnect sequences, including any required control characters. This capability is also found in a time-out feature that enhances system efficiency by automatically disconnecting idle ports after a predefined period of inactivity.

When the data switch is equipped with a logging port, the network administrator can obtain a complete record of port connects and disconnects to aid in maintaining security. When used with the optional security call back, this feature provides a precise audit trail of connections, as well as the users who made them. In addition, all alarms may be logged for output to a PC or designated port on the host to aid in network analysis and management.

The network administrator may decide to group ports with similar characteristics and capabilities under a class name to permit a more efficient utilization of shared resources. Password-protected classes also can be used to restrict access and enhance system security. The CoS may designate the use of particular host ports for general access, while other classes may designate a high-speed printer, a modem pool, or a gateway to a packet network.

Summary

Data switches were advocated as an economical alternative to LANs when host connectivity needs were relatively simple and LANs were still expensive. They also made a suitable LAN server when networking needs grew more complex and the cost of LANs became more affordable. Later gateway functions were added to data switches through plug-in cards that provided the appropriate protocol converters. This permitted asynchronous or synchronous terminals from one vendor to communicate with the mainframes and network nodes provided by other vendors. It also permitted the data switch to connect to various types of LANs. Although data switches are still available, advances in hub technology and, more recently, LAN switches have overtaken them to the point that they are rarely considered for new data centers.

See also

Hubs

LAN Switching

DIGITAL SUBSCRIBER LINE TECHNOLOGIES

Digital Subscriber Line (DSL) is a category of local loop technologies that turns an existing twisted-pair line, normally used for plain old telephone service (POTS), into a high-speed digital line for Internet access. The electronics at both ends of the local loop compensate for impairments that normally would impede high-speed data transmission over ordinary twisted-pair copper lines. This enables Incumbent Local Exchange Carriers (ILECs) and their competitors to offer high-speed connections over which a variety of advanced broadband data services and value-added applications can be offered, including Web surfing, news feeds, virtual private network (VPN) access, and in some cases, videoconferencing and telephone calls.

DSL provides an economical way to satisfy surging demand among businesses and consumers for huge amounts of bandwidth, which is especially needed for Internet access. In leveraging existing copper local loops, DSL obviates the need for a huge capital investment to bring fiber to the customer premises or to the curb in order to offer broadband services. Instead, idle twisted pairs to the customer premises can be provisioned to support high-speed data services. Users are added simply by installing DSL access products at the customer premises and connecting the DSL line to the appropriate voice or data network switch via a DSL concentrator at the central office (CO) or a serving wire center (SWC) where the data and voice are split out for distribution to the appropriate network (Figure D-2).

There are about a dozen DSL technologies currently available—each optimized for a given level of performance relative to the distance of the customer premises from the CO or serving wire center SWC. The farther away the customer is from the CO or SWC, the lower is the speed of the DSL in both the upstream (toward the network) and downstream (toward the user) directions. The closer the customer location is from the CO or SWC, the greater is the speed in both directions.

Common Characteristics of DSL

Regardless of the specific type of technology used to implement DSL, the different varieties share some common characteristics:

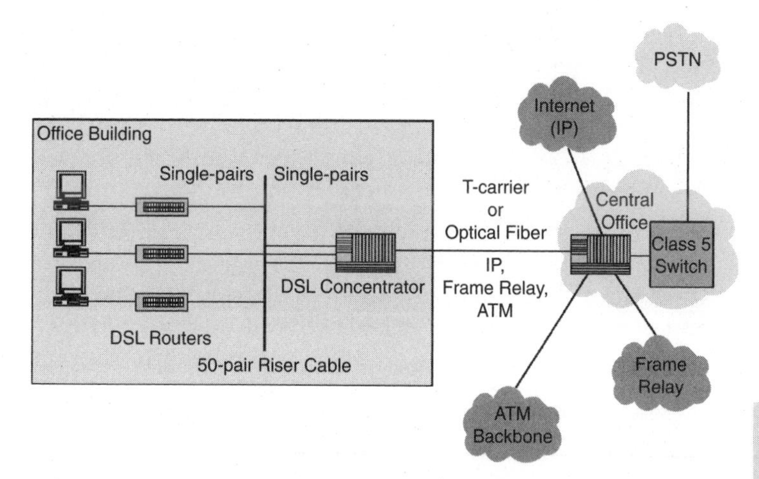

Figure D-2 Each customer location is equipped with one or more DSL modems. DSL lines from multiple customers are concentrated at a DSL Access Multiplexer (DSLAM) that can be positioned in a building's equipment room or at the central office where voice and data are split out for delivery to appropriate networks.

- All DSL services are provisioned over the same unshielded twisted-pair copper wiring that is commonly used for POTS. Thus no special connection must be installed to the customer premises to obtain this type of service.
- All the DSL varieties offer a means to turn a low-quality voice-grade POTS line into a high-quality broadband data line. Electronic equipment at both ends of the POTS lines adapts or compensates for impairments that normally would corrupt high-speed data transmission over ordinary twisted-pair copper lines. The method by which this "line conditioning" is achieved differs by equipment vendor, which helps account for the slight variance in maximum line speeds for the same DSL service at comparable distances. Other factors that account for these variances include the specific gauge of the wire [e.g., 24 American Wire Gauge (AWG) versus 26 AWG] over which the DSL service is provisioned.

- All the DSL technologies conform to a similar configuration of user and service provider equipment. At a telecommuter's home, a branch office, or other corporate facility, DSL access requires a copper phone line that is connected to a DSL modem, which is actually a router. At the service provider's CO or SWC, concentration/server equipment connects multiple users and passes the transmissions to their respective voice and data networks.
- Once the DSL modem or router is connected to the DSL and is put into service by the carrier, it remains continuously available to the user without the need to dial up every time access is required—the service is "always on," just like a LAN. To use the service to access the Internet, for example, users merely open their Web browser.
- DSL access concentrators at the CO or SWC help carriers relieve congestion in their voice switching systems. This equipment partitions voice and data traffic, directing data onto a separate packet, frame, or cell-based data network, and directing voice onto the Public Switched Telephone Network (PSTN).
- DSL is inherently more secure than other access technologies. DSL provides a dedicated point-to-point connection to the network that cannot be accessed by others, except by physically tapping into the line. On cable TV networks, for instance, many subscribers share the same cable for Internet access, so there is always the possibility of an intrusion by hackers. To guard against this, cable customers must purchase their own firewall software and have the technical expertise to configure and manage it.
- DSL is also very flexible in the types of traffic data formats it can accommodate. In addition to voice, DSL can be used to transport IP, Frame Relay, and Asynchronous Transfer Mode (ATM) traffic from the customer premises through the local loop and to the appropriate network, including the PSTN, the Internet, or a corporate VPN.

When it comes to price, DSL is far more economical than other digital technologies, such as T1. Like T1, DSL is priced at a flat rate per month with unlimited hours of access. While T1 access circuits cost anywhere from $150 to $700 per month or more, depending on market, DSL can cost much less for comparable bandwidth—as low as $29.95 a month for 256 kbps to $250 a month for 7 Mbps.

Asymmetric versus Symmetric

The bandwidth available over the DSL is carved up in a variety of ways to meet the needs of particular applications. When the upstream and downstream speeds are different, the DSL is referred to as "asymmetric"; that is, much greater bandwidth is available in the downstream direction than the upstream direction, as in the case of Asymmetric Digital Subscriber Line (ADSL).

ADSL runs at up to 8 Mbps in the downstream direction and up to 640 kbps in the upstream direction, with the actual speeds depending on the distance of the customer location to the CO or SWC. This would meet the needs of Internet users who want to retrieve multimedia Web content very quickly without waiting an indeterminate period for the pages to be loaded to their computer, as is the case with dial-up modem connections. The lower upstream speed of ADSL is more than adequate for issuing information requests from the computer to the network, since simple queries usually do not require more than 16 kbps. This leaves enough bandwidth capacity to handle multiple voice channels as well as data.

Asymmetric operation is fine for such activities as Web surfing, but it may not be appropriate for applications like server mirroring, where huge amounts of data must be able to flow in both directions. Symmetrical Digital Subscriber Line (SDSL) service, on the other hand, offers the same amount of bandwidth in both the upstream and downstream directions—160 kbps to about 2 Mbps is available in each direction, depending on the distance of the user's location to the CO or SWC. SDSL and other symmetric DSL services, such as High Bit Rate DSL (HDSL), are more suited to applications that once required a T1/E1 line. Among the popular symmetric applications are videoconferencing, interactive distance learning, and telecommuting.

Two versus Four Wire

Traditionally, voice has been handled in the local loop via two wires (i.e., a "twisted pair"). This continues to be a very economical way to provide millions of residential customers with POTS. Businesses, however, require better-quality local loops for high-speed digital communication. This is achieved by providing them with four-wire connections for such services as T1. Businesses pay a premium price for these connections.

But with the growing popularity of the Internet, even residential customers have a need for more bandwidth than can be obtained with 56-kbps modems and basic rate Integrated Services Digital Network (ISDN). Where the infrastructure is almost exclusively two-wire, new DSL technologies have been developed that are capable of providing bandwidth in the multimegabit-per-second range without requiring expensive upgrades of the local loop.

Increasingly, two-wire DSL solutions are moving into the business sector, and their performance now exceeds the previous performance levels of traditional four-wire solutions. This means that carriers can offer services to twice the number of businesses or double the transmission speed to the same businesses—without incurring major local loop upgrade costs.

There are other advantages to offering two-wire DSL solutions to businesses. Provisioning T1 service requires the installation of repeaters every 4000 to 6000 feet to boost signal strength. This is an expensive, time-consuming task for the carrier, which inflates the costs of T1 service. Some two-wire DSL technologies, such as HSDSL two-wire service (HDSL2), match or exceed the performance of T1 without the need for repeaters.

Rate Adjustment

A version of ADSL is available that adjusts dynamically to varying lengths and qualities of twisted-pair local access lines. Like ADSL, Rate Adaptive DSL (RADSL) delivers a high-capacity downstream channel and a lower-speed upstream channel while simultaneously providing POTS over standard copper loops. Unlike ADSL, which does not tune itself to changing line conditions, RADSL adjusts data rates up or down in much the same way ordinary modems do.

With RADSL, it is also possible to connect over different lines at varying speeds. Connection speed can be determined when the line synchs up, while the connection is active, or as the result of a signal from the CO.

Inverse Multiplexing

Multiple DSL lines can be bonded together to provide users with higher-speed services. For example, two-wire SDSL, which tops out at 2 Mbps, can be bonded with another two-wire SDSL to offer an access speed of up to 4 Mbps. With this much bandwidth, small to medium-sized companies can get the transmission capacity they need without resorting to more expensive four-wire T1 access lines.

The bonding arrangement for DSL requires the user to have multiple telephone lines over which the higher-speed service can be run. No additional hardware is required; a simple software change to the inverse multiplexers in the service provider's network implements the bonding process. Once the DSL lines are bonded together in the service provider's network, the user's data load is balanced across the active lines.

Service Provisioning

Most DSL service providers offer customers the means to check for local DSL availability via forms posted on their Web sites, which prompt for the address and phone number of the DSL location (Figure D-3). If DSL service is available,

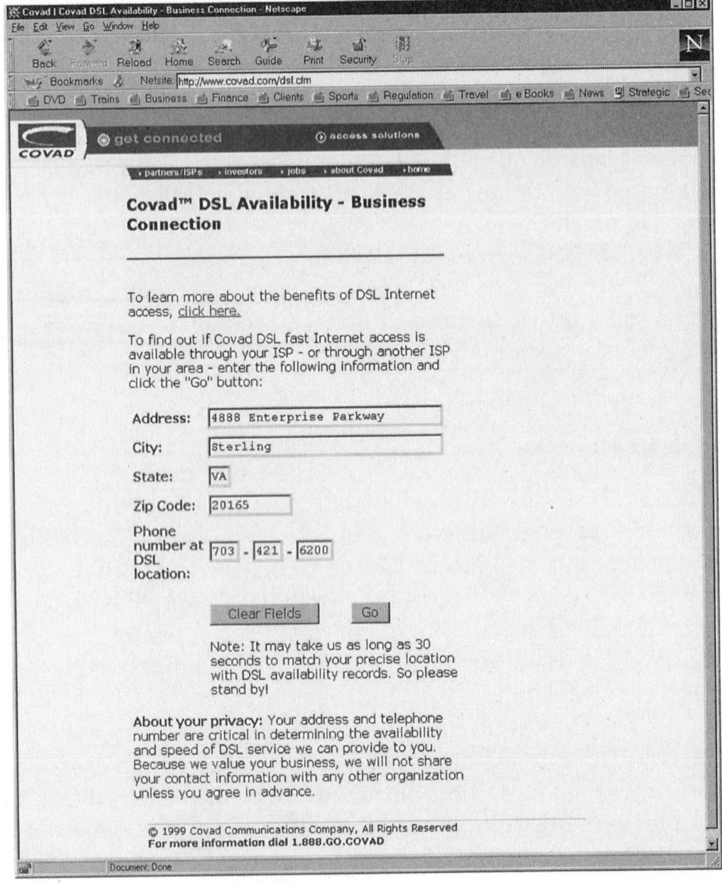

Figure D-3 Covad Communications is among the many companies that provide a Web form to allow consumers to check on the availability and speed of DSL in their service area.

the database application notifies the user of the type of service available as well as the speed of the connection. Since this type of service is only 70 percent accurate, a quirk of DSL services has been that the actual speed of the connection will not be known positively until the service is actually provisioned. Customers in high-rise office buildings have to factor in the vertical distance as part of the total circuit distance. Now there are online services users can call that conduct live tests of their line to determine if it is suitable for DSL and at what speed. These services can even determine the existence of a load coil on the line, which would preclude DSL operation.

Once DSL service is ordered, the service provider will arrange with the local exchange carrier to connect a line to the network interface outside the customer premises or share the existing POTS line for data. Either way, the DSL service provider (or a local agent) visits the customer premises and connects the line to a DSL modem.[1] The DSL modem is typically leased from the service provider, but users may now purchase them from a retail source and configure them through a Web browser (Figure D-4). By connecting the DSL modem to a hub, multiple LAN users can share the available bandwidth to access the Internet for such applications as e-mail, Web browsing, and news group discussions.

For large installations, such as a multitenant building, an appropriately sized DSL concentrator usually will be installed on the premises to aggregate the wires from individual DSL modems in each office. The building manager can have its own technician install the modems at each workstation and connect the wires to the concentrator or have the service provider do it for a fee. The concentrator usually will be scalable in terms of port density to accommo-

[1]So-called DSL modems are actually routers. They are called "modems" because most consumers have become familiar with modems for dialing into the Internet or sending faxes from their home computers. Manufacturers stayed with this familiar term rather than risk confusing consumers with the unfamiliar term "router."

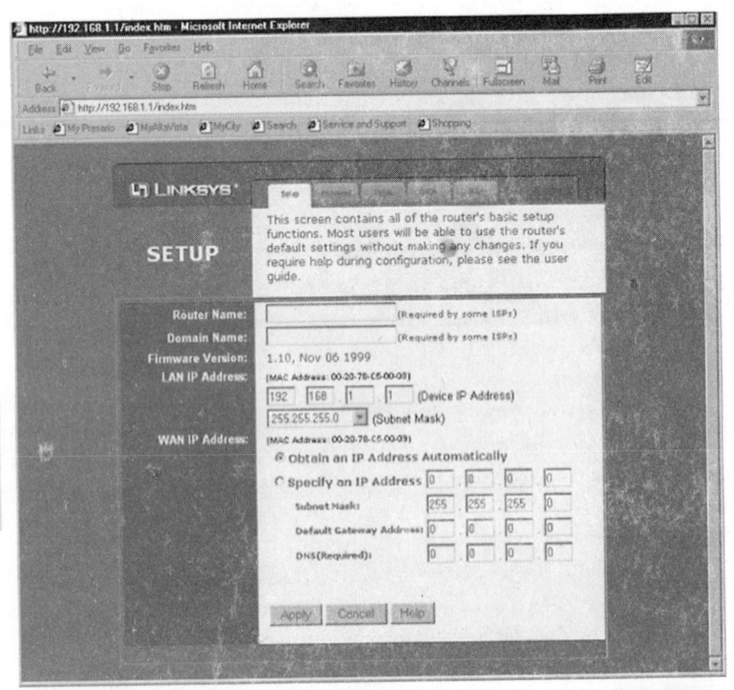

Figure D-4 The Linksys Instant Broadband EtherFast Cable/DSL Router is one of a growing number of off-the-shelf products that is easily configured by the user through a Web browser interface.

date future growth and support multiple types of DSLs to meet the varying needs of users.

Several methods may be used to send aggregated traffic from the on-premises DSL concentrator to the service provider's location. Depending on the amount of traffic involved, leased T1 lines, Frame Relay over DS3, and ATM over SONET may be used. At the service provider's location, another concentrator splits out individual user sessions and tunnels for distribution to other networks.

Depending on the switching platform used, value-added services can be offered over DSL. Cisco's BPX 8650 switching platform, for example, lets service providers offer DSL on an IP-over-ATM backbone. This platform will let the carrier offer quality of service (QoS) through either ATM or multi-protocol label switching (MPLS), thus permitting the delivery of value-added services such as voice over DSL.

Provisioning Obstacles

In addition to distance limitations from CO/SWC to customer premises and the presence of fiber in the loop, there are several other obstacles that may stand in the way of provisioning DSL.

Load Coils DSL service cannot be provisioned over lines that have load coils. These are used on persistently noisy phone lines to improve voice quality. The load coils are inserted at

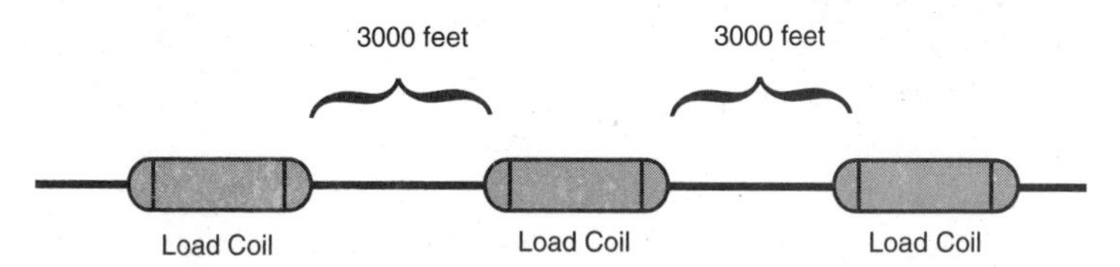

Figure D-5 Load coils improve line quality by limiting the audio band to about 4 kHz to filter out noise, but they also render the line useless for digital services like DSL, which use the higher ranges of the audio band.

3000-foot intervals along the line (Figure D-5). They filter out noise by limiting the usable audio band to about 4 kHz, which is just enough to allow voice conversation. But in limiting the audio band to 4 kHz, the upper frequency ranges of the audio band are no longer available to provision DSL. Although telephone company maps can pinpoint the locations of load coils, the companies will not remove them from the line. Not only would this open the audio band to noise, it also might introduce cross-talk into adjacent pairs. Where load coils are encountered, an unused pair must be selected for DSL.

Bridged Taps A bridged tap and lateral is a circuit that has been used in provisioning analog telephone service for many years. The bridged tap itself is a splicing mechanism for attaching an additional circuit to the normal distribution cable (Figure D-6). One leg allows the normal distribution path to continue farther along, and the other is attached to an unused "lateral." A lateral or "spur" is any portion of a cable pair that is not in the direct path between the customer and the CO.

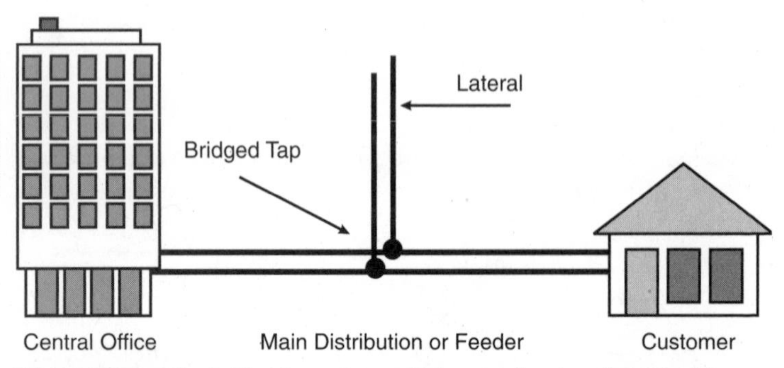

Figure D-6 The bridged tap is a splicing mechanism for attaching an additional circuit to the normal distribution cable. This renders both the lateral and main circuits useless for digital services like DSL.

Laterals create problems on voice circuits that have been converted to digital. In addition to the main circuit, a lateral creates a second path for a digital signal, which weakens the signal on both paths. If the digital signal travels down a lateral that is open (i.e., not terminated), it is reflected back into the main circuit, where it can mix with the "good" digital signals. These echoes effectively render the data useless. In order for the digital circuit to operate properly, the bridged taps must be removed.

Pair-Gain Equipment DSL service cannot be provisioned over lines that attach to digital pair-gain systems (Figure D-7).

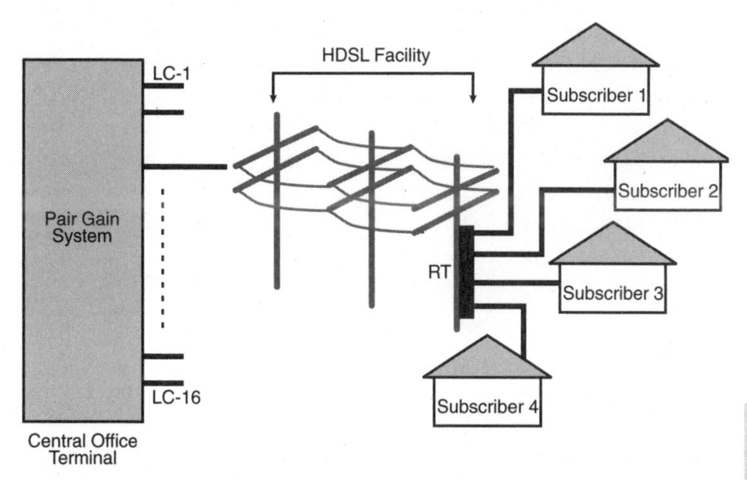

Figure D-7 When a carrier already uses a DSL technology to minimize its remote equipment requirements, the affected subscriber lines cannot be provisioned for DSL services.

These systems typically use HDSL technology to consolidate four, eight, or twelve 64-kbps channels over a single twisted copper pair that is connected to a line card of the pair-gain system. All subscriber circuits are completely independent of each other, and different applications can be mixed within the available channels of the system, including voice, fax, and modem data. These pair-gain systems provide an immediate, low-cost solution to subscriber loop shortages. They also provide the telephone company with flexibility in network planning and conserve existing investments in outside plant. However, DSL services cannot be provisioned over the individual subscriber lines. Instead, HDSL is used by the telephone company to consolidate multiple 64-kbps subscriber lines at a remote terminal for connection to a line card at the pair gain system.

Digital Loop Carrier Systems This type of outside voice-centric plant equipment concentrates local loop lines onto a shared T-carrier backbone, which backhauls them to the CO. While the DLCS enables telephone services to be extended to outlying locations that are beyond the normal reach of the CO, shared backhaul prevents the lines from being provisioned for DSL. This is so because DSL requires a dedicated pair of copper wires from the customer location to the DSL access multiplexer in the CO. About 65 million of the 250 voice lines in the United States are served by DLCS. Next-generation DLCS and upgrades to legacy DLCS overcome this limitation by supporting DSL and voice on the same plug-in cards.

Security

While DSL is certainly more secure than cable or wireless, it is not entirely immune to the problem. Although the access line is certainly more secure than cable because it is dedicated rather than shared, the security problem begins on the other side of the DSLAM, where traffic from multiple copper loops is aggregated for transport over a high-capacity fiber link to the Internet. Since the DSL connection is always on, it is possible for hackers on the Internet to find their way into the DSLAM and, from there, to individual computers on the other side. Since the DSLAM is not equipped with firewall capabilities, the CPE must be equipped to provide security.

Several vendors have addressed the need for security in their DSL products. Netopia, for example, offers a built-in firewall in its SDSL routers. The units come with preconfigured firewalls to disallow all inbound traffic originating from the Internet. They also filter packets on a per-connection profile basis for source/destination address, service, and protocol. Up to 255 rules are available in up to eight filter sets. The routers also support secure VPN access to corporate intranets and extranets via the Point-to-Point Tunneling Protocol (PPTP), plus 56-bit Data Encryption Standard (DES) extensions for added protection. Security is further enhanced with a Network Address Translation (NAT), which hides all IP addresses on the LAN behind a single statically or dynamically assigned IP address on the Internet.

Management

Like any other service, the management of DSL has carrier and customer components. Initially, DSL services were hard for the carriers to provision, which stalled service delivery. But new tools have become available to enable carriers to streamline the rollout, setup, and ongoing management of DSL. There are now tools that can be used by customers, enabling them to change services, add bandwidth, and monitor carrier performance for compliance with service level agreements (SLAs).

Paradyne's Service Level Agreement Reporter, for example, can be used by both a service provider, to offer SLAs on its DSL service, and corporate network managers, to ensure that such SLAs are being met. The SLA Reporter supports Paradyne's multiservice DSLAM that carriers use to offer a variety of DSL services. The SLA Reporter enables service providers to obtain factual, network-based statistics and operational information to verify the quality of services being provided to their customers.

Graphical charts provide the performance information and are delivered to customers via a secure Web site. Using familiar Web browser software, customers can view, and even interact with, the data provided in the SLA Reporter output. This allows corporate network planners, architects, and authorized usersto obtain their own local views of network performance, throughput, capacity planning, and line quality.

Another company, Syndesis, offers service-provisioning software for DSL. Its NetProvision Creator streamlines carrier setup of DSL lines and extends service-ordering functions to the customer. The company's NetProvision Activator issues the commands that configure the individual network devices, including the customer modem, the DSL access multiplexer in the carrier switching office, and core switches and routers. By eliminating paper-based provisioning requests, the process is not only speeded up, but the number of order errors is reduced.

NetProvision Creator allows DSL service profiles to be defined in plain language, such as the speed of a line. This feature makes it easy for sales representatives to use the software to take customer orders for DSL services, and it allows customers to modify their DSL-based services via the Web. For example, a customer with a 256-kbps DSL service to an Internet service provider (ISP) could increase that bandwidth to 384 kbps via NetProvision Creator installed on a Web site. NetProvision Activator implements the requested changes.

Service Provider Selection

The choice of DSL service provider typically will hinge on the following key factors:

- Broadness of coverage
- Range of DSL services offered and their price points
- Type of equipment offered and its security features
- Service level agreement
- Availability of technical support
- Field service infrastructure
- Track record of timely installations
- Pace of DSL service rollouts to new areas
- Availability of value-added services
- Financial stability of the service provider

Another factor that enters into the choice of provider is the quality of the business partnerships the DSL service provider has in place. For example, partnering with a big Interexchange Carrier (IXC) in the launch of enterprise services could help the DSL provider differentiate itself from the competition and accelerate the deployment of value-added services such as voice over DSL and secure VPNs. Partnering with a Web hosting company could allow the DSL provider to further differentiate itself by offering customers a total e-commerce solution that includes site development and trend-reporting tools and collocation space for the servers.

Summary

DSL is a local loop technology that provides a high-speed digital service over an unused portion of the audio band on an ordinary twisted-pair copper line. This turns the POTS line into an economical platform for multiservice networking, enabling ILECs and their competitors (i.e., CLECs, DLECs, and ISPs) to offer high-speed connections over which a variety of broadband data services can be offered, including Internet access, corporate connectivity via VPNs, and extra voice channels. Although fiber is faster and more reliable, it does not extend into enough customer locations. Twisted-pair copper lines, however, reach into every home and business and can be leveraged to provide advanced services and new revenue streams. By some estimates, the U.S. market for DSL is expected to reach $5 billion with 20 million subscribers in 2005. Although some major industry players are struggling financially, even declaring bankruptcy, demand for DSL remains strong.

See also

Cable Television Networks
T1 Lines

DOWNSIZING

From a computing perspective, *downsizing* refers to the process among large companies of easing the load of the central mainframe by distributing appropriate processing and information resources to LANs, specifically, the servers and clients located throughout the organization.[2] This arrangement can yield optimal results to all users: It provides PC users with ready access to the information they need, and in a friendly format, while permitting mission-critical applications and large databases to remain on the mainframe where security and access privileges can best be applied.

Benefits of Downsizing

In moving applications and information closer to departments and individuals, knowledge workers at all levels in the organization are empowered to do their jobs, improving the quality and timeliness of decision making and allowing a more effective corporate response to customer issues and market dynamics. In addition,

- Downsized applications may run as fast or faster on PCs and servers than on mainframes, at only a fraction of the cost.
- Even if downsizing does not improve response time, it can improve response-time consistency.
- In the process of rewriting applications to run in the downsized client-server environment, there is the opportunity to realize greater functionality and efficiency from the software.
- With greater reliance on off-the-shelf applications, the number of programmers and analysts needed to support the organization is minimized.
- The lag time between applications development and business processes can be shortened substantially, enabling significant competitive advantage to be realized.

[2]Downsizing also has been used to describe how organizations restructure themselves to become more competitive by trimming the number of employees, outsourcing specialized tasks, and spinning off operations unrelated to the core business.

- The organization need not be locked into a particular vendor, as is typically the case with centralized host-based architectures.

Distributed Environment

The distributed computing environment brought about by downsizing can take many forms. For example, dedicated file servers can be used on the LAN to control access to application software and to prevent users from modifying or deleting certain types of files. Several file servers can be deployed throughout the network, each supporting a single application (e.g., electronic mail, facsimile, graphics, or specific types of databases). Metering tools can be included on the server to monitor usage and prevent unauthorized copying, which might violate various software licenses. Metering can even qualify the organization for a limited-use network license, whereas the absence of metering might force the organization to buy a more expensive unlimited-use network license.

When connected to a LAN, an appropriately equipped PC can act as a server. With the client-server approach, an application program is divided into two parts on the network. The client portion of the program, or front end, executes on the user's workstation, enabling such tasks as querying databases, producing printed reports, or entering new records. The server portion of the program, or back end, is resident on a computer (i.e., server) that is configured to support multiple clients. This setup offers users shared access to numerous application programs as well as to printers, file storage, database management, communications, and other capabilities. Automated tools ensure data concurrency, implement security, and ease software maintenance tasks.

Mainframe as Server

The mainframe is still valued for its unequaled processing and security capabilities. The mainframe can continue supporting traditional applications that require processing power and access to legacy applications but also act as a server in the distributed computing environment. In the case of IBM, for example, the addition of special software allows the mainframe to act as a server:

- *LAN Resource Extension and Services (LANCES)* A server-based software product that is used with NetWare LANs.
- *Data Facility Storage Management Subsystem (DFSMS)* A suite of software programs that automate storage management.
- *Network File Server (NFS)* Originally designed to operate on LANs, this software has been tuned to operate with multiple virtual system (MVS) on the mainframe.
- *File Transfer Protocol (FTP)* The FTP server application can be used as part of the native TCP/IP stack under the Virtual Telecommunications Access Method (VTAM) or as a single third-party FTP server application also running as a VTAM application.

The mainframe also can play the role of master server in a hierarchical arrangement, backing up, restoring, and archiving data from multiple LAN servers.

Summary

Determining the benefits of downsizing, the extent to which it will be implemented, who in the organization will have responsibility for accomplishing it, and the justification of its upfront costs are highly subjective activities. The answers will not appear in the form of cookie-cutter solutions from market savvy vendors. In fact, the likelihood of success will be improved greatly if the downsizing effort is approached within the context of business process reengineering.

See also

Client-Server Networks

Peer-to-Peer Networks

E

ELECTRONIC MAIL

The popularity of electronic mail (e-mail) in recent years has paralleled the growth of the Internet. More mail is now delivered in electronic form over the Internet than is delivered by the U.S. Postal Service. That amounts to tens of billions of messages per month. Message delivery usually takes only minutes over the Internet instead of days or weeks, as is typical with postal services in many countries.

Ray Tomlinson (Figure E-1) invented e-mail in 1971 while working at Bolt, Beranek, and Newman (BBN), a company that had a government contract to work on the Advanced Research Projects Agency Network (ARPANET), the precursor of the Internet. An MIT graduate, Tomlinson was part of a team building an operating system when he came up with a "Send Message" program. At first, it worked only on a local system, but he developed it further into cross-ARPANET mail. Tomlinson needed a character to separate a name from a place so that computers sending messages would not confuse the two. He immediately came up with the @ sign, the only prepositional character on the keyboard.

Figure E-1 Ray Tomlinson invented e-mail in 1971.

Advantages of E-Mail

Automating information delivery and processing with electronic mail can dramatically reduce the cost of doing business, since the manual tasks of sorting, matching, filing, reconciling, and mailing paper files are virtually eliminated. There are also attendant cost savings on overnight delivery services, supplies, file storage, and clerical personnel.

Users can even send e-mail from within any application that supports Microsoft's Messaging Applications Programming Interface (MAPI) specification. Thus a file done in a MAPI-compliant word processor or spreadsheet application, for example, can be e-mailed as an attachment without having to leave that application.

Through mail gateways, e-mail can be sent to people who may subscribe to some other type of service such as America Online or Microsoft Network (MSN). Some gateway services even transport messages from the Internet to wireless devices such as personal digital assistants (PDAs) and portable computers equipped with radio modems. And with more paging services supporting short-text messaging, e-mail can even be sent over the Internet to alphanumeric pagers.

The servers and gateways on the Internet take care of message routing and delivery. The e-mail address contains all the information necessary to route the message. If a message cannot be delivered because of some problem on the Internet, an error message is returned that explains the reason and estimates the time of delivery.

Internet Protocols

Currently, the dominant mail protocols used on the Internet are the Simple Mail Transfer Protocol (SMTP) and Post Office Protocol 3 (POP3). When installed on a server, SMTP gives it the capability to send and route messages over the Internet. POP3 is also installed on a server, giving it the capability to hold incoming e-mail until the recipient is ready to download it to his or her own computer. Once downloaded, the e-mail message can be opened, edited for reply, cut and pasted into another application, saved as a document, filed for future reference, forwarded, or deleted (Figure E-2).

POP was designed to support offline message access. Once downloaded, messages can be opened at any time and then marked for deletion from the mail server on next log-on. This mode of access is not compatible with access from multiple computers because it tends to distribute messages across all the computers used for mail access.

A newer protocol, Internet Mail Access Protocol (IMAP4), allows users to access messages on the server as if they were

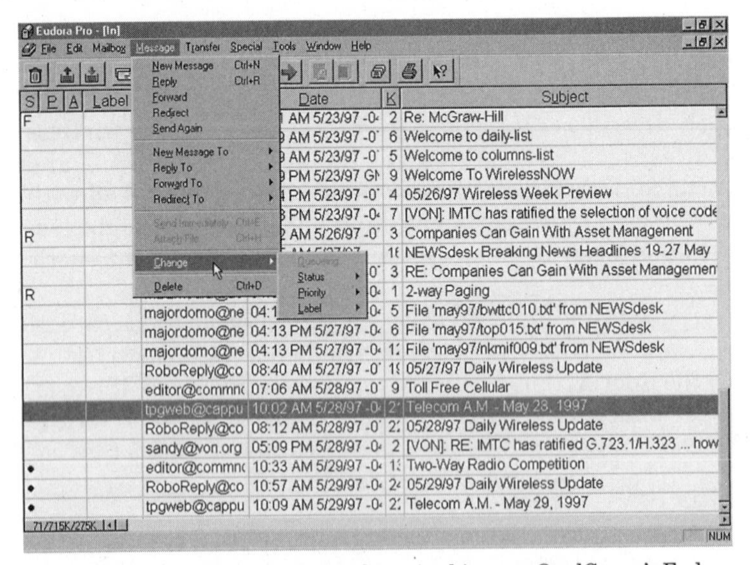

Figure E-2 A typical e-mail interface—in this case, QualComm's Eudora Pro.

local. For example, e-mail stored on an IMAP server can be manipulated from a desktop computer at home, a workstation at the office, and a notebook computer while traveling—without the need to transfer messages or files back and forth between them.

IMAP's ability to access messages (both new and saved) from more than one computer has become extremely important as reliance on electronic messaging and the use of multiple computers among mobile professionals increase. The key advantages of IMAP include

- Full compatibility with other Internet messaging standards, such as MIME[1]
- Message access and management from more than one computer

[1] MIME stands for Multipurpose Internet Mail Extensions. It is a technique for encoding text, graphics, audio, and video files as attachments to SMTP-compatible Internet mail messages.

- Message access without reliance on less efficient file access protocols
- Support for "online," "offline," and "disconnected" access modes
- Support for concurrent access to shared mailboxes
- Client software needs no knowledge about the server's file store format

IMAP supports operations for creating, deleting, and renaming mailboxes; checking for new messages; permanently removing messages; setting and clearing flags; server-based MIME parsing (relieving clients of this burden) and searching; and selective fetching of message attributes and text. In certain circumstances, IMAP also allows sent messages to be recalled.

When a user has second thoughts about sending a message, it can be recalled with another message to save embarrassment (Figure E-3). Alternatively, the original message can be replaced with a new message. The status of the recall attempt, whether it succeeded or failed, is reported back to the originator as a message in the Inbox. However, the recall/replace feature works only if the recipient has not yet opened the message.

Outsourcing Arrangements

Many businesses find it difficult and time-consuming to run their own e-mail systems. Businesses with up to 1000 employees may find outsourcing their e-mail a more attractive option. By subscribing to a carrier-provided e-mail service, these companies can save from 50 to 75 percent on the cost of buying and maintaining an in-house system.

Carriers even offer guaranteed levels of service that will minimize downtime and maintain 24 × 7 support for this mission-critical application. An interface allows the subscribing company to partition and administer accounts. Companies can even create multiple domains for divisions or

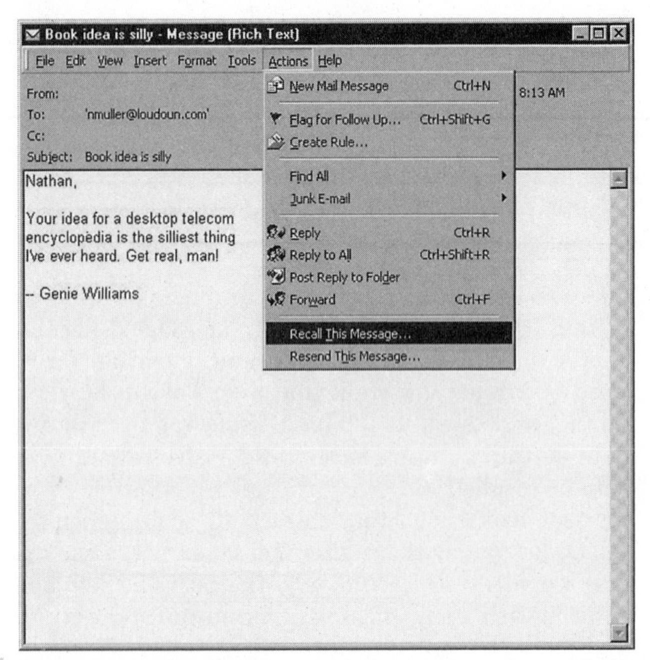

Figure E-3 When Microsoft Outlook is configured as an IMAP client, it can recall or replace a sent message, provided that the recipient has not yet opened it.

contractors. Administration is all done through the service provider's Web site. Changes are easy and intuitive—no programming experience is necessary—and the changes take place immediately. Many services offer spam controls to virtually eliminate junk mail.

Outsourcing can level the playing field for companies competing in the Internet economy. Businesses no longer need to budget for endless rounds of software upgrades. A carrier-provided service also can include value-added applications such as fax, collaboration, calendaring, and unified messaging services.

For companies that are not ready to entrust their entire e-mail operation to a third-party service provider, there are customized solutions that allow them to selectively outsource certain aspects and functions. For example, a company can choose to have its e-mail system hosted on the service provider's server. Such "midsourcing" can reduce the costs of administration and support and provide improved performance. In addition, the company can add new functionality easily, without incurring a costly upgrade.

Summary

Once considered by most companies to be just a fad, electronic mail is valued for its role in supporting daily business operations. In fact, the popularity of e-mail is now so great that many companies are being forced to upgrade the capacity of their communications links and systems to accommodate the growing traffic load. Steps are also being taken to minimize unnecessary traffic, such as by storing only one copy of e-mail attachments on a server rather than allowing the same attachment to be duplicated to all recipients of the message. Other companies are considering outsourcing their e-mail operations to carriers and Internet service providers (ISPs). This is part of the relatively new trend of applications outsourcing.

See also

Unified Messaging

ELECTRONIC SOFTWARE DISTRIBUTION

With a growing population of PC and workstation users deployed across widely dispersed geographic locations—each potentially using different combinations of operating systems, applications, databases, and network protocols—software has become more complex and difficult to install, maintain, and meter. The ability to perform these tasks over a network from a central administration point can leverage investments in software, enforce vendor license agreements, qualify the organization for discounts on network licenses, and greatly reduce network support costs.

Industry experts estimate that the average 5-year cost of managing a single desktop PC exceeds $45,000 and that the 5-year cost of deployment and managing changes to new client-server applications averages an additional $45,000 to $55,000 per user. Automating the distribution and maintenance of software can cut support costs in half. Electronic software distribution (ESD) tools also can provide useful reports that can aid in problem resolution and determining the need for license upgrades.

Automating File Distribution

The complexity of managing the distribution and implementation of software at the desktop requires that network administrators make use of automated file distribution tools. By assisting a network administrator with tasks like packaging applications, checking for dependencies, and offering links to event and fault management platforms, these tools reduce installation time, lower costs, and speed problem resolution.

One of these tools is a programmable file distribution agent. It is used to automate the process of distributing files to particular groups or workstations. A file distribution job can be defined as software installations and upgrades, startup file updates, or file deletions. Using a file distribution agent, these types of changes can be applied to each workstation or group automatically.

The agent can be set up to collect file distribution status information. The network administrator can view this information at the console to determine if files were distributed successfully. The console allows the administrator to review status data, such as which workstations are set up for file distributions, the stations to which files have been distributed, and the number of stations waiting for distributions.

Because users can be authorized to log onto the network at one or more workstations, the file distribution agent determines where to distribute files based on the primary user (owner) of the workstation. The owner is established the first time that a hardware/software inventory is taken of the workstation. Before automated file distributions are run, the hardware inventory agent is usually run to check for resource availability, including memory and hard disk space. Distributions are made only if the required resources are available to run the software.

Via scripts, the network administrator can define distribution criteria, including the group or station to receive files and the day or days on which the files are to be distributed. Scripts usually identify the files for distribution and the hardware requirements needed to run the file distribution job successfully. Many vendors provide templates to ease script creation. The templates are displayed as a preset list of common file tasks. Using the templates, the network administrator can outline a file distribution script and then use the outline to actually generate the script.

To help network administrators prepare for a major software distribution, some products offer routines called "wizards" that walk administrators through the steps required to assemble a "package." A package is a complete set of scripts, files, and recipients necessary to successfully complete a distribution and install the software. To reduce network traffic associated with software distributions, some products automatically compress packages before they are sent to another server or workstation. At the destination, the package is automatically decompressed when accessed.

When a file distribution job is about to run, target users receive a message indicating that files are about to be sent and requesting that they choose either to continue or to cancel the job or postpone it for a more convenient time.

Managing Installed Software

Maintaining a software inventory allows the network administrator to quickly determine what operating systems and applications are installed on various servers and clients. In addition to knowing what software components are installed and how many are in use on the network (Figure E-4), the network administrator can track application usage to ensure compliance with vendor license agreements.

The ESD tool creates and maintains a software inventory by scanning all the disk drives on the network. Usually software management tools come with preset lists of software packages they can identify during a scan of all disk drives on

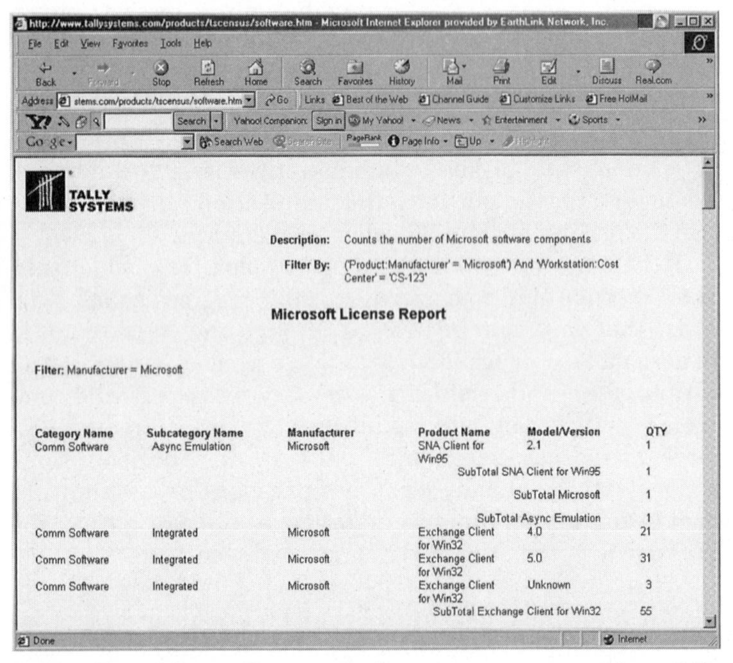

Figure E-4 Software license compliance summary report from Tally Systems' TS.Census. With the Microsoft filter applied, the report shows what Microsoft software components are installed on the network.

the network. Some tools can recognize several thousand software packages. Software that cannot be identified during a scan is tagged for further inquiry and manual data entry. The next time a scan is done, the added software packages will be identified properly.

Whether the initial software inventory is established by drive scanning or manual entry, the following information about the various software packages usually can be added or updated at any time:

- *Package information* The number of available software licenses, the product manufacturers, and the project code to log application use to a particular project.
- *Software availability* When the software is available for user access. The software package can be closed during the upgrade process and a message displayed to indicate when the application becomes available. A startup message can be displayed when users open particular applications. For example, the network administrator can post a message telling users that this is a beta copy of an application or announce the date a new release of the software will be installed. Being able to relay such information to users helps prevent unnecessary support calls to the help desk.
- *Files in package* Executable files associated with the software and files for which to verify integrity—for example, the presence of a computer virus or unauthorized file access.
- *Additional information about the application* Information such as which departments in the organization have access to the application or the vendor's technical support phone number.
- *Optional information* Information such as whether the software is a Windows or OS/2 application and the directory to which it should be added.

Applications also can be restricted according to user, workgroup, department, or project such that certain financial software, for example, can be restricted to accounting personnel. Personnel management software can be restricted to the human resources department. CAD/CAM software can be restricted to an engineering workgroup. Locking out unauthorized users prevents inadvertent data loss or malicious damage to files.

Metering Software Usage

The ability to track software usage helps the network administrator ensure that the organization complies with software license agreements while making sure that users have access to required applications. Tracking software usage also helps reduce software acquisition costs, since accurate usage information can be used to determine which applications are run most before deciding on upgrades and how many copies to buy.

Metering allows the network administrator to control the number of concurrent users of each application. The network administrator also can choose to be notified of the times when users are denied access to particular applications because all available copies are in use. This may identify the need to purchase additional copies of the software or pay an additional license charge to the vendor so more users can access the application.

Before users are granted access to metered software, the software inventory is checked to determine whether there are copies available. If no copies are available, a status message is issued, indicating that there are no copies available. The user waits in a queue until a copy becomes available.

Metering software can save money on software purchases and ensure compliance with software copyright laws. For example, if there are 100 users of Microsoft Word on an enterprise network and only half that number use it concurrently, the software metering tool's load balancing feature automatically handles the transfer of software licenses from one server to another on a temporary or permanent basis.

Load balancing helps network administrators purchase licenses based on need rather than on the number of potential users at a given location.

License management capabilities are important to have because it is a felony under U.S. federal law to copy and use (or sell) software. Companies found guilty of copyright infringement face civil penalties of up to $150,000 for each work infringed on. In criminal cases, the maximum penalty for copyright infringement is up to $250,000 and jail term of up to 5 years.

The Software & Information Industry Association (formerly the Software Publishers Association) runs a toll-free hotline and receives about 40 calls a day from whistle-blowers. It sponsors an average of 250 lawsuits a year against companies suspected of software copyright violations. Having a license management capability can help the network administrator track down illegal copies of software and eliminate a company's exposure to litigation and financial risk.

Some software metering packages allow application usage to be tracked by department, project, workgroup, and individual for chargeback purposes. Charges can be assigned on the basis of general network use, such as time spent logged onto the network or disk space consumed. Reports and graphs of user groups or department charges can be printed out or exported to other programs, such as an accounting application.

Although companies may not require departments or divisions to pay for application or network usage, chargeback capabilities can still be a valuable tool for breaking down operations costs and planning for budget increases.

Distribution over the Internet

Since 1996, corporations have had the option of subscribing to Internet-based software distribution services offered by application service providers (ASPs). Such services are usually hosted on a Web server and provide managed Internet-based delivery and tracking of business-critical software applications and documents. Available via a link from any Web or extranet site, such services offer all the functionality of an enterprise-class software management system, with the convenience and flexibility of a hosted Web service.

Such services are designed for large and midsized companies that are using the Web to extend their businesses and to lower the cost of customer support and product distribution. As these companies make broader use of their extranets, there is an increasing need to distribute proprietary digital files to support the online business service. On an outsource basis, service providers offer confirmed delivery and tracking of these software and document packages securely and cost effectively.

The entire service is available from a standard browser, offering users the flexibility and ease of use of an intuitive Web interface. Businesses seeking to outsource their software distribution operations simply create authorized user lists and packages software, documents, and digital files that will be available for download. The service provider's management server authenticates users, and maintains a database to track distribution transactions, package versions and customer accounts, providing comprehensive reports to support business operations.

Once set up, users are assigned an authorization code and are directed to a secure, fully customized download portal site hosted by the service provider, where they can browse through available software, check their download histories, and select packages by name, version, or category. The download manager can be a Java applet that loads dynamically when a customer requests an update package. The applet executes a managed download, confirming receipt of the files on the customer's computer. As additional packages are made available, the service provides automatic e-mail notification of the updates, providing a way to instantly notify authorized users and groups.

The service also may include a block-level restart feature, which automatically resumes the file transfer if a download is interrupted and synchronizes from the point of failure. All download activities are secure, using multilevel encryption and authentication features.

In addition to a completely managed service, some service providers also offer a lower-cost "self-drive" option, enabling information technology (IT) and Web site managers to run the service from a browser, with all distribution and tracking activity operated from the service provider's distribution center.

Summary

ESD tools are essential for managing software assets. They can help trim support costs by permitting software to be distributed and installed from a central administration point; ensure compliance with vendor license agreements, thereby eliminating exposure to lawsuits for copyright infringement; and help companies manage software to minimize their investments while meeting the needs of all users.

See also

Asset Management

ETHERNET

Ethernet is a type of local area network (LAN) that uses a contention-based method of access to allow computers to share resources, send files, print documents, and transfer messages. The Ethernet LAN originated as a result of the experimental work done by Xerox Corporation at its Palo Alto Research Center (PARC) in the mid-1970s. However, Robert Metcalfe (Figure E-5) is the individual generally credited with the development work that led to Ethernet.

Once developed, Ethernet quickly became a de facto standard with the backing of DEC and Intel. Xerox licensed Ethernet to other companies that developed products based on the specification issued by Xerox, Intel, and DEC. Much of the original Ethernet design was incorporated into the IEEE 802.3 Standard adopted in 1980 by the Institute of Electrical and Electronics Engineers (IEEE).

Figure E-5 Robert M. Metcalfe began working for Xerox Corporation at its Palo Alto Research Center (PARC) in 1972 while working on his Ph.D. at Harvard. It was at PARC, in 1973, that Dr. Metcalfe and D. R. Boggs invented Ethernet. In 1979, Metcalfe founded 3Com Corp., a computer networking company in Santa Clara, California.

Ethernet is contention-based, meaning that stations compete with each other for access to the network, a process that is controlled by a statistical arbitration scheme. Each station “listens” to the network to determine if it is idle. On sensing that no traffic is currently on the line, the station is free to transmit. If the network is already in use, the station backs off and tries again. If multiple stations sense that the network is idle and transmit at the same time, a “collision” occurs, and each station backs off to try again at staggered intervals. This media access control scheme is known as Carrier Sense Multiple Access with Collision Detection (CSMA/CD).

Frame Format

The IEEE 802.3 Standard defines a multifield frame format, which differs only slightly from that of the original version of Ethernet, known as "pure" Ethernet (Figure E-6).

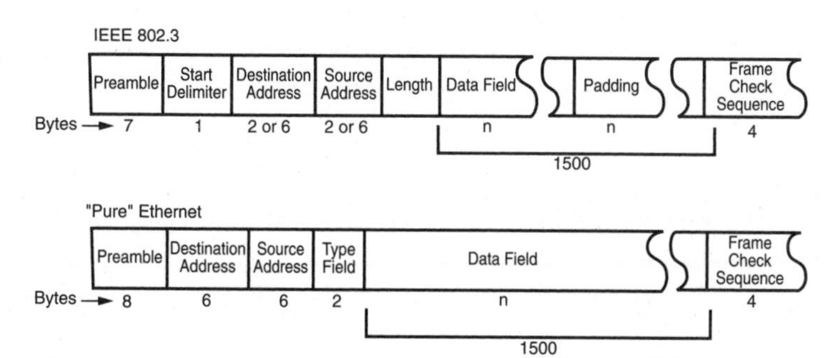

Figure E-6 Comparison of Ethernet frame formats: IEEE 802.3 and "pure" Ethernet.

Preamble The frame begins with an 8-byte field called a "preamble," which consists of 56 bits having alternating 1 and 0 values. These are used for synchronization and to mark the start of the frame. The same bit pattern used in the pure Ethernet preamble is used in the IEEE 802.3 preamble, which includes the 1-byte start-frame delimiter field.

Start-Frame Delimiter The IEEE 802.3 Standard specifies a start-frame delimiter field, which is really a part of the preamble. This is used to indicate the start of a frame.

Address Fields The destination address field identifies the station or stations that are to receive the frame. The source address field identifies the station that sent the frame. If addresses are locally assigned, the address field can be either 2 bytes (16 bits) or 6 bytes (48 bits) in length. A destination address can refer to one station, a group of stations, or all stations. The original Ethernet specifies the use of 48-bit addresses, while IEEE 802.3 permits either 16- or 48-bit addresses.

Length Count The length of the data field is indicated by the 2-byte count field. This IEEE 802.3–specified field is used to determine the length of the information field when a pad field is included in the frame.

Pad Field To detect collisions properly, the frame that is transmitted must contain a certain number of bytes. The IEEE 802.3 Standard specifies that if a frame being assembled for transmission does not meet this minimum length, a pad field must be added to bring it up to that length.

Type Field Pure Ethernet does not support length and pad fields, as does IEEE 802.3. Instead, 2 bytes are used for a type field. The value specified in the type field is only meaningful to the higher network layers and was not defined in the original Ethernet specification.

Data Field The data field of a frame is passed by the client layer to the data link layer in the form of 8-bit bytes. The minimum frame size is 72 bytes, while the maximum frame size is 1526 bytes, including the preamble. If the data to be sent use a frame that is smaller than 72 bytes, the pad field is used to stuff the frame with extra bytes. In defining a minimum frame size, there are fewer problems to contend with in handling collisions. If the data to be sent use a frame that is larger than 1526 bytes, it is the responsibility of the higher layers to break it into individual packets in a procedure called "fragmentation." The maximum frame size reflects practical considerations related to adapter card buffer sizes and the need to limit the length of time the medium is tied up in transmitting a single frame.

Frame Check Sequence A properly formatted frame ends with a frame check sequence, which provides the means to check for errors. When the sending station assembles a frame, it performs a cyclic redundancy check (CRC) calculation on the bits in the frame. The sending station stores the results of the calculation in the 4-byte frame check sequence field before sending the frame. At the receiving station, an identical CRC calculation is performed and a comparison made with the original value in the frame check sequence field. If the two values do not match, the receiving station assumes that a transmission error has occurred and requests that the frame be retransmitted. In pure Ethernet, there is no provision for error correction; if the two values do not match, notification that an error has occurred is simply passed to the client layer.

Media Access Control

Several key processes are involved in transmitting data across the network, among them data encapsulation/decapsulation and media access management, which are performed by the Media Access Control (MAC) sublayer of OSI's Data Link Layer.

Data Encapsulation/Decapsulation Data encapsulation is performed at the sending station. This process entails adding information to the beginning and end of the data unit to be transmitted. The data unit is received by the MAC sublayer from the logical link control (LLC) sublayer. The added information is used to perform the following tasks:

- Synchronize the receiving station with the signal
- Indicate the start and end of the frame
- Identify the addresses of sending and receiving stations
- Detect transmission errors

The data encapsulation function is responsible for constructing a transmission frame in the proper format. The destination address, source address, type, and information fields are passed to the Data Link Layer by the client layer in the form of a packet. Control information necessary for transmission is encapsulated into the offered packet. The CRC value for the frame check sequence field is calculated, and the frame is constructed.

When a frame is received, the data decapsulation function performed at the receiving station is responsible for recognizing the destination address, performing error checking, and then removing the control information that was added by the data encapsulation function at the sending station. If no errors are detected, the frame is passed up to the LLC sublayer.

Specific types of errors are checked in the decapsulation process, including whether the frame is a multiple of 8 bits or exceeds the maximum packet length. The address is also checked to determine whether the frame should be accepted and processed further. If it is, a CRC value is calculated and checked against the value in the frame check sequence field. If the values match, the destination address, source address, type, and data fields are passed to the client layer. What is passed to the station is the packet in its original form.

Media Access Management The method used to control access to the transmission medium is known as "media access management" in IEEE terms but is called "link management" in Ethernet parlance. Link management is responsible for several functions, starting with collision avoidance and collision handling, which are defined by the IEEE 802.3 Standard for contention networks.

Collision Avoidance Collision avoidance entails monitoring the line for the presence or absence of a signal (carrier). This is the "carrier sense" portion of CSMA/CD. The absence of a signal indicates that the channel is not being used and that it is safe to begin transmission. Detection of a signal indicates that the channel is already in use and that transmission must be withheld. If no collision is detected during the period of time known as the "collision window," the station acquires the channel and can complete the transmission without risking a collision.

Collision Handling When two or more frames are offered for transmission at the same time, a collision occurs, which triggers the transmission of a sequence of bits called a "jam." This is the means whereby all stations on the network recognize that a collision has occurred. At that point, all transmissions in progress are terminated. Retransmissions are attempted at calculated intervals. If there are repeated collisions, link management uses a process called "backing off," which involves increasing the retransmission wait time following each successive collision.

On the receiving side, link management is responsible for recognizing and filtering out fragments of frames that resulted from a transmission that was interrupted by a collision. Any frame that is less than the minimum size is assumed to be a collision fragment and is not reported to the client layer as an error.

Methods have been developed to improve the performance of Ethernet by reducing or totally eliminating the chance for collisions without having to segment the LAN into smaller subnetworks. Special algorithms sense when frames are on a collision course and will temporarily block one frame while allowing the other to pass. This is called "collision avoidance."

Summary

Ethernet is the most popular type of LAN. Its success has spawned continued innovations leading to higher speeds and overcoming distance limitations. 10BaseT Ethernet, for example, enables the LAN to operate over ubiquitous unshielded twisted-pair (UTP) wiring instead of thick or thin coaxial cable. For those who find 10 Mbps inadequate for supporting large file transfers and graphic-intensive applications, there are higher-speed versions of Ethernet, including Fast Ethernet at 100 Mbps, Gigabit Ethernet, and 10× Gigabit Ethernet—all of which can operate over various grades of UTP wiring within a building. There are also versions of Ethernet that run over optical fiber for metropolitan area or wide area connectivity.

See also

Ethernet (10BaseT)
Ethernet (100BaseT)
Ethernet (1000BaseT)

ETHERNET (10BASET)

10BaseT is the IEEE standard for providing 10-Mbps Ethernet performance and functionality over ubiquitously available unshielded twisted-pair wiring. This standard is noteworthy in that it specifies a star topology, unlike traditional 10Base-2 and 10Base-5, which use coaxial cabling arranged in a bus or ring topology. The star topology permits centralized network monitoring, which enhances fault isolation and bandwidth management. For new installations, twisted-pair wire is substantially less expensive as well as easier to install and maintain than Ethernet's original "thick" coaxial cable (10Base5) or the "thin" coaxial alternative (10Base2).

Performance

Traditionally, Ethernet has relied on coaxial cable with multidrop connections to a LAN segment. Repeaters between the segments keep the signal strength at a consistent level across the Ethernet. The key disadvantage of this bus topology is that any disturbance to the continuity of the cable at any point renders the entire LAN inoperable.

In contrast, the star topology relies on a hub or switch to support a dedicated link to each user. If an individual link, port, or workstation were to fail, there would be no impact on the rest of the LAN. This is precisely the benefit promised by 10BaseT: Because the network can tolerate a malfunction of any end-user device or its physical link, the rest of the network will not be affected, resulting in improved network availability. This is made possible by the link test and autopartition logic inherent in the port-level circuitry of a 10BaseT hub or switch.

A network management system can make problems even easier to identify. When a predefined error threshold is exceeded, for example, the network management system will alert the LAN administrator, enabling a technician to be dispatched before the user even realizes that a problem exists. Alarm notification is usually by a screen message or visual indicator; some systems can even notify the LAN administrator via pager.

10BaseT and traditional bus Ethernet LANs have different distance limitations: 10BaseT operates reliably over cable segments not exceeding 100 meters (330 feet), whereas traditional Ethernet LANs using thick coaxial cable operate reliably over segments of up to 500 meters (1650 feet) in length and 10Base2 Ethernets are effective at 200 meters (660 feet) using thin coax. The cable limitation of 10BaseT aside, its bit error rate performance is at least as good as 10Base2 and 10Base5 systems. The 10BaseT specification allows for bit error rates of no more than 1 in 100 million bits. The data-encoding scheme used for 10BaseT systems is the same as that used for coaxial-based Ethernets—self-clocking Manchester Encoding.

10BaseT LANs are designed to support the same applications as traditional Ethernet LANs. The relatively short distance over which 10BaseT LANs operate is rarely a factor in their ability to support these applications. During the standardization process, a survey by AT&T revealed that over 99 percent of employee desktops are located within 100 meters of a telephone wire closet where all the connections meet at a patch panel.

Media

Most installed telephone wiring is of the type known as 24 American Wire Gauge (AWG). Even if existing telephone wiring follows the required star configuration and is within the roughly 100-meter distance limitation, it may still not be suitable to handle the 10-Mbps data rate. This is so because no particular wire gauge or type is specified in the 10BaseT standard, although 24 AWG is what most equipment vendors have used in conformance tests to confirm that their products transmit reliably at up to 100 meters.

More important than gauge, however, are the attenuation (signal loss), impedance (resistance), delay, and cross-talk characteristics of the wiring. Minimally acceptable levels for all these are spelled out in the 10BaseT specifications. Generally, the inside wiring installed in the last 20 years for telephone connections meets the 10BaseT specifications at cabling runs of up to 100 meters. Older wiring may not meet the standards, in which case poor or erratic LAN performance could result and maximum transmission distances could be considerably less than 100 meters.

The 10BaseT standard was designed to eliminate the requirement for shielded wiring. It relies on the twists in twisted-pair wire to hold down frequency loss, which in turn improves the integrity of the signal. The twists minimize the effects of this loss by preventing high-frequency signal energy from radiating to and corrupting the signal being carried over nearby twisted pairs (cross-talk). Even the individual wire pairs in 25-pair telephone cable are twisted. This cable is used widely in large installations; it enables as many as twelve 10BaseT segments to be neatly carried, separated, and patched to a panel for distribution to the hub.

Adapter Cards

Adapter cards, also known as network interface cards (NICs), are boards that insert into an expansion slot of PCs and servers. The adapter card connects the device's bus directly to the LAN segment, eliminating the need for a separate transceiver. Depending on the type of adapter, it permits connections to thick or thin coaxial cable, as well as to unshielded twisted pairs. An adapter with an Attachment Unit Interface (AUI) port will permit connection to a transceiver, allowing the card to be used with thick and thin coaxial cables and fiberoptic cables.

10BaseT adapters vary considerably in their capabilities and features for reporting link status. Some adapters notify the user of a miswired connection to the LAN. Many have light-emitting diodes (LEDs) that indicate link status after the connection is made to the hub, such as link, collision, transmit, and receive. Others provide minimal information, such as if the link is correctly wired and working, while some light only when something is wrong. Still other cards have no LEDs to display link information. Some cards feature a menu-driven diagnostic program to help the user isolate problems with the cards, such as finding out if the adapter is capable of responding to commands from the software.

Transceivers

Transceivers connect PCs and peripherals already equipped with legacy Ethernet cards to 10BaseT wiring. Typically, a transceiver consists of a small external box with an RJ45 jack at one end for connecting to the unshielded twisted-pair LAN segment and an AUI port at the other end for connecting to the Ethernet adapter card.

Medium Access Unit

Medium access units (MAUs) terminate each end of the 10BaseT link. The MAU accommodates two wire pairs: one pair for transmitting the Ethernet signal and the other pair for receiving the signal. The 10BaseT standard describes seven basic functions performed by the MAU: Transmit, receive, collision detection, and loop-back functions direct data transfer through the MAU, and the jabber detect, signal quality error test, and link integrity functions define ancillary services provided by the MAU.

The jabber function removes equipment from the network whenever it continuously transmits for periods significantly longer than required for a maximum-length packet, indicating a possible problem with the NIC. The signal quality error test detects silent failures in the circuitry, while the link integrity signal detects breaks in the wire pairs. Both assist in fault isolation.

The Hub

All 10BaseT stations are connected to the hub via two twisted-pair wire pairs—one two-wire path for transmitting and the other two-wire path for receiving—over a point-to-point link. In essence, the hub acts as a multiport repeater. It contains the circuitry to retime and regenerate the signal received from any of the wire segments that connect at the hub to each of the other segments. However, a 10BaseT hub is more than a simple repeater; it serves as an active filter that rejects damaged packets.

The multiport repeater provides packet steering, fragment extension, and automatic partitioning. The packet steering function broadcasts copies of packets received at one repeater port to all its other ports. Fragment extension ensures that partially filled packets are sent to their proper destination. Automatic partitioning isolates a faulty or misconnected 10BaseT link to prevent it from disrupting traffic on the rest of the network.

There are several types of 10BaseT hubs, the most common being chassis-based solutions and stackable 10BaseT hubs that can be cascaded with appropriate cable connections. At the low end, hubs come in managed or unmanaged versions.

The chassis models are high-end devices that provide a variety of connections to the wide area network (WAN), support multiple LAN topologies and media, and offer advanced network management, with SNMP typically included as well. This type of 10BaseT hub is also the most expensive. Aside from enhancing their products with more interfaces and network management features to further differentiate them from competitive offerings, vendors are continually increasing the port capacity of these devices to bring down the price per port, making them more appealing to larger users.

Depending on port capacity, stackable hubs are used to link members of a workgroup or department. With a pass-through or crossover cable, additional units can be added in a cascading arrangement to meet growth requirements or to connect to a high-end hub.

Management

The promise of 10BaseT networks was not only that Ethernet would run over economical unshielded twisted-pair wiring and offer unprecedented configuration flexibility but also that it would offer a superior approach to LAN management. Some of the most important management capabilities available through the hub include

- Support for the IEEE 802.3 Repeater Management Standard and the Internet's Repeater Management Information Base (MIB).
- Remote site manageability from a central management station.
- Provision of performance statistics not only at the port level where such information traditionally has been available but also at the module and hub level.
- Autopartitioning, which entails the ability to automatically remove disruptive ports from the network.
- The ability to set performance thresholds that notify managers of a problem or automatically take action to address the problem.

- Port address association features that connect the Ethernet media access control address of a device with the port to which it is attached.
- Source/destination address information, which aids network redesign, traffic redistribution, troubleshooting, and security.

Summary

Overall, 10BaseT LANs offer a sound technical solution for most routine applications. In the office environment, packet throughput and error rates over twisted-pair wiring are the same as with coaxial systems, but the former is limited to shorter distances. The standard offers protection against equipment and media faults that can potentially disrupt the network, and the signaling method used is reasonably immune from most sources of electromagnetic interference commonly found in the office environment. As organizations have added more applications, more users, and more workstations, 10BaseT has been displaced by 100-Mbps (100BaseT) and 1000-Mbps (1000BaseT) Ethernet LANs. To ease the migration, vendors support 10/100BaseT from the same network interface card, and some support 10/100/1000BaseT from the same card.

See also

Ethernet
Ethernet (100BaseT)
Ethernet (1000BaseT)

ETHERNET (100BASET)

100BaseT is the IEEE standard for providing 100-Mbps Ethernet performance and functionality over ubiquitously available unshielded twisted-pair (UTP) wiring. Like 10BaseT Ethernet, this standard specifies a star topology. The need for 100 Mbps came about as a result of the emergence of data-intensive applications and technologies such as multimedia, groupware, imaging, and the explosive growth of high-performance database software packages on PC platforms. All tax today's client-server environments and demand even greater bandwidth for improved response time.

Compatibility

Also known as Fast Ethernet, 100BaseT uses the same contention-based media access control method (MAC)—Carrier Sense Multiple Access with Collision Detection, or CSMA/CD—that is at the core of IEEE 802.3 Ethernet. The Fast Ethernet MAC specification simply reduces the "bit time"—the time duration of each bit transmitted—by a factor of 10, enabling a tenfold boost in speed over 10BaseT. Fast Ethernet's scaled CSMA/CD MAC leaves the remainder of the MAC unchanged. The packet format, packet length, error control, and management information in 100BaseT are all identical to those used in 10BaseT.

Since no protocol translation is required, data can pass between 10BaseT and 100BaseT stations via a hub equipped with a 10/100-Mbps bridge module. Both technologies are also full-duplex capable, meaning that data can be sent and received at the same time. This compatibility enables existing LANs to be inexpensively upgraded to the higher speed as demand warrants.

Media Choices

To ease the migration from 10BaseT to 100BaseT, Fast Ethernet can run over Category 3, 4, or 5 UTP cable while preserving the critical 100-meter (330-foot) segment length between hubs and end stations. The use of fiber allows even more flexibility with regard to distance. For example, the maximum distance from a 100BaseT repeater to a fiberoptic bridge, router, or switch using fiber optic cable is 225 meters (742 feet). The maximum fiber distance between bridges, routers, or switches is 450 meters (1485 feet). The maximum distance between a fiber bridge, router, or switch—when the network is configured for half-duplex—is 2 kilometers (1.2 miles). By interconnecting repeaters with other internetworking devices, large, well-structured networks can be easily created with 100BaseT. The types of media used to implement 100-Mbps Ethernets are summarized as follows:

- *100BaseTX* A two-pair system for data-grade (EIA 568 Category 5) unshielded twisted-pair (UTP) and shielded twisted-pair (STP) cabling
- *100BaseT4* A four-pair system for both voice and data-grade (Category 3, 4, or 5) UTP cabling
- *100BaseFX* A multimode two-strand fiber system

Together, the 100BaseTX and 100BaseT4 media specifications cover all cable types currently in use in 10BaseT networks. Since 100BaseTX, 100BaseT4, and 100BaseFX systems can be mixed and interconnected through a hub, users can retain their existing cabling infrastructure while migrating to Fast Ethernet.

100BaseT also includes a Media-Independent Interface (MII) specification, which is similar to the 10-Mbps Attachment Unit Interface (AUI). The MII provides a single interface that can support external transceivers for any of the 100BaseT media specifications.

Summary

Unlike other high-speed technologies, Ethernet has been installed for over 20 years in business, government, and educational networks. The migration to 100-Mbps Ethernet is made easier by the compatibility of 10BaseT and 100BaseT technologies, making it unnecessary to alter existing applications for transport at the higher speed. This compatibility allows 10BaseT and 100BaseT segments to be combined in both shared and switched architectures, allowing network administrators to apply the right amount of bandwidth easily, precisely, and cost-effectively. Fast Ethernet is managed with the same tools as 10BaseT networks, and no changes to current applications are required to run them over the higher-speed 100BaseT network.

See also

Ethernet

Ethernet (10BaseT)

Ethernet (1000BaseT)

ETHERNET (1000BASET)

Ethernet is a highly scalable LAN technology. It has long been available in two versions—10-Mbps Ethernet and 100-Mbps Fast Ethernet—and a third version has been standardized by the IEEE, offering another order of magnitude increase in bandwidth. Offering a raw data rate of 1000 Mbps or 1 Gbps, so-called Gigabit Ethernet uses the same frame format and size as previous Ethernet technologies. It also maintains full compatibility with the huge installed base of Ethernet nodes through the use of LAN hubs, switches, and routers.

Gigabit Ethernet supports full-duplex operating modes for switch-to-switch and switch-to-end-station connections and half-duplex operating modes for shared connections using repeaters and the CSMA/CD access method. Figure E-7 illustrates the functional elements of Gigabit Ethernet.

The initial efforts in the IEEE 802.3z standards process drew heavily on the use of Fibre Channel and other high-speed networking components. Fibre Channel encoding/

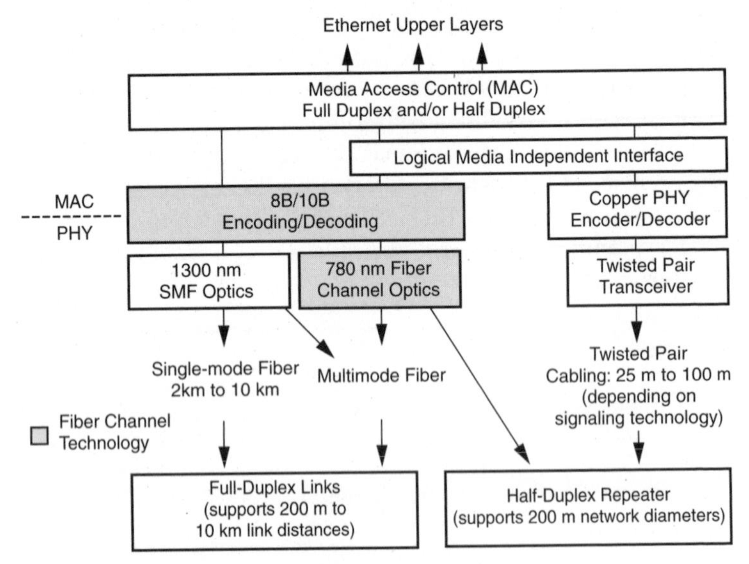

Figure E-7 Functional elements of Gigabit Ethernet.

decoding integrated circuits and optical components were readily available and are specified and optimized for high performance at relatively low costs. The first implementations of Gigabit Ethernet employed Fibre Channel's high-speed, 780-nanometer (short wavelength) optical components for signaling over optical fiber and 8B/10B encoding/decoding schemes for serialization and deserialization. Fibre Channel technology operating at 1.063 Gbps was enhanced to run at 1.250 Gbps, thus providing the full 1000-Mbps data rate for Gigabit Ethernet. Link distances—up to 2 kilometers over single-mode fiber and up to 550 meters over 62.5-micrometer multimode fiber were specified as well.

In mid-1999, the IEEE and the Gigabit Ethernet Alliance formally ratified the standard for Gigabit Ethernet over copper. The IEEE 802.3ab Standard defines Gigabit Ethernet operation over distances of up to 100 meters (330 feet) using four pairs of Category 5 balanced copper cabling. The standard adds a Gigabit Ethernet physical layer to the original IEEE 802.3 Standard, allowing for the higher speed over the existing base of Category 5 unshielded twisted pair wiring. It also allows for autonegotiation between 100- and 1000-Mbps equipment. Table E-1 summarizes Gigabit Ethernet standards for various media.

The initial applications for Gigabit Ethernet will be for campuses or buildings requiring greater bandwidth between routers, switches, hubs and repeaters, and servers. Examples include switch-to-router, switch-to-switch, switch-to-server, and repeater-to-switch links.

At this writing, a draft specification for Ethernet at the Synchronous Optical Network (SONET) OC-192 rate of 10 Gbps is in the works. This technology will be used not for building high-speed campus and building backbones but as a service offering from carriers for metropolitan area networks (MAN) or as a fat pipe to access the Internet. Service providers are looking for a familiar, low-cost, high-speed

Table E-1 A Summary of Gigabit Ethernet Standards for Various Media

Specification	Transmission Facility	Purpose
1000BaseLX	Long-wavelength laser transceivers	Support links of up to 550 meters of multimode fiber or 3000 meters of single-mode fiber
1000BaseSX	Short-wavelength laser transceivers operating on multimode fiber	Support links of up to 300 meters using 62.5-micrometer multimode fiber or links of up to 550 meters using 50-micrometer multimode fiber
1000BaseCX	Shielded twisted-pair (STP) cable spanning no more than 25 meters	Support links among devices located within a single room or equipment rack
1000BaseT	Unshielded twisted-pair (UTP) cable	Support links of up to 100 meters using four-pair Category 5 UTP

SOURCE: IEEE 802.3z Gigabit Task Force.

technology to support networked applications, such as virtual private networks, Internet Protocol (IP) telephony, transparent LAN services, and e-commerce. Ethernet running at 10 Gbps offers all that and more. It reduces operational costs because it is the one technology that can be used from LAN to MAN to WAN. Because there is no need for ancillary equipment, such as protocol translators, equipment and operational costs can be greatly reduced.

Summary

The seamless connectivity to the installed base of 10- and 100-Mbps equipment, combined with Ethernet's scalability and flexibility to handle new applications and data types over a variety of media, makes Gigabit Ethernet a practical choice for high-speed, high-bandwidth networking. 1000BaseT enables the deployment of Gigabit Ethernet into the large installed base of Category 5 cabling, preserving the investment organizations have made in existing cable infrastructure.

See also

Ethernet
Ethernet (10BaseT)
Ethernet (100BaseT)
Fibre Channel

F

FIBER DISTRIBUTED DATA INTERFACE

The Fiber Distributed Data Interface (FDDI) is a 100-Mbps token-passing network that employs a dual counterrotating ring topology for fault tolerance. Originally conceived to operate over multimode fiberoptic cable, the standard has evolved to embrace single-mode fiberoptic cable, shielded twisted-pair copper, and even unshielded twisted-pair copper wiring. It is designed to provide high-bandwidth, general-purpose interconnection between computers and peripherals, including the interconnection of local area networks (LANs) (Figure F-1) and other networks, within a building or campus environment.

FDDI Operation

A timed token-passing access protocol is used to pass frames of up to 4500 bytes in size, supporting up to 1000 connections over a maximum multimode fiber path of 200 kilometers (124 miles) in length. Each station along the path serves as the means for attaching and identifying devices on the network, regenerating and repeating frames sent to it. Unlike other types of LANs, FDDI allows both asynchronous (time-

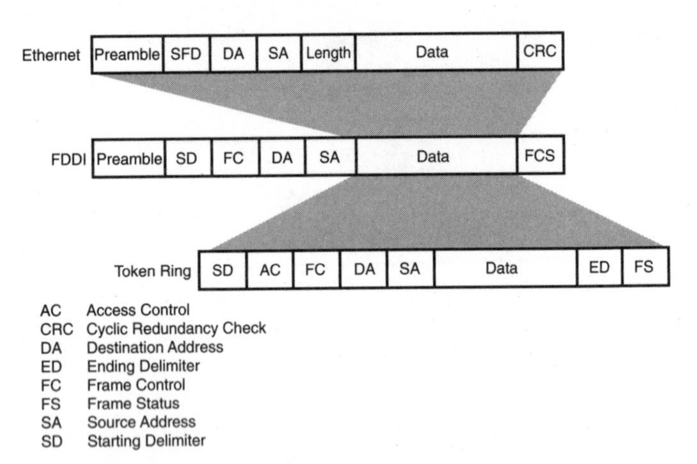

Figure F-1 FDDI can carry Ethernet and Token Ring frames as data, providing a multiprotocol backbone network.

insensitive) and synchronous (time-sensitive) devices to share the network. Synchronous services (e.g., voice and video) are intolerant of delays and must be guaranteed a fixed bandwidth or time slot. Synchronous traffic is therefore given priority over asynchronous traffic, which is better able to withstand delay. FDDI stresses reliability, and its architecture includes integral management capabilities, including automatic failure detection and network reconfiguration.

Any change in the network status—such as power-up or the addition of a new station—leads to a "claim" process during which all stations on the network bid for the right to initialize the network. Every station indicates how often it must see the token to support its synchronous service. The lowest bid represents the station that must see the token most frequently. That request is stored as the Target Token Rotation Time (TTRT). Every station is guaranteed to see the token within 2 × TTRT seconds of its last appearance.

This process is completed when a station receives its own claim token. The winning station issues the first unrestricted token, initializing the network on the first rotation. On the second rotation, synchronous devices may start transmitting. On the third and subsequent rotations, asynchronous devices may transmit if there is available bandwidth. Errors are corrected automatically via a beacon-and-recovery process during which the individual stations seek to correct the situation.

FDDI Architecture

These processes are defined in a set of standards sanctioned by the American National Standards Institute (ANSI). The standards address four functional areas of the FDDI architecture (Figure F-2).

Physical Media–Dependent Sublayer Data are transmitted between stations after converting the data bits into a series of optical pulses. The pulses are then transmitted over the cable linking the various stations. The Physical Media–Dependent (PMD) sublayer describes the optical transceivers—specifically, the minimum optical power and sensitivity levels over the optical data link. This layer also defines the connectors and media characteristics for point-to-point communications between stations on the FDDI network. The PMD sublayer is a subset of the physical layer of the OSI reference model, defining all the services needed to transport a bit stream from station to station. It also specifies the cabling requirements for FDDI-compliant cable plant, including worst-case jitter and variations in cable attenuation.

Physical Layer The Physical Layer (PHY) protocol defines those portions of the physical layer that are media-independent, describing data encoding/decoding, establishing clock synchronization, and defining the handshaking sequence used between adjacent stations to test link integrity. It also provides the synchronization of incoming and outgoing code-bit clocks and delineates octet boundaries as required for the

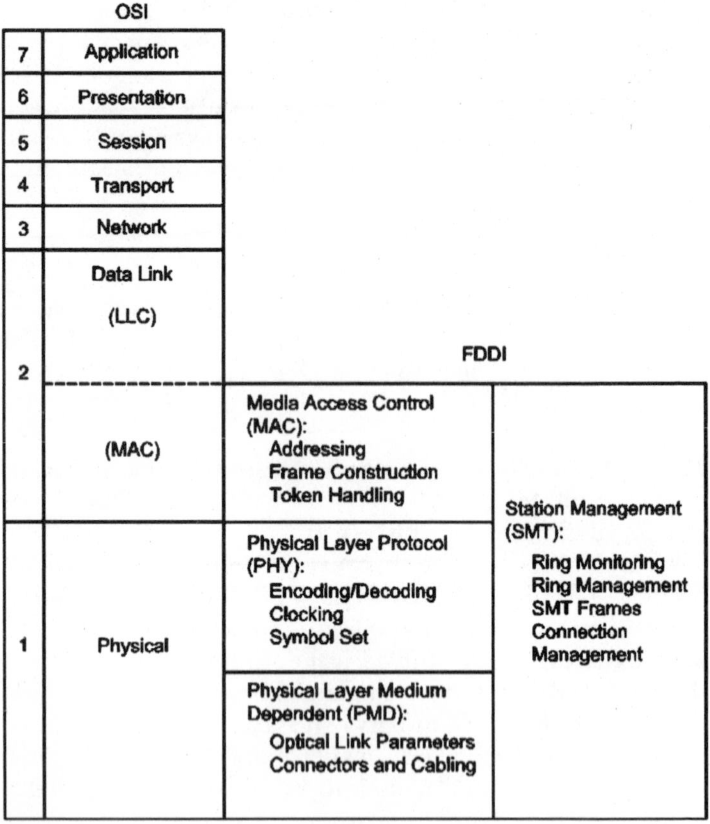

OSI		FDDI	
7	Application		
6	Presentation		
5	Session		
4	Transport		
3	Network		
2	Data Link (LLC)		
	(MAC)	Media Access Control (MAC): Addressing, Frame Construction, Token Handling	Station Management (SMT): Ring Monitoring, Ring Management, SMT Frames, Connection Management
1	Physical	Physical Layer Protocol (PHY): Encoding/Decoding, Clocking, Symbol Set	
		Physical Layer Medium Dependent (PMD): Optical Link Parameters, Connectors and Cabling	

Figure F-2 FDDI layers and their relationship to the seven-layer OSI reference model.

transmission of information to or from higher layers. These processes allow the receiving station to synchronize its clock to the transmitting station.

Media Access Control FDDI's Data Link Layer is divided into two sublayers. The Media Access Control (MAC) sublayer governs access to the medium. It describes the frame format, interprets frame content, generates and repeats frames, issues and captures tokens, controls timers, monitors the ring, and interfaces with station management.

The Logical Link Control (LLC) sublayer is required for proper ring operation and is part of the IEEE 802.2 Standard. In keeping with the IEEE model, the FDDI MAC is fully compatible with the IEEE 802.2 Logical Link Control Standard. Applications that interface to the LLC and operate over existing LANs, such as IEEE 802.3 CSMA/CD or IEEE 802.5 Token Ring, are able to operate over an FDDI network.

The FDDI MAC, like the IEEE 802.5 Token Ring MAC, has two types of protocol data units: a frame and a token. Frames are used to carry data (such as LLC frames), while tokens are used to control a station's access to the network. At the MAC layer, data are transmitted in 4-bit blocks called "4B/5B symbols." The symbol coding is such that 4 bits of data are converted to a 5-bit pattern; thus the 100-Mbps FDDI rate is provided at 125 million signals per second on the medium. This signaling type is employed to maintain signal synchronization on the fiber.

Station Management The Station Management (SMT) facility provides the system management services for the FDDI protocol suite, detailing control requirements for the proper operation and interoperability of stations on the FDDI ring. It acts in concert with the PMD, PHY, and MAC layers. The SMT facility is used to manage connections, configurations, and interfaces. It defines such services as ring and station initialization, fault isolation and recovery, and error control. SMT is also used for statistics gathering, address administration, and ring partitioning.

FDDI Topology

FDDI is a token-passing ring network. Like all rings, it consists of a set of stations connected by point-to-point links to form a closed loop. Each station receives signals on its input side and regenerates them for transmission on the output side. Any number of stations, theoretically, can be attached to the network, although default values in the FDDI standard assume no more than 1000 physical attachments and a 200-kilometer path.

FDDI uses two counterrotating rings: a primary ring and a secondary ring. Data traffic usually travels on the primary ring. The secondary ring operates in the opposite direction and is available for fault tolerance. If appropriately configured, stations may transmit simultaneously on both rings, thereby doubling the bandwidth of the network to 200 Mbps.

Three classes of equipment are used in the FDDI environment: single attached stations (SASs), dual attached stations (DASs), and concentrators (CONs).

A DAS physically connects to both rings, while a SAS connects only to the primary ring via a wiring concentrator. In the case of a link failure, the internal circuitry of a DAS can heal the network using a combination of the primary and secondary rings. If a link failure occurs between a concentrator and a SAS, the SAS becomes isolated from the network.

These equipment types may be arranged in any of three topologies: dual ring, tree, and dual ring of trees (Figure F-3). In the dual-ring topology, DASs form a physical loop, in which case all the stations are dual attached. In a tree topology, remote SASs are linked to a concentrator, which is connected to another concentrator on the main ring.

Any DAS connected to a concentrator performs as a SAS. Concentrators may be used to create a network hierarchy that is known as a "dual ring of trees." This topology offers a flexible hierarchical system design that is efficient and economical. Devices requiring highly reliable communications attach to the main ring, while those less critical attach to branches off the main ring. Thus SAS devices can communicate with the main ring but without the added cost of equipping them with a dual-ring interface or a loop-around capability that would otherwise be required to ensure the reliability of the ring in the event of a station failure.

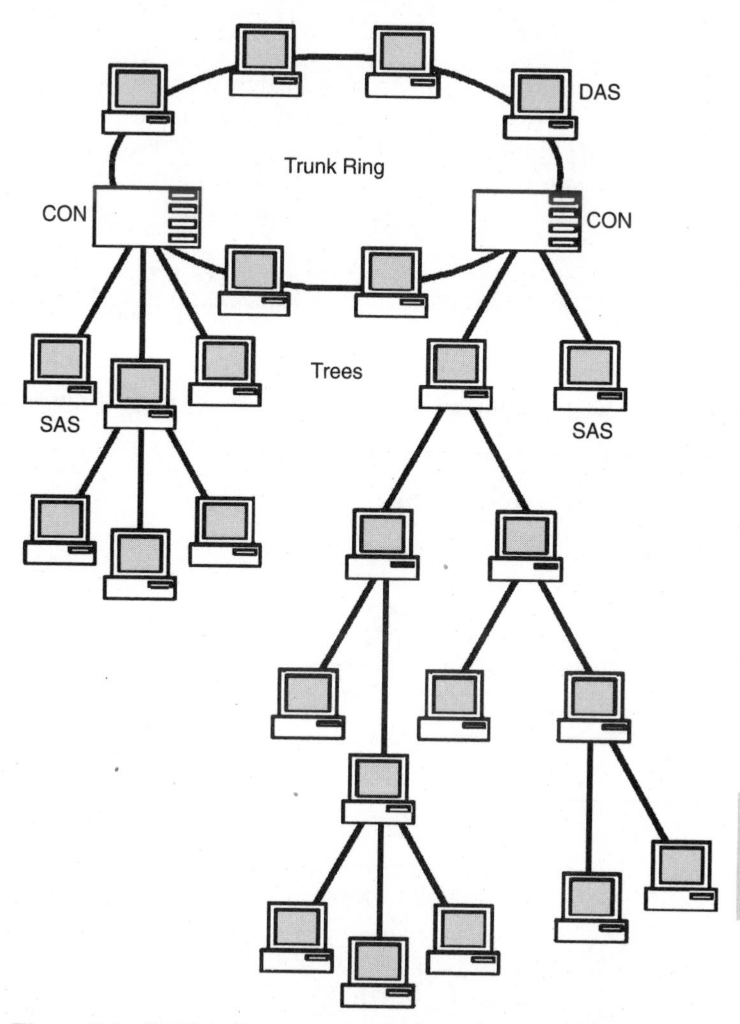

Figure F-3 FDDI dual-ring topology with three types of interconnecting devices.

Failure Protection

FDDI provides an optional bypass switch at each node to overcome a failure anywhere on the network. In the event of a node failure, it is bypassed optically and thus removed from the network. Up to three nodes in sequence may be bypassed; enough optical power will remain to support the operable portions of the network.

In the event of a cable break, the dual counterrotating ring topology of FDDI allows use of the redundant cable to handle normal 100-Mbps traffic. If both the primary and secondary cables fail, the stations adjacent to the failures automatically loop the data around and between rings (Figure F-4), thus forming a new C-shaped ring from the operational portions of the original two rings. When the fault is healed, the network will reconfigure itself again.

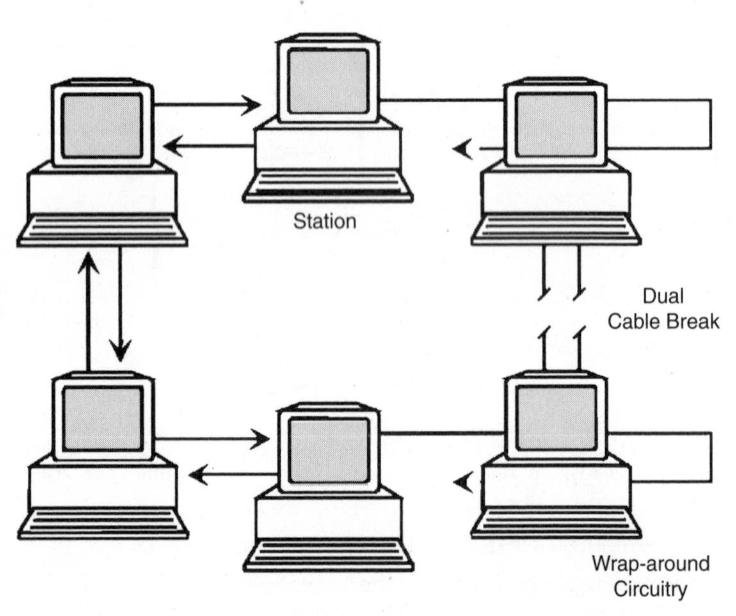

Figure F-4 Self-healing capability of FDDI's dual-ring topology.

Normally, FDDI concentrators offer two buses that correspond to the two FDDI backbone rings. Fault tolerance is also provided for stations that are connected to the ring via a concentrator because the concentrator provides the looparound function for attached stations.

Summary

An extension of FDDI, called FDDI-2, uses portions of its 100-Mbps bandwidth to carry voice and video. FDDI Full Duplex Technology (FFDT) uses the same network infrastructure but can potentially support data rates up to 200 Mbps. However, neither technology has been widely accepted. Furthermore, FDDI is limited by distance and is expensive to implement. Although FDDI offers high reliability and fault tolerance for mission-critical applications, it has been overtaken by more scalable technologies such as Gigabit Ethernet and Fibre Channel.

See also

Ethernet
Fibre Channel

FIBRE CHANNEL

Fibre Channel is a high-performance interconnect standard designed for bidirectional point-to-point serial data channels between desktop workstations, mass storage subsystems, peripherals, and host systems. Serialization of the data permits much greater distances to be covered than in parallel communications. Unlike networks where each node must share the bandwidth capacity of the media, Fibre Channel devices are connected through a flexible circuit/packet switch capable of providing the full bandwidth to all connections simultaneously.

Advantages

The key advantage of Fibre Channel is speed—it is 10 to 250 times faster than typical LAN speeds. Fibre Channel started out by offering a transmission rate of 100 Mbps (200 Mbps in full-duplex mode), the equivalent of 60,000 pages of text per second. Such speeds are achieved simply by transferring data between one buffer at the source device and another buffer at the destination device without regard for how it is formatted—cells, packets, or frames. It is inconsequential what the individual protocols do with the data before or after they are in the buffer—Fibre Channel only provides complete control over the transfer and offers simple error checking.

Unlike many of today's interfaces—including the Small Computer Systems Interface (SCSI)—Fibre Channel is bidirectional, achieving 100 Mbps in both directions simultaneously. Thus it provides a 200-Mbps channel if usage is balanced in both directions. Fibre Channel also overcomes the restrictions on the number of devices that can be connected—up to 126, versus 15 for SCSI.

Fibre Channel overcomes the distance limitations of today's interfaces. A fast SCSI parallel link from a disk drive to a workstation, for example, can transmit data at 20 Mbps, but it is restricted in length to about 20 meters. In contrast, a quarter-speed Fibre Channel link transmits information at 25 Mbps over a single, compact optical cable pair at up to 10 kilometers in length. This allows disk drives to be placed almost anywhere and enables more flexible site planning.

Applications

The high-speed, low-latency connections that can be established using Fibre Channel make it ideal for a variety of data-intensive applications, including

- *Backbones* Fibre Channel provides the parallelism, high bandwidth, and fault tolerance needed for high-speed backbones. It is the ideal solution for mission-critical internetworking. The scalability of Fibre Channel makes it practical to create backbones that grow as one's needs grow—from a few servers to an entire enterprise network.

- *Workstation clusters* Fibre Channel is a natural choice to enable supercomputer-power processing at workstation costs.
- *Imaging* Fibre Channel provides the "bandwidth on demand" needed for high-resolution medical, scientific, and prepress imaging applications, among others.
- *Scientific/engineering* Fibre Channel delivers the needed throughput for today's new breed of visualization, simulation, CAD/CAM, and other scientific, engineering, and manufacturing applications that demand megabytes of bandwidth per node.
- *Mass storage* Current mass storage access is limited in rate, distance, and addressability. Fibre Channel provides mass storage attachments at distances of up to several kilometers. Fibre Channel also interfaces with current SCSI, high performance parallel interface (HIPPI), and IPI-3 connections, among others.
- *Multimedia* Fibre Channel's bandwidth supports real-time videoconferencing and document collaboration between several workstation users and is capable of delivering multimedia applications containing voice, music, animation, and video.

Topology

Fibre Channel uses a flexible circuit/packet-switched topology to connect devices. Through the switch, Fibre Channel is able to establish multiple simultaneous point-to-point connections. Devices attached to the switch do not have to contend for the transmission medium as they do in a network. Through its intrinsic flow control and acknowledgment capabilities, Fibre Channel also supports connectionless traffic without suffering the congestion of the shared transmission media used in traditional networks.

The fabric relieves each Fibre Channel port of the responsibility for station management. All that a Fibre Channel port has to do is manage a simple point-to-point connection between itself and the fabric. If an invalid connection is attempted, the fabric rejects it. If there is a congestion problem en route, the fabric responds with a busy signal, and the calling port tries again.

Fibre Channel Layers

Fibre Channel employs a five-layer stack that defines the physical media and transmission rates, encoding scheme, framing protocol and flow control, common services, and the upper-level applications interfaces.

- FC-0, the lowest layer, specifies the physical characteristics of the media, transmitters, receivers, and connectors that can be used with Fibre Channel, including electrical and optical characteristics, transmission rates, and other physical components of the standard.
- FC-1 defines the 8B/10B encoding/decoding scheme used to integrate the data with the clock information required by serial transmission techniques. Fibre Channel uses 10 bits to represent each 8 bits of "real" data, requiring it to operate at a speed sufficient to accommodate this 25 percent overhead. The 2 extra bits are used for error detection and correction, known as "disparity control."
- FC-2 defines the rules for framing the data to be transferred between ports, the different mechanisms for using Fibre Channel's circuit- and packet-switched service classes (discussed below), and the means of managing the sequence of data transfer. All frames belonging to a single transfer are uniquely identified by sequential numbering from 0 through *n*, enabling the receiver to determine if a frame is missing, as well as which one.
- FC-3 provides the common services required for advanced features such as striping (to multiply bandwidth) and

hunt groups (the ability for more than one port to respond to the same alias address). A hunt group can be likened to a business that has 10 phone lines but requires only a single number to be dialed. Whichever line is free will ring.

- FC-4 provides seamless integration of legacy standards, including FDDI, HIPPI, IPI, SCSI, and IP, as well as IBM's Single Byte Command Code Set (SBCCS) of the Block Multiplexer Channel (BMC), Ethernet, Token Ring, and Asynchronous Transfer Mode (ATM).

Classes of Service

To accommodate a wide range of communications needs, Fibre Channel provides different classes of service at the FC-2 layer.

- *Class 1* This class of service provides exclusive use of the connection for the duration of a session, much like a dedicated physical channel. It is used for time-critical, nonbursty traffic such as a link between two supercomputers.
- *Class 2* A connectionless, frame-switched link that provides guaranteed delivery and confirms receipt of traffic. No dedicated connection is established between ports as in Class 1; instead, each frame is sent to its destination over any available route. This service is typically used for data transfers to and from a shared mass-storage system physically located at some distance from several individual workstations.
- *Class 3* A one-to-many connectionless service that allows data to be sent rapidly to multiple devices attached to the fabric. This service is used for real-time broadcasts and any other application that can tolerate lost packets. Since no confirmation of receipt is given, this service is faster than that provided by the Class 2 level of service.
- *Class 4* A connection-based service that offers guaranteed fractional bandwidth and guaranteed latency levels. This is achieved by allowing users to lock down a physical path through the Fibre Channel switch fabric.

The same switching matrix can support multiple classes of service simultaneously, according to the requirements of each application (Figure F-5).

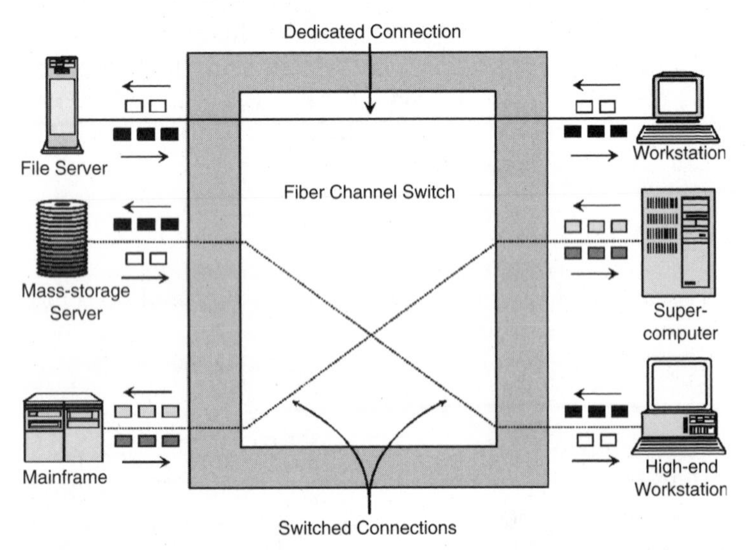

Figure F-5 A switching matrix supporting Fibre Channel's dedicated (Class 1) and switched (Class 2) connections.

Fibre Connection

Fibre Connection (FICON) is a high-speed input-output (I/O) interface for connecting mainframes to storage devices. The channel link speed of FICON is 100 Mbps, more than five times faster than the 17 Mbps of IBM's Enterprise Systems Connection (ESCON), which was the previous standard for transferring data to and from a mainframe over an optical channel connection.

FICON supports full-duplex data transfers—meaning that data can be read and written over the same link at the same time—while ESCON operates just in half-duplex mode. FICON uses a mapping layer that is based on technology developed for Fibre Channel. Because it uses Fibre

Channel's multiplexing capabilities, small data transfers are multiplexed on the link alongside larger transfers. That way, small data transfers, typical for transactions, do not have to wait for large data transfers to complete. However, for users who want to mix and match FICON and ESCON, IBM offers a FICON director that supports both technologies.

Increased Speed

The demands on networks and systems for moving and managing data are increasing exponentially, and improvements in performance across the infrastructure are required to enable users to move and manage their data efficiently and reliably. The current specification for Fibre Channel allow for 1-Gbps speeds and higher. Companies like Brocade, Gadzoox, Qlogic, and Vixel now offer 2-Gbps Fibre Channel products that are based on the FC-SW-2 Open Fabric standard.

FC-SW-2 establishes the foundation for building interoperable, multivendor switch fabrics. Users can connect existing 1-Gbps products with the newer 2-Gbps technology and, through standards-based autonegotiation, extend their current storage area network (SAN) installations instead of having to replace them. Not only does this technology provide a clear path for the industry, but it also holds great promise for new and revolutionary products that greatly extend the capabilities of SANs.

The Fibre Channel Industry Association (FCIA) has introduced a proposal for 10-Gbps Fibre Channel that supports LAN and wide area network (WAN) devices over distances ranging from 15 meters (about 50 feet) to 10 kilometers (about 6 miles). The standard also supports bridging SANs over metropolitan area networks (MANs) through Dense Wave Division Multiplexing and SONET. The 10-Gbps draft specification requires backward compatibility with 1- and 2-Gbps devices; 10-Gbps devices also will be able to use the same cable, connectors, and transceivers used in Ethernet and Infiniband.

At the systems level, SAN architectures also make use of a number of underlying bus technologies, including all variants of SCSI and PCI. The latest innovation in bus technology is Infiniband, a channel-based, switched-fabric architecture that provides scalable performance from 500 Mbps to 6 Gbps, meeting a range of needs from entry-level to high-end enterprise systems. Supported by the computing industry's leading companies, it is anticipated that this new high-speed bus technology will soon replace the current PCI bus standard, since it overcomes I/O bottlenecks for substantial improvements in link speeds between servers and storage, as well as overall data throughput in the server systems themselves.

Summary

Fibre Channel is not a telecommunications solution for the wide area but a cost-effective, high-speed technology for transporting large volumes of data in the local area, where link distances do not exceed 10 kilometers. Fibre Channel's ability to transfer data at high speeds securely and bidirectionally makes it an effective connectivity option for distributed computing environments, particularly those involving mass storage and server clusters. While there is some overlap of capability between ATM and Fibre Channel, each can do something the other cannot. With ATM, it is the ability to span the local and wide areas. With Fibre Channel, it is the ability to attach CPU and peripheral devices directly to a very high-speed network infrastructure. ATM and Fibre Channel are complementary technologies.

See also

Asynchronous Transfer Mode
Ethernet
Infiniband
Synchronous Optical Network

FIREWALLS

Firewalls occupy a strategic position between a "trusted" corporate network and an "untrusted" network, such as the Internet (Figure F-6). They implement perimeter security by monitoring all traffic to and from the enterprise network to determine which packets can pass and which cannot pass. A firewall can identify suspected break-in events and issue appropriate alarms to a management station and then invoke a predefined action to head off the attack. Firewalls also can be used to trace attempted intrusions from the Internet through logging and auditing functions.

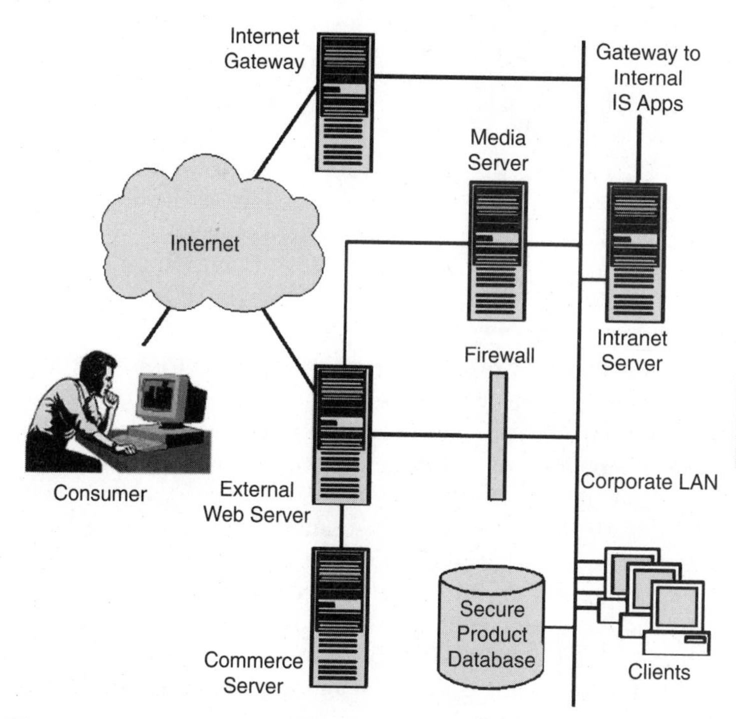

Figure F-6 Firewalls guard resources on the corporate network from access by unauthorized persons on the greater Internet.

Firewalls can be stand-alone devices that are dedicated to safeguarding the enterprise network. Similar functionality can be added to routers, in which case the security features are programmed through the router's operating system. Internet appliances can have firewall capabilities as well, such a Digital Subscriber Line (DSL) routers and cable modems. In addition, there is firewall software that can be loaded into desktop computers that gives users personal control over security (Figure F-7).

Operation

A packet-filtering firewall examines all the packets passed to it and then forwards them or drops them according to predefined rules (Figure F-8). The network administrator can control how packet filtering is performed, permitting or denying connections using criteria based on the source and destination host or network and the type of network service (Figure F-9).

In addition to packet filtering, a firewall offers other useful security features, such as

- *Stateful packet inspection* State information is derived from past communications and other applications to make the control decision for new communication attempts. With this method of security, the packet is intercepted by an inspection engine, which extracts state-related information. It maintains this information in dynamic state tables for evaluating subsequent connection attempts. Packets are allowed to pass only when the inspection engine examines the list and verifies that the attempt is in response to a valid request. The list of connections is maintained dynamically, so only the required ports are opened. As soon as the session is closed, the ports are locked, ensuring maximum security.

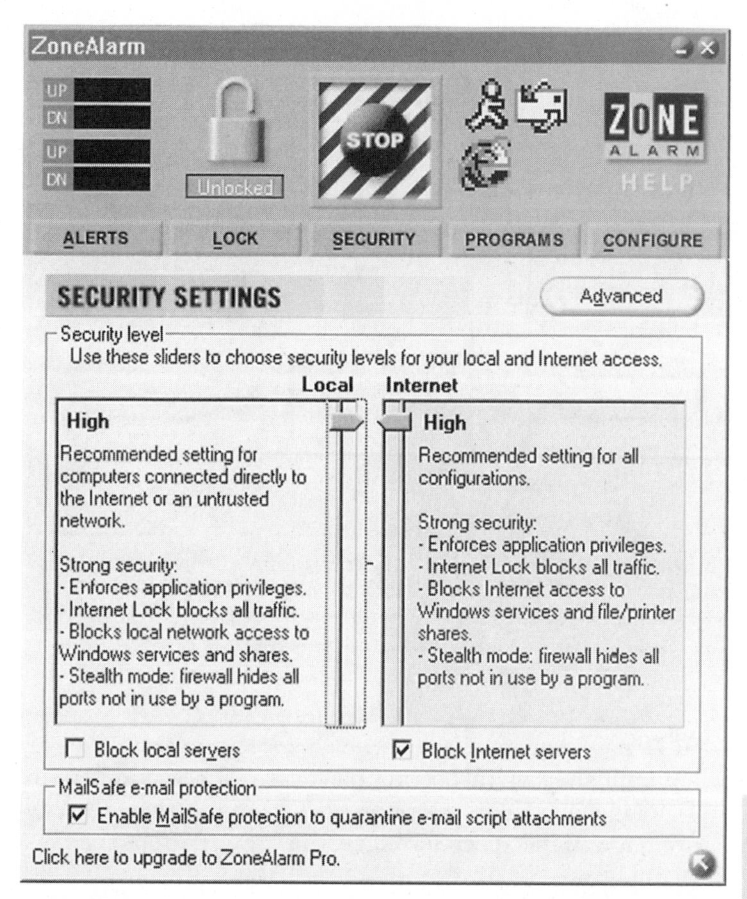

Figure F-7 Firewall software for individual computers allows users to control their own level of security. Shown is ZoneAlarm from Zone Labs, which is available free at the company's Web site for personal use.

- *Network Address Translation (NAT)* Hides internal Internet Protocol (IP) addresses from public view, preventing them from being used for "spoofing"—a technique for impersonating authorized users by using a valid IP address to gain access to an internal network.

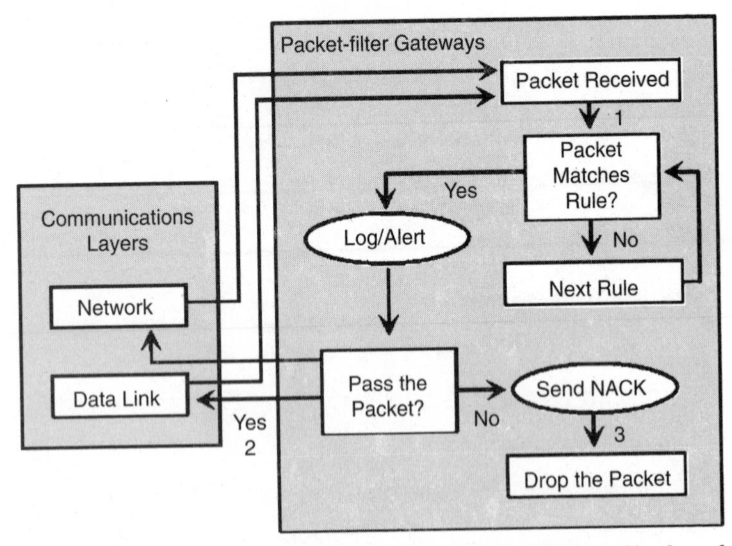

Figure F-8 Operation of a packet-filtering firewall: (1) inbound/outbound packets are examined for compliance with company-defined security rules; (2) packets found to be in compliance are allowed to pass into the network; (3) packets that are not in compliance are dropped.

- *Denial-of-service detection* Defends the network against SYN flooding,[1] port scans, and packet injection. This is accomplished by inspecting packet sequence numbers in Transmission Control Protocol (TCP) connections. If they are not within expected ranges, the firewall drops them as

---

[1] SYN flooding involves sending a continuous stream of bogus messages to a targeted computer, keeping it busy and locking out legitimate users. This method of attack exploits the synchronization (SYN) feature of the Transmission Control Protocol (TCP). When users connect to a Web site, they are actually asking the computer to send back requested information. That request initiates an interaction, called a "handshake," between the computer looking for data and the computer sending data. When the first computer begins to "talk" to the second, it sends a message that essentially means "Hello." The second computer answers with the equivalent of, "Hello, how are you?" On answering back, both computers have established that they are listening, or synchronized. They exchange confirmation of the connection so that the transfer of information can begin. In a SYN-flood attack, hackers send a series of forged messages that do not contain real return addresses to which the Web site's computer can send its response. As a result, the computer under attack waits a long time—as long as a minute—for the second computer to respond. Soon its storage buffers fill up as it struggles to complete the connections, preventing new requests from legitimate users from being answered while it attempts to deal with the congestion problem.

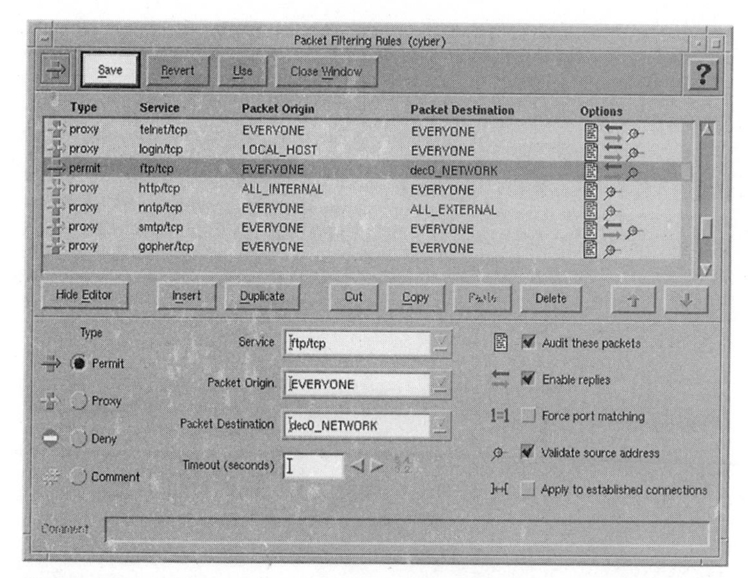

Figure F-9 Like other products, CyberGuard Corp.'s Firewall 3 allows the network administrator to control packet filtering, which permits or denies connections using criteria based on the source and destination host or network and the type of network service.

suspicious packets. When the firewall detects unusually high rates of new connections, it issues an alert message so that appropriate action can be taken.

- *Virus scanning* In addition to hiding viruses, Java applets and ActiveX controls can be used to hide intelligent agents that can give intruders access to corporate resources once they get inside the enterprise network. With the increasing use of Java applets and ActiveX controls on Web sites, more firewalls offer the means to either deny corporate users access to Web pages that contain these elements or to filter such content from the Web pages when they are downloaded.
- *Probe detection* Firewalls offer alarms that are activated when port probing is detected. The alarm system can be configured to watch for TCP or User Datagram Protocol

(UDP) probes from either external or internal networks. Alarms can be configured to trigger e-mail, pop-up windows, or output messages to a local printer.

- *Event logging* Automatically logs system error messages to a console terminal or system log server, allowing administrators to track potential security breaches or other nonstandard activities on a real-time basis (Figure F-10).

Automated Intrusion Detection

Many vendors now offer automated intrusion detection tools for their firewalls. These real-time tools monitor the audit trails of distributed systems for "footprints" that signal suspicious or unauthorized activity on all major operating systems, Web servers, firewalls, routers, applications, databases, and

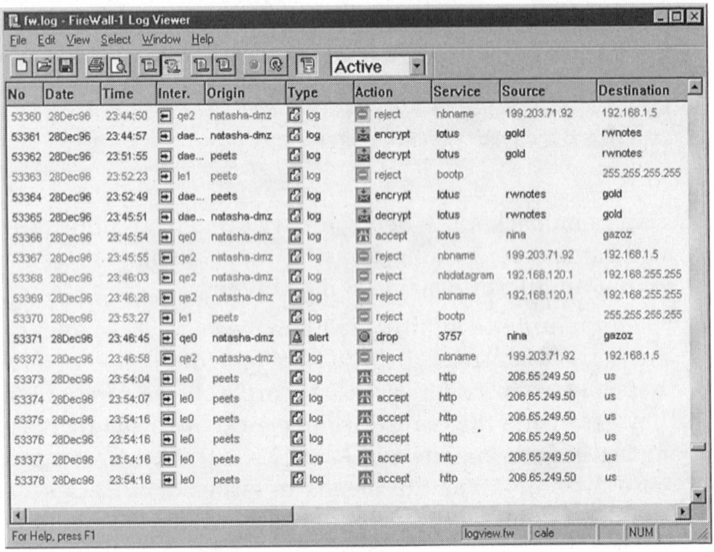
fw.log - FireWall-1 Log Viewer
File Edit View Select Window Help
Active
No Date Time Inter. Origin Type Action Service Source Destination
53360 28Dec96 23:44:50 qe2 natasha-dmz log reject nbname 199.203.71.92 192.168.1.5
53361 28Dec96 23:44:57 dae... natasha-dmz log encrypt lotus gold rwnotes
53362 28Dec96 23:51:55 dae... peets log decrypt lotus gold rwnotes
53363 28Dec96 23:52:23 le1 peets log reject bootp 255.255.255.255
53364 28Dec96 23:52:49 dae... peets log encrypt lotus rwnotes gold
53365 28Dec96 23:45:51 dae... natasha-dmz log decrypt lotus rwnotes gold
53366 28Dec96 23:45:54 qe0 natasha-dmz log accept lotus nina gazoz
53367 28Dec96 23:45:55 qe2 natasha-dmz log reject nbname 199.203.71.92 192.168.1.5
53368 28Dec96 23:46:03 qe2 natasha-dmz log reject nbdatagram 192.168.120.1 192.168.255.255
53369 28Dec96 23:46:28 qe2 natasha-dmz log reject nbname 192.168.120.1 192.168.255.255
53370 28Dec96 23:53:27 le1 peets log reject bootp 255.255.255.255
53371 28Dec96 23:46:45 qe0 natasha-dmz alert drop 3757 nina gazoz
53372 28Dec96 23:46:58 qe2 natasha-dmz log reject nbname 199.203.71.92 192.168.1.5
53373 28Dec96 23:54:04 le0 peets log accept http 206.65.249.50 us
53374 28Dec96 23:54:07 le0 peets log accept http 206.65.249.50 us
53375 28Dec96 23:54:16 le0 peets log accept http 206.65.249.50 us
53376 28Dec96 23:54:16 le0 peets log accept http 206.65.249.50 us
53377 28Dec96 23:54:16 le0 peets log accept http 206.65.249.50 us
53378 28Dec96 23:54:16 le0 peets log accept http 206.65.249.50 us
For Help, press F1
logview.fw
cale
NUM

Figure F-10 Check Point Software Technologies' FireWall-1 includes the Live Connections Monitor, which gives network administrators the ability to view all currently active connections. The live connections are stored and handled in the same way as ordinary log records but in a special file that is continuously updated as connections start and end.

SNMP traps from other network devices. Unlike other intrusion detection tools, which typically report suspicious activity hours or even days after it occurs, the new breed of real-time tools instantly takes action to alert network administrators, shut systems down, terminate offending sessions, execute commands, and take other actions to stop intrusions before they damage critical systems.

As new security threats emerge, network administrators can quickly protect their systems by loading new drop-and-detect scenarios into their firewalls. These are available from the firewall vendors, whose computer security staff focus on hacking techniques and the latest computer security threats. These new scenarios, which can be downloaded from the vendor's Web site and installed enterprise-wide, make it easy for network administrators to keep systems safe from evolving threats.

From a single management workstation, network administrators can quickly drag new security policies and attack scenarios to different enterprise domains, implementing additional protection for hundreds or thousands of systems in a matter of minutes. The management console also provides a correlated, graphical view of security trends, letting network administrators view graphs that illustrate real-time security trends and drill-down to additional details on activity.

Security Appliances

For small remote offices or remote users who cannot justify the expense of an enterprise firewall, there are appliances that combine firewall and virtual private network (VPN) security capabilities. Security is particularly important when DSL or cable is used for Internet access. One drawback of these "always on" services is that they are vulnerable to security attacks.

NetScreen Technologies, for example, offers its NetScreen-5 security appliance for use in small offices and telecommuter sites with DSL or cable modem access. Its stateful-inspection firewall prevents hacker attacks, and the

VPN delivers secure remote access to a corporate network through encrypted tunnels. NetScreen-5 unit is installed between the PC and cable/DSL modem. Since small remote offices typically do not have staff with technical expertise, the box can be configured by a network administrator and shipped out to a user for plug-and-play installation. The network administrator can then centrally manage and reconfigure dispersed units through the NetScreen Global Manager or a Web browser.

Administrators can check the status of multiple NetScreen appliances, monitor performance, troubleshoot existing configurations, or add remote sites to the network from one location. All such activities are conducted via VPN tunnels for the highest level of security.

Risk Assessment

Even after a firewall solution is implemented, it is recommended that a comprehensive security risk assessment be conducted periodically. This helps network administrators identify and resolve security breaches before they are discovered and exploited by hackers and cause serious problems later. There are a number of risk assessment tools available.

Among these tools is bv-Control for Internet Security (formerly known as Hacker-Shield) from BindView, a provider of network scanning and response software. The tool scans and detects networks for potential security holes and offers the user patches or corrective actions to fix the breaches before they become a threat. The tool identifies and resolves security vulnerabilities at both the operating system level and the network level, protecting against both internal and external threats. It also monitors key system files for unauthorized changes and, by referencing a 1 million-word dictionary, identifies vulnerable user passwords through a variety of password-cracking techniques.

A detailed report provides network administrators with a description of each vulnerability and corrective action as well as a ranking of vulnerabilities by the risk they pose to a site's security. Network administrators are also presented with a high-level overview of the vulnerability and its solution with an option to link to a more detailed explanation and reference materials.

Employing an implementation model similar to antivirus products, BindView provides ongoing security updates via the Internet to ensure that users are protected from the latest threats. BindView uses secure push technology to broadcast the vulnerability updates. Users are not required to reinstall the software in order to integrate the updates.

Load Balancing

While the value of firewalls is undisputed, they can degrade performance and create a single point of network failure. This is so because such tasks as stateful packet inspection, encryption, and virus scanning require significant amounts of processing. As traffic load increases, the firewall can become bogged down. And since the firewall sits directly in the data path, it constitutes a single point of failure. If the firewall cannot keep up with filtering all of the packets coming through, it will go down and isolate the whole network behind it.

A solution to this problem is to add a firewall for redundancy and put dynamic load-balancing switches on each side of them. That way, the switches can distribute incoming requests to the firewalls according to their availability. This configuration eliminates the firewalls as single points of failure, dramatically simplifies configuration, and increases end-user performance by balancing traffic among multiple firewalls.

The switches monitor the health of attached firewalls through automatic, periodic health checks. They also monitor physical link status of the switch ports connected to each firewall. Since the firewalls are no longer directly in line and traffic is evenly distributed among them, the end-user experience is improved. The switches automatically recognize failed firewalls and redirect entire sessions through other available firewalls, while maintaining the state of each session.

Managed Firewall Services

For small and medium-sized companies, maintaining a full-time staff of network security professionals at a reasonable cost is virtually impossible. Knowing this, hackers have been known to use the unprotected networks of smaller companies to launch denial-of-service attacks against the e-commerce servers of larger companies. Now smaller companies can implement effective security and mitigate their exposure to risk by subscribing to a managed firewall service.

The service provider performs an initial security assessment, which includes reviewing the customer's security requirements, configuring any required equipment, and managing the remote turnup of management services. Continuous remote monitoring is performed via encrypted channels over the Internet. In scenarios where the firewall resides in a router, that device is managed as well. Performance reports may be accessed from the service provider's secure Web site.

As part of the managed firewall service, periodic reports are furnished that focus on "hot spots" or anomalies in the firewall. Such reports include a performance analysis and recommendations for modifications that will improve throughput and close potential breach points. These recommendations might include software changes, hardware upgrades or changes, or topologic or transport changes.

Summary

Although firewalls can provide a formidable defense against many kinds of attacks, they are not a panacea for all network security problems, particularly those which originate from the corporate side. For example, even if virus scanning is provided at the firewall, it will protect only against viruses that come from the Internet. It does nothing to guard against more likely sources such as floppy disks brought into the company by employees who upload the contents to a desktop computer and inadvertently (or deliberately) spread the virus throughout the network. Companies must take appropriate internal security measures to safeguard mission-critical resources.

See also

Network Security

FRAME RELAY

Frame Relay is a stripped-down version of the X.25 protocol for packet data networking that was designed to transport data very reliably over noisy analog-line networks. In running over cleaner digital-line networks, however, Frame Relay could eliminate many of the functions of X.25, including node-to-node error correction. In stripping away unnecessary functions and relegating error correction to the Frame Relay access devices (FRADs) or routers at the edge of the network, Frame Relay could transport data at much higher speeds than X.25, making it suited for interconnecting LANs over a WAN.

About X.25

In the 1970s, WANs were built using low-speed analog facilities that were used primarily for voice traffic. For reliable data transmission, however, private and public packet-switched networks had to employ the X.25 suite of protocols to overcome noise and other impairments that made the transmission of data difficult. X.25 was endowed with substantial error-correction capabilities so that any node on the network could request a retransmission of errored data from the node that sent them. Errors had to be detected and corrected within the network, since the user's equipment typically did not have the intelligence and spare processing power to devote to this task.

However, the error-correction and flow-control capabilities of the X.25 protocol, plus its many other functions, entail an overhead burden that limits network throughput. This, in turn, limited X.25 to niche applications, such as point-of-sale transaction processing, where the reliable transmission of credit card numbers and other financial information—not speed—was the overriding concern. At the same time, LANs were becoming popular in the 1980s, and there was a growing need to interconnect them over the WAN. Point-to-point T-carrier lines were becoming commercially available, but they were cost prohibitive for all but the largest companies. Frame Relay was developed specifically to provide LAN interconnectivity as a service, eliminating costly leased line charges based on distance.

With the increasing use of digital facilities, there is less need for error protection. At the same time, the end devices increased in intelligence, processing power, and storage capacity, making them better at handling error control and diverse protocols. Consequently, the communications protocols used over the network may be scaled down to its bare essentials to greatly increase throughput. This is the idea behind Frame Relay, which can support voice traffic, as well as data, packaged in variable-sized frames of up to 4000 bytes in length.

Frame Relay was introduced commercially in May 1992. It initially gained acceptance as a method for providing end users with a solution for data connectivity requirements, such as LAN-to-LAN connections. Frame Relay provided both an efficient and flexible data transport mechanism and also allowed for a cheaper bandwidth cost associated with connecting legacy Systems Network Architecture (SNA) networks.

Whereas X.25 operates at the bottom three layers of the Open Systems Interconnection (OSI) reference model, Frame Relay operates at the first layer and the lower half of the second layer. This cuts the amount of processing by as much as 50 percent, improving network throughput. Although the Frame Relay network can detect errors, it does not correct them. Bad frames are simply discarded. When the receiving device detects corrupt or missing frames, it can request a retransmission from the originating device, whereupon the appropriate frames are sent again.

Advantages of Frame Relay

The most compelling advantages of a carrier-provided Frame Relay service include

- *Improved throughput/low delay* Frame Relay service uses high-quality digital circuits end to end, making it possible to eliminate the multiple levels of error checking and error control. The result is higher throughput and less delay compared to legacy packet-switched networks like X.25.
- *Any-to-any connectivity* Any node connected to the Frame Relay service can communicate with any other node via predefined permanent virtual circuits (PVCs) or dynamically via switched virtual circuits (SVCs).
- *No long-distance charges* Since Frame Relay is offered as a service over a shared network, the need for a highly meshed private-line network is eliminated, for substantial cost savings. There are no distance-sensitive charges with Frame Relay, as there is with private lines.
- *Oversubscription* Multiple PVCs can share one access link, even exceeding the port speed of the Frame Relay switch. In oversubscribing the port, multiple users can access the Frame Relay network—but not all at the same time—eliminating the cost of multiple private-line circuits and their associated customer premises equipment (CPE), for further cost savings.
- *Higher speeds* Whereas X.25 tops out at 56 kbps, Frame Relay service supports transmission speeds up to 44.736 Mbps. If the Frame Relay switches in the network support

Frame Relay Forum Implementation Agreement 14 (FRF 14), speeds at the OC-3 rate of 155 Mbps and the OC-12 rate of 622 Mbps over fiber backbones are possible.

- *Simplified network management* Customers have fewer circuits and less equipment to monitor. In addition, the carrier provides proactive monitoring and network maintenance 24 hours a day.
- *Intercarrier connectivity* Frame Relay service is compatible between the networks of various carriers, through network-to-network interfaces (NNIs), enabling data to reach locations not served by the primary service provider.
- *Customer-controlled network management* Allows customers to obtain network management information via in-band SNMP queries and pings launched from their own network management stations.
- *Performance reports* Enables customers to manage their Frame Relay service to maximum advantage. Available network reports, accessible on the carrier's secure Web site, include those for utilization, errors, health, trending, and exceptions.
- *Service-level guarantees* Frame Relay service providers offer customers service-level agreements (SLAs) that specify availability as a percentage of uptime, round-trip delay expressed in milliseconds, and throughput in terms of the committed information rate (CIR). If the carrier cannot meet the SLA, it credits the customer's invoice accordingly.

Types of Circuits

Packet networks make use of virtual circuits, sometimes referred to as "logical channels." The two primary types of virtual circuits supported by Frame Relay are switched virtual circuits (SVCs) and permanent virtual circuits (PVCs). SVCs are analogous to dial-up connections that require path setup and tear-down. A key advantage of SVCs is that they permit any-to-any connectivity between devices connected to the Frame Relay network. PVCs are more like dedicated private lines; once set up, the predefined logical connections between sites attached to the Frame Relay network stay in place. This allows logical channels to be dedicated to specific terminals. The SVC requires fewer logical channels at the host because the terminals contend for a lesser number of logical channels. Of course, it is assumed that not everyone will require access to the host at the same time.

Another type of virtual circuit is the multicast virtual circuit (MVC), which is used to broadcast the same data to a group of users over a reserved data link connection in the Frame Relay network. This type of virtual circuit might be useful for expediting communications among members of a single workgroup dispersed over multiple locations or to facilitate interdepartmental collaboration on a major project. It also can be used for broadcast faxing, news feeds, and "push" applications.

The same Frame Relay interface can be used to set up SVCs, PVCs, and MVCs. All three may share the same digital facility. In supporting multiple types of virtual circuits, Frame Relay networks provide a high degree of configuration flexibility, as well as more efficient utilization of the available bandwidth.

The virtual circuits have a committed information rate (CIR), which is the minimum amount of bandwidth the carrier agrees to provide for each virtual circuit. If some users are not accessing the Frame Relay network at any given time, extra bandwidth becomes available to users who are online. The CIR of their virtual circuit can burst up to the full port speed. As other users come online, however, the virtual circuits that were bursting beyond their CIR must back down to the assigned CIRs.

Congestion Control

Real-time congestion control must accomplish the following critical objectives in a frame relay network:

- Maintain high throughput by minimizing timeouts and out-of-sequence frame deliveries.
- Prevent session disconnects, unless required for congestion control.
- Protect against unfair users who attempt to hog the available network resources by exceeding their CIR or established burst size.
- Prevent the spread of congestion to other parts of the network.
- Provide delays consistent with application requirements and service objectives.

In the Frame Relay network, congestion can be avoided through control mechanisms that provide backward explicit congestion notification (BECN) and forward explicit congestion notification (FECN), which are depicted in Figure F-11.

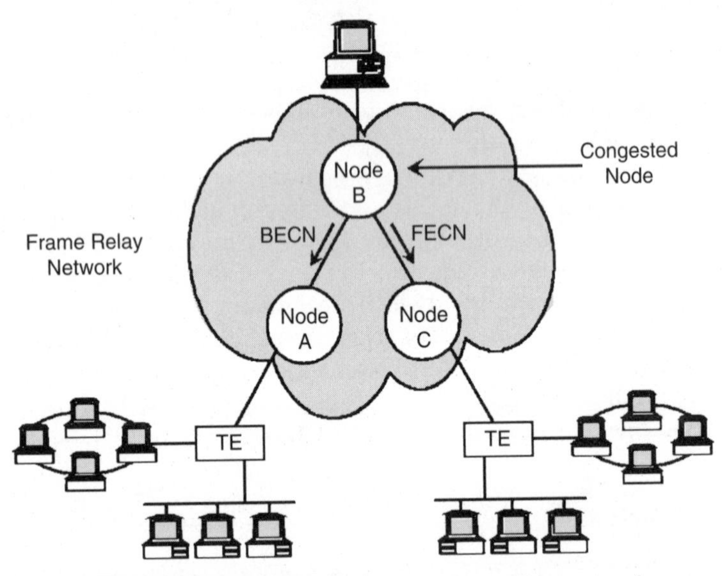

Figure F-11 Congestion notification on the Frame Relay network.

BECN is indicated by a bit set in the data frame by the network to notify the user's equipment that congestion-avoidance procedures should be initiated for traffic in the opposite direction of the received frame. FECN is indicated by a bit set in the data frame by the network to notify the user that congestion-avoidance procedures should be initiated for traffic in the direction of the received frame. On receiving either indication, the end point (i.e., bridge, router, or other internetworking device) takes appropriate action to ease congestion.

The response to congestion notification depends on the protocols and flow-control mechanism employed by the end point. The BECN bit typically would be used by protocols capable of controlling traffic flow at the source. The FECN bit typically would be used by protocols implementing flow control at the destination.

On receipt of a frame with the BECN bit set, the end point must reduce its offered rate to the CIR for that Frame Relay connection. If consecutive data frames are received with the BECN bit set, the end point must reduce its rate to the next "step" rate below the current offered rate. The step rates are 0.675, 0.50, and 0.25 of the current rate. After the end point has reduced its offered rate in response to receipt of BECN, it may increase its rate by a factor of 0.125 times the current rate after receiving two consecutive frames with the BECN bit clear.

If the end point does not respond to the congestion notification, or the user's data flow into the network is not significantly reduced as a result of the response to the congestion notification, or an end point is experiencing a problem that exacerbates the congestion problem, the network switches collaborate in implementing congestion-recovery procedures. These procedures include discarding frames, in which case, the end-to-end protocols employed by the end points are responsible for detecting and requesting the retransmission of missing frames.

Frame discard can be done on a priority basis; that is, a decision is made on whether certain frames should be discarded in preference to other frames in a congestion situation based on predetermined criteria. Frames are discarded on the basis of their "discard eligibility" setting of 1 or 0, as specified in the data frame. A setting of 1 indicates that the frame should be discarded during congestion, while a setting of 0 indicates that the frame should not be discarded unless there are no alternatives.

The discard eligibility may be determined in several ways. The user can declare whether the frames are eligible for discard by setting the discard eligibility bit in the data frame to 1. Or the network access interface may be configured to set the discard eligibility bit to 1 when the user's data have exceeded the CIR, in which case the data are considered excess and subject to discard. For users who subscribe to CIR = 0, which moves data through the Frame Relay network on a best-effort basis subject to bandwidth availability, all traffic is discard-eligible.

Frame Relay Charges

Frame Relay charges differ by carrier and may differ further by configuration. Accordingly, Frame Relay service charges may include

- Port charge for access to the nearest Frame Relay switch, which is applied to every user location attached to the Frame Relay network.
- Local loop charge, which is the monthly cost of the facility providing access to the Frame Relay network. This charge may not apply if the customer's building is directly connected to the carrier's metropolitan fiber ring, in which case the customer is charged only a one-time setup fee.
- Charges for the PVCs and SVCs, which are determined according to the CIR assigned to each virtual circuit.
- Burst capability, usually determined by the burst excess size. Most carriers do not specifically charge customers for bursting beyond the CIR.

- Customer premises equipment (CPE), which includes the Frame Relay or internetworking access equipment optionally leased from the service provider and bundled into the cost of the service.
- Intra–local access and transport area (LATA)/interLATA service. Usually, there is one price for "local" Frame Relay service and another price for "national" Frame Relay service. Neither is distance sensitive, however.

Voice over Frame Relay

Voice over Frame Relay (VoFR) is receiving growing attention. Most data-oriented Frame Relay access devices (FRADs) and routers use the first-in, first-out (FIFO) method of handling traffic. In order to achieve the best voice quality, however, voice frames cannot be allowed to accumulate behind a long queue of data frames. Voice FRADs and routers, therefore, employ traffic prioritization schemes to minimize delay for voice traffic.

Traffic prioritization schemes ensure that voice packets have preference over data. During times of network congestion, one of the easiest prioritization methods is to simply discard frames. In such cases, data rather than voice frames will be discarded first, giving voice a better chance of making it through the network.

Some service providers offer prioritization of PVCs within the Frame Relay network. Prioritization features on both the CPE and the Frame Relay network can result in better voice application performance. The CPE ensures that higher-priority traffic is sent to the network first, while PVC prioritization within the network ensures that higher-priority traffic is delivered to its destination first.

VoFR equipment compresses the voice signal from 64 to at least 32 kbps. In most cases, compression to 16 or even 8 kbps is possible. Some equipment vendors support dynamic compression options. When bandwidth is available, a higher voice quality is achieved using 32 kbps, but as other calls are placed or other traffic requires bandwidth, the 16- or 8-kbps compression algorithm is implemented. Most voice FRADs also support fax traffic. A fax can take up as little as 9.6 kbps of bandwidth for each active line.

VoFR usually allows a company to use its existing phones and numbering plan. In most cases, an internal dialing plan can be set up that allows users to dial fewer digits to connect to internal locations.

A persistent myth about VoFR is that voice calls can be carried free on an existing Frame Relay network. In fact, VoFR requires special CPE and entails an increase in the port speed and possibly an increase of the CIR—all of which have a cost.

Standards

Formed in 1991, the Frame Relay Forum is an association of vendors, carriers, users, and consultants committed to the implementation of Frame Relay in accordance with national and international standards. The forum's technical committee takes existing standards, which are necessary but not sufficient for full interoperability, and creates Implementation Agreements. These documents represent an agreement by all members of the Frame Relay community as to the specific manner in which standards will be applied, thus helping to ensure interoperability. As of mid-2001, the forum had issued the following Implementation Agreements:

- FRF.1.2 User-to-Network (UNI) Implementation Agreement, April 2000
- FRF.2.1 Frame Relay Network-to-Network Interface (NNI) Implementation Agreement, July 1995
- FRF.3.2 Multiprotocol Encapsulation Implementation Agreement (MEI), April 2000
- FRF.4.1 SVC User-to-Network Interface (UNI) Implementation Agreement, January 2000

- FRF.5 Frame Relay/ATM Network Interworking Implementation, December 1994
- FRF.6 Frame Relay Service Customer Network Management Implementation Agreement (MIB), March 1994
- FRF.7 Frame Relay PVC Multicast Service and Protocol Description, October 1994
- FRF.8.1 Frame Relay/ATM PVC Service Interworking Implementation Agreement, February 2000
- FRF.9 Data Compression Over Frame Relay Implementation Agreement, January 1996
- FRF.10.1 Frame Relay Network-to-Network SVC Implementation Agreement, September 1996
- FRF.11.1 Voice over Frame Relay Implementation Agreement, May 1997; Annex J added March 1999
- FRF.12 Frame Relay Fragmentation Implementation Agreement, December 1997
- FRF.13 Service Level Definitions Implementation Agreement, August 1998
- FRF.14 Physical Layer Interface Implementation Agreement, December 1998
- FRF.15 End-to-End Multilink Frame Relay Implementation Agreement, August 1999
- FRF.16 Multilink Frame Relay UNI/NNI Implementation Agreement, August 1999
- FRF.17 Frame Relay Privacy Implementation Agreement, January 2000
- FRF.18 Network-to-Network FR/ATM SVC Service Interworking Implementation Agreement, April 2000
- FRF.19 Frame Relay Operations, Administration and Maintenance Implementation Agreement, March 2001
- FRF.20 Frame Relay IP Header Compression Implementation Agreement, June 2001

Summary

The need for Frame Relay arose largely out of the need to interconnect LANs at different locations. Since Frame Relay was designed to operate over digital networks, which are faster and less prone to transmission errors than older analog lines, there was less need for the network to perform error correction. This could be effectively handled by the CPE at the edges of the network. The X.25 protocol overcame the limitations of analog lines but did so with a significant performance penalty, due mainly to its extensive error-checking and flow-control capabilities. In being able to do without these and other functions, Frame Relay offers higher throughput, less delay, and more efficient utilization of the available bandwidth.

See also

Asynchronous Transfer Mode

Switched Multimegabit Data Service

G

GATEWAYS

Gateways are used to interconnect dissimilar networks or applications. Gateways operate at the highest layer of the Open Systems Interconnection (OSI) reference model—the Application Layer (Figure G-1). A gateway consists of protocol conversion software that usually resides in a server, minicomputer, mainframe, or front-end device. One application of gateways is to interconnect disparate networks or media by processing the various protocols used by each so that information from the sender is intelligible to the receiver despite differences in network protocols or computing platforms.

For example, when an Systems Network Architecture (SNA) gateway is used to connect an asynchronous PC to a synchronous IBM SNA mainframe, the gateway acts as both a conduit through which the computers communicate and as a translator between the various protocol layers. The translation process consumes considerable processing power, resulting in relatively slow transmission rates when compared with other interconnection methods—hundreds of packets per second for a gateway versus tens of thousands of packets per second for a bridge.

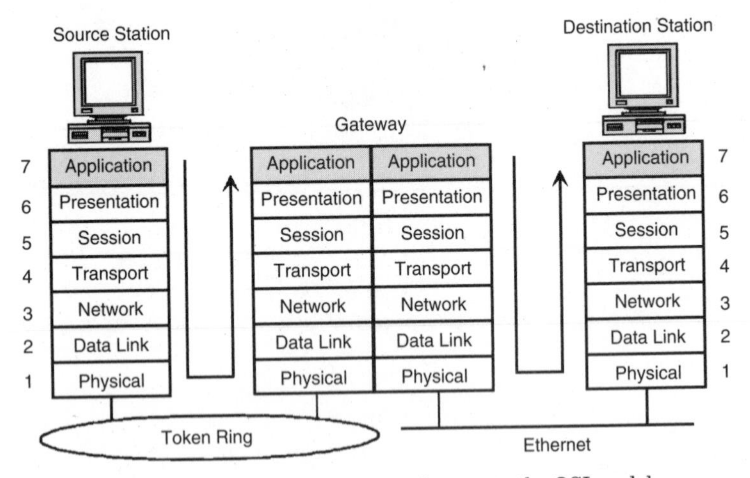

Figure G-1 Gateway functionality in reference to the OSI model.

In addition to its translation capabilities, a gateway can check on the various protocols being used, ensuring that there is enough protocol processing power available for any given application. It also can ensure that the network links maintain a level of reliability for handling applications in conformance with predefined error rate thresholds.

Gateway Applications

Gateways have a variety of applications. In addition to facilitating local area network (LAN) workstation connections to various host environments, such as IBM's SNA systems and midrange systems, they facilitate connections to X.25 packet-switching networks. Other applications of gateways include the interconnection of various electronic-mail systems, enabling mail to be exchanged between normally incompatible formats. The gateway function is actually provided by servers equipped with the X.400 international messaging protocol.

In some cases, gateways can be used to consolidate hardware and software. An SNA 3270 gateway shared among multiple networked PCs, for example, can be used in place of IBM's 3270 Information Display System or many individual 3270 emulation products. Although the IBM systems offer a standard means of achieving the PC-host connection, it is expensive when used to attach a large number of stand-alone PCs. The relatively high connection cost per computer discourages host access for occasional users and limits the central control of information.

If the PCs are attached to a LAN, however, one gateway can emulate a cluster controller and thereby provide all workstations with host access at a very low cost. Cluster controller emulators use an RS-232C or compatible serial interface to a host adapter or communications controller, such as an IBM 3720 or 3745. They can support up to 254 simultaneous sessions.

IP-PSTN Gateways

A relatively new type of gateway provides connections between the Internet and the Public Switched Telephone Network (PSTN), enabling users to place phone calls from their multimedia PCs or conventional telephones over the Internet or a carrier's managed Internet protocol (IP) data network and vice versa. This arrangement allows users to save on long-distance and international call charges.

The IP-PSTN gateways perform the translations between the two types of networks. When a standard voice call is received at a near-end gateway, the analog voice signal is digitized, compressed, and packetized for transmission over an IP network. At the far-end gateway, the process is reversed, with the packets decompressed and returned to their original digital form for delivery to the nearest Class 5 central office.

The gateways support one or more of the internationally recognized G.7xx voice codec specifications for toll-quality voice. The most commonly supported codec specifications are as follows:

- G.711 describes the requirements for a codec using Pulse Code Modulation (PCM) of voice frequencies to achieve 64 kbps, providing toll-quality voice on managed IP networks with sufficient available bandwidth.
- G.723.1 describes the requirements for a dual-rate speech codec for multimedia communications (e.g., videoconferencing) transmitting at 5.3 and 6.3 kbps. This codec provides near-toll-quality voice on managed IP networks.[1]
- G.729A describes the requirements for a low-complexity codec that transmits digitized and compressed voice at 8 kbps. This codec provides toll-quality voice on managed IP networks.

The specific codec to be used is negotiated on a call-by-call basis between the gateways using the H.245 control protocol. Among other things, the H.245 protocol provides for capability exchange, enabling the gateways to implement the same codec at the time the call is placed. The gateways may be configured to implement a specific codec at the time the call is established, based on predefined criteria, such as the following:

- Use G.711 only, in which case the G.711 codec will be used for all calls.
- Use G.729 (A) only, in which case the G.729 (A) codec will be used for all calls.
- Use highest-common-bit-rate codec, in which case the codec that will provide the best voice quality is selected.
- Use lowest-common-bit-rate codec, in which case the codec that will provide the lowest packet bandwidth requirement is selected.

This capability exchange feature provides carriers and Internet service providers (ISPs) with the flexibility to offer

[1] The mean opinion score (MOS) used to rate the quality of speech of codecs measures toll-quality voice as having a top score of 4.0. With G.723.1, voice quality is rated at 3.98, which is only 2 percent less than that of analog telephone.

different quality voice services at different price points. It also allows corporate customers to specify a preferred proprietary codec to support voice or a voice-enabled application through an intranet or IP-based virtual private network (VPN).

Summary

Gateways are available as software or may be dedicated hardware systems equipped with appropriate software to make the translations between different applications, networked devices, or different types of networks. Gateways can even be used to reconcile the differences between network management systems or operations support systems (OSSs), enabling them to interoperate with the systems of other vendors.

See also

Bridges
Open Systems Interconnection (OSI)
Repeaters
Routers

HELP DESKS

With the proliferation of hardware and software throughout an organization, information technology (IT) and telecom managers are faced with the task of providing troubleshooting assistance to branch offices, telecommuters, and mobile professionals, as well as internal users, workgroups, and departments. The consequences of not providing adequate levels of assistance are too compelling to ignore: lost corporate productivity, slowed responses to competitive pressures, and eventual loss of market share. One way to efficiently and economically service the needs of a growing population of computer and communications users is to set up a help desk.

Briefly, the help desk acts a central clearinghouse for support issues and is manned by a technical staff that addresses support problems and attempts to solve them in-house before calling in vendors or carriers. The help desk operator logs every call and, if possible, attempts to isolate the cause of the problem. Help desk operators are usually able to answer from 50 to 70 percent of all calls without having to pass them to another authority. If the problem cannot be solved over the phone, the operator transfers the problem to someone with more expertise or dispatches a technician and monitors progress to a satisfactory conclusion before closing out the transaction.

Although the concept of the help desk originated in the mainframe environment, with increased corporate reliance on local area networks (LANs) and the Internet, the role of the help desk has expanded into realms not ventured into by many mainframe professionals. Fortunately, there are now software packages available that assist with help desk administration. There are also problem-determination tools based on expert systems that can quickly bring untrained help desk personnel up to competency. Aside from handling calls from users, help desks can provide such services as order and delivery tracking, asset and inventory tracking, preventive maintenance, and vendor performance monitoring.

Help desk support has been extended to the Internet. With the ability to update transactions via the Web, field technicians now can update information pertaining to outstanding requests from any location where they have Web browser access. Users can perform a range of actions such as authorizing a change request and adding new information to an existing trouble ticket. Help desk applications and data generate Web forms (schemas) and hyperlinks dynamically (Figure H-1). This lets help desk managers focus on serving customers and improving business processes instead of constantly maintaining static Web pages.

Benefits

Establishing a centralized help desk to coordinate the resolution of system problems offers a number of benefits. Users have a single number to remember, support personnel are assured of an orderly, controlled flow of tasks and assignments, and managers are provided with an effective means of tracking problems and solutions. The help desk provides users with a "warm and fuzzy" level of support. Knowing that someone is available to solve any problem or even to help them find their way through unfriendly documentation

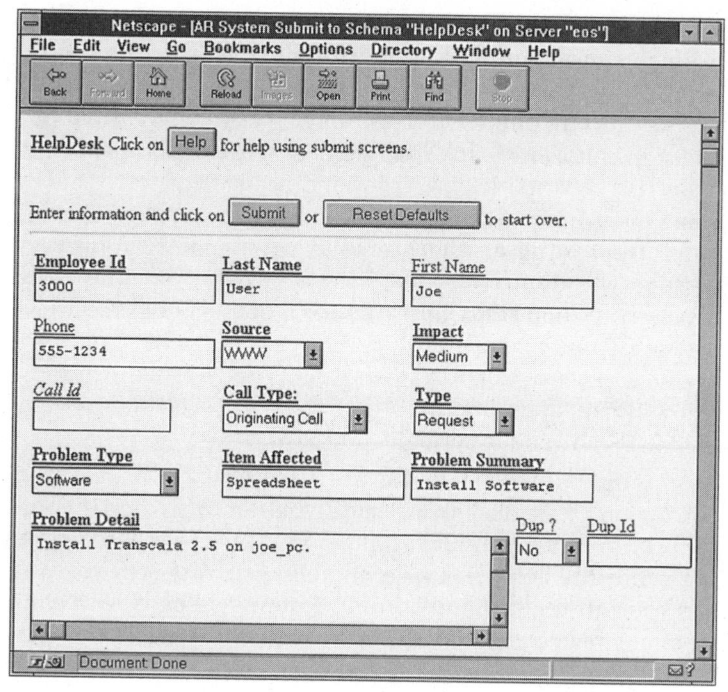

Figure H-1 Remedy Corporation's Action Request System has extensions to the Web. The company's ARWeb client lets organizations create a Web-based help desk that can be accessed by anyone with a Web browser. In this case, a user can access a Web form for reporting a problem to the help desk.

adds to an individual's confidence and willingness to learn new applications and office technologies.

Although a help desk costs money, it can pay for itself in ways that, unfortunately, can be hard to quantify. The fact is that most companies have millions of dollars invested in computer and communications systems. They also have millions of dollars invested in people. To ensure that both are utilized to optimal advantage, there should be an entity in place that is capable of solving the many and varied problems that inevitably arise.

Determining what level of support the help desk should provide can be difficult. One way to determine the proper level of support is to have a system that automatically tracks calls to the help desk and records what problems users encounter most. That way, people with appropriate expertise can be identified to lend assistance or recruited from outside the organization to fill in any knowledge gaps. Depending on the problems, in-house training may provide the dual benefits of helping users become more productive and reducing the support burden of IT and telecom staff.

Internal versus External Support

Assuming that these benefits are attractive, the next step in setting up a help desk is to decide whether to provide the service internally or rely on an outside vendor. Another option is to offer the help desk as a billable corporate service. This can prevent the help desk from becoming deluged with calls from users who can just as easily look up the information in manuals or use the online help facilities that come with many application packages.

While it is important that the help desk staff have technical skills, "people" skills are much more useful. When a user calls with a problem, the support person must extract information from someone who does not know the technical jargon needed to reach a solution or how to use seemingly arcane procedures to isolate and solve problems.

The help desk operator can carry out all the maintenance activities that once required the dispatch of an on-site technician to accomplish. When users had a problem before, the help staff or technician had to go to users' desks and physically look over their shoulder. With a remote-control product, the help desk operator can stay put, hit a few keys, and instantly see a user's screen and control the keyboard. By taking control of the user's computer, the help desk operator can often assess the problem immediately.

An alternative to internally staffed help desks is to subscribe to a commercial service. Subscribers typically call an 800 number to get help in the use of Windows-based PCs, software programs, and peripherals, for example. A number of pricing schemes are available. An annual subscription, for instance, may entitle the company to unlimited advice and consultation. Per-call pricing is also available. The cost of basic services is usually $10 to $15 per call, depending on the volume of support calls. Depending on the service provider, calls are usually limited to 15 or 20 minutes. If the problem cannot be resolved within that time, there may even be no charge for the call.

Resolution can be anything from talking the caller through a system reboot to helping the user ascertain that a PC-to-mainframe link is out of order and what must be done to get it back into service. Callers also can obtain solutions to common software problems, from how to recover an erased file, to importing files from one package to another, to modifying system configuration files when adding new applications packages or hardware.

Some hotline services even offer advice on the selection of software and hardware products, provide software installation support, and guide users through maintenance and troubleshooting procedures. Other service providers specialize in LANs of one type or another.

Since many callers are not familiar with the configuration details of their hardware and software, this information is compiled into a subscriber profile for easy access by online technicians. With this information readily available, the time spent with any single caller is greatly reduced. This helps to keep the cost per call low, which is what attracts new customers. The cost of a customer profile varies greatly, depending on the size of the database that must be compiled. It may be based on the number of potential callers or entail a flat yearly charge for the entire organization.

The operations of some service providers are becoming quite sophisticated. To service their clients, staff may access a shared knowledge base that may contain tens of thousands of questions and answers. Sometimes called "expert systems," the initial knowledge base is compiled by technical experts in their respective fields. As online technicians encounter new problems and devise solutions, this information is added to the knowledge base for immediate access when the same problem is encountered at a later time.

Even users with access to vendors' free help line services can benefit from a third-party help line. This is so because the third-party service may cover situations where more than one application or hardware platform is being used, whereas vendor help line services provide assistance only for their products. Moreover, most users rarely bother to phone the vendors because of constantly busy lines. So rather than supplanting vendor services and in-house support desks (which may be overburdened), third-party services may be used to complement them.

In addition to delivering a variety of help services, for an extra charge, third-party service providers can issue call-tracking and accounting reports to help clients keep a lid on expenses for this kind of service and allocate expenses appropriately among departments and other internal cost centers.

Summary

The help desk can improve the quality of service while decreasing service costs. In addition, it frees up experts to work in other areas and provides consistent answers to questions. Some expert systems use the database to generate graphics and text reports on what types of hardware and software cause the most problems. This information can be used to guide product purchases and ascertain the response times of vendors. The growth of distributed computing via the Internet, as well as LANs and wide area networks (WANs), plus the increasing complexity of hardware and software, have expanded the role of the traditional help desk in many organizations.

See also

Asset Management
Electronic Software Distribution
Network Management Systems

HERTZ

The term *hertz* is a measure of frequency, or the speed of transmission. The frequency of electromagnetic waves generated by radio transmitters is measured in cycles per second (cps), but this designation was officially changed to hertz (Hz) in 1960.

An electromagnetic wave is composed of complete cycles. The number of cycles that occur each second gives radio waves their frequency, while the peak-to-peak distance of the waveform gives the amplitude of the signal (Figure H-2).

The frequency of standard speech is between 3000 cps, or 3 kilohertz (kHz), and 4000 cps, or 4 kHz. Some radio waves may have frequencies of many millions of hertz (megahertz, or

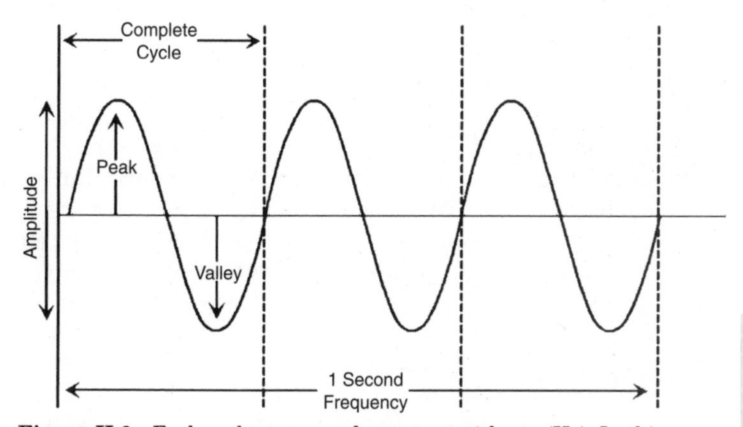

Figure H-2 Each cycle per second equates to 1 hertz (Hz). In this case, 3 cycles occur in 1 second, which equates to 3 Hz.

MHz) and even billions of hertz (gigahertz, or GHz). Table H-1 provides the range of frequencies and their band classification.

The term *hertz* was adopted in 1960 by an international group of scientists and engineers at the General Conference of Weights and Measures in honor of Heinrich R. Hertz (1857–1894), a German physicist (Figure H-3). Hertz is best known for proving the existence of electromagnetic waves, which had been predicted by British scientist James Clerk Maxwell in 1864.

Figure H-3 German physicist Heinrich R. Hertz (1857–1894) proved the existence of electromagnetic waves, which led to the development of radio, microwave, radar, and other forms of wireless communication.

Hertz used a rapidly oscillating electric spark to produce ultrahigh-frequency waves. These waves caused similar electrical oscillations in a distant wire loop. The discovery of electromagnetic waves and how they could be manipulated paved the way for the development of radio, microwave, radar, and other forms of wireless communication.

Summary

As interest in electromagnetic waves grew in the nineteenth century, a physical model to describe it was proposed. It was

Table H-1 List of Frequency Ranges and Corresponding Band Classifications

Frequency	Band Classification
Less than 30 kHz	Very low frequency (VLF)
30–300 kHz	Low frequency (LF)
300 kHz to 3 MHz	Medium frequency (MF)
3–30 MHz	High frequency (HF)
30–300 MHz	Very high frequency (VHF)
300 MHz to 3 GHz	Ultrahigh frequency (UHF)
3–30 GHz	Superhigh frequency (SHF)
More than 30 GHz	Extremely high frequency (EHF)

suggested that electromagnetic waves, including light, were like sound waves but that they propagated through some previously unknown medium called the "luminiferous ether," which filled all unoccupied space throughout the universe. The experiments of Albert A. Michelson and Edward W. Morley in 1887 proved that the ether did not exist. Albert Einstein's theory of relativity, proposed in 1905, eliminated the need for a light-transmitting medium, so today the term *ether* is used only in a historical context, as in the term *Ethernet.*

See also

Bluetooth

HomeRF

Wireless Fidelity

HIERARCHICAL STORAGE MANAGEMENT

Hierarchical storage management (HSM) uses two or more levels of storage (three is typical: online, offline, and nearline) to provide a cost-effective and efficient solution to meet the demand for increased storage space and appropriate data-retrieval response time. HSM came about because of the need to move low-volume and infrequently accessed files from disk, thus freeing up valuable online storage space.

Although more disks can be added to file servers to keep up with storage needs, budget constraints often limit the long-term viability of this solution. In an HSM scheme, data can be categorized according to their frequency of usage and stored appropriately: online, near-line, or offline. Different storage media come into play for each of these categories, and migration operations are under control of an HSM management system (Figure H-4).

Frequently used files are stored online on local disk drives installed in a server or workstation. Occasionally used files are stored near-line on secondary storage devices such as optical disks installed in a server-like device called an "autochanger" or "jukebox." Infrequently used files are usually migrated offline to tape cartridges that are stored in a tape jukebox or a

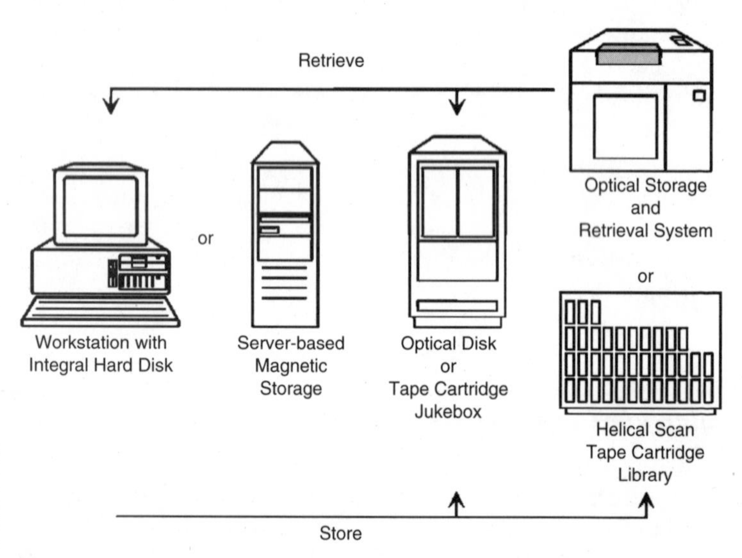

Figure H-4 Hierarchical data storage spanning magnetic disk, optical disk, and tape options.

library facility capable of holding hundreds or thousands of tapes. The library facility uses sophisticated robotics to retrieve individual bar-coded cartridges and inserts them into a tape drive so the data can be moved to local storage media. The exchange time can be several minutes.

For organizations with mixed needs for online and offline storage, a near-line automated tape library offers the best compromise between price and performance. These systems bridge the gap between fast, expensive online disk storage and slow, high-capacity offline tape libraries. The exchange time is about 30 seconds.

A management system determines when a file should be transferred or retrieved, initiates the transfer, and keeps track of its new location. As files are moved from one type of medium to another, they are put into the proper directory for user access. The management system automatically optimizes storage utilization across different media types by removing files from one to the other until they are permanently archived in the most economical way, usually a tape library. At the same time, individual files or whole directories can be excluded from migration.

Data migration can be controlled according to such criteria as file size and last date of access. Files also can be migrated when the hard disk reaches a specified capacity threshold. When a disk reaches the threshold of 80 percent full, for example, files are migrated automatically to tape storage, freeing space on the hard disk until it reaches another specified threshold, say 60 percent. When files are no longer needed but must be stored permanently, they can be migrated off the tape drive to optical media.

Summary

Huge amounts of data—hundreds of gigabytes or even terabytes—cannot be managed efficiently without the help of HSM solutions. HSM balances the cost, capacity, and efficiency of different storage media according to the frequency of data usage. This provides all clients and servers with expanded online storage through the migration of files between hard disks, magnetic tape, and optical media, according to predefined rules.

See also

Redundant Array of Inexpensive Disks

HOME PHONE-LINE NETWORKING

Home phone-line networking refers to the ability of consumers to implement a local area network (LAN) in the home over ordinary telephone wire, enabling them to connect computers and peripherals at speeds of up to 10 Mbps and share the same Internet access connection. A home network permits file and application sharing, messaging, and multiuser gaming among family members or with others via the Internet.

The number of multiple-PC homes is growing faster than the number of single-PC homes. Currently, there are about 35 million homes in the United States that have at least two PCs. The need for a network in the home to share printers, modems for Internet access, CD-ROMs, hard drives, and other equipment becomes essential, if only to save money on expensive resources, including multiple phone lines and Internet access accounts. Beyond cost savings, a home network offers convenience. Users no longer have to waste time looking all over the house to check each computer for the documents, spreadsheets, images, or software they need for office work, school work, tax preparation, or to play their favorite games. Further, a network in the home provides the opportunity to implement special features such as a family message board and voice intercom.

The cost of installing and configuring a basic two-node network in the home is less than $100 (U.S.), and several vendors offer kits that include all the necessary components. Third-party software is usually required for features such as internal messaging and voice intercom. More computers can be added to the network by purchasing extra adapter cards to make the phone-line connections. Such networks can be built and configured at a leisurely pace, with professional results achievable within a Saturday afternoon.

The first home phone-line products were introduced in 1998 but offered a top transmission speed of only 1 Mbps. Today's phone-line networking products offer a top transmission speed of 10 Mbps. In either case, the existing phone wiring within the home provides the connectivity among computers, including shared Internet access, without the need for a hub device. In addition, the newer 10-Mbps products are backward-compatible with the older 1-Mbps products.

Data run over the home's existing phone wiring without disrupting normal phone service. This is accomplished through the use of Frequency Division Multiplexing (FDM), which essentially divides the data traveling over the phone lines into separate frequencies—one for voice, one for high-bandwidth net access such as Digital Subscriber Line (DSL), and one for the network data. These frequencies can coexist on the same telephone line without interfering with each other. The technology is designed to operate with computers and devices up to 500 feet apart, making it suitable for homes of up to 10,000 square feet.

Components

For each computer that will be networked in the home, a network interface card (NIC) is required, just as at the office. This is an adapter that plugs into a vacant slot of the computer. For phone-line networking, the NIC will have two RJ11 ports into which a segment of phone wire will be connected—one to the phone and the other to the wall jack. Many NICs are optimized to work in Windows environments, which provide the software drivers for most brands of NICs, making installation and configuration essentially a plug-and-play affair.

With the computer turned off, the user opens the case and inserts the NIC into the appropriate type of card slot. The short white slots are for Peripheral Component Interconnect (PCI) cards, and the longer black slots are for the older Industry Standard Architecture (ISA) cards. Because of the limited number of vacant card slots in each computer—usually only three to five—it is best to inventory each computer before actually buying the NICs or a network kit. Whenever possible, PCI-type cards should be used for their higher performance.

Once the card is installed, the computer can be restarted. Windows will recognize the new hardware and configure it automatically. If the driver or a driver component is not found, however, it may be necessary to get out the Windows CD-ROM or the NIC vendor's 3.5-inch installation disk so that Windows can find the components it needs.

After Windows has recognized and configured the NICs, it is time to connect each computer. One segment of phone line is inserted into the wall jack and the other end into an RJ11 port of the computer's NIC. Another phone line segment plugs into the phone and terminates at the other RJ11 port on the NIC. This allows the phone to be used even while a file transfer is in progress between computers.

Instead of using one of the two RJ11 ports for connection to a telephone, the extra RJ11 port can be used for daisy-chaining computers together over the same phone line. Some vendors also include an RJ45 port on the NIC, which lets users migrate to 100 Mbps using Category 5 LAN cabling.

To actually share files and peripherals over the phone-line connections, each computer must be configured with the right protocols and be set up for file and printer sharing. Once this is done, a modem also can be shared. Up to 25 users can share that modem for Internet access, eliminating the need for separate telephone lines, modems, and accounts with an Internet service provider (ISP). If Internet access is provided over a broadband service such as cable or DSL, the computers connected over the phone-line network can share these resources as well.

Configuring the Network

As noted, after installing a NIC and powering up the computer, Windows will recognize the new hardware and automatically install the appropriate network-card drivers. If the drivers are not already available on the system, Windows will prompt the user to insert the manufacturer's disk containing the drivers, and they will be installed automatically.

Next, the user must select the client type. Since this is a Microsoft peer-to-peer network that is being created, the user must add "Client for Microsoft Networks" as the primary network log-on (Figure H-5). Since the main advantage of networking is resource sharing, it is important to enable the sharing of both printers and files, which is done by clicking on the "File and Print Sharing" button and choosing one or both of these capabilities (refer again to Figure H-5).

Identification and security are the next steps in the configuration process. From the "Identification" tab of the dialog box, the user must select a unique name for the computer and the workgroup to which it belongs, as well as a brief description of the computer (Figure H-6). This information will be visible to others when they use Network Neighborhood to browse the network.

From the "Access Control" tab of the dialog box, the user selects the security type. For a small peer-to-peer network, share-level access is adequate (Figure H-7). It allows printers, drives, directories, and CD-ROM drives to be shared and enables the user to establish password access for each of these resources. In addition, read-only access allows users to view (not modify) a file or directory.

To allow (or disallow) disk drives to be shared, the user double-clicks on the "My Computer" icon on the desktop and then right-clicks on the drive to be shared. Next, the user selects "Sharing" from the pick list (Figure H-8). The type of access can be set as Read-Only, Full, or Depends on Password. This is done for each drive the user wants to share, including CD-ROM, CD-RW, CD-R, and DVD optical disk drives. Individual directories can be set for no sharing in the same way.

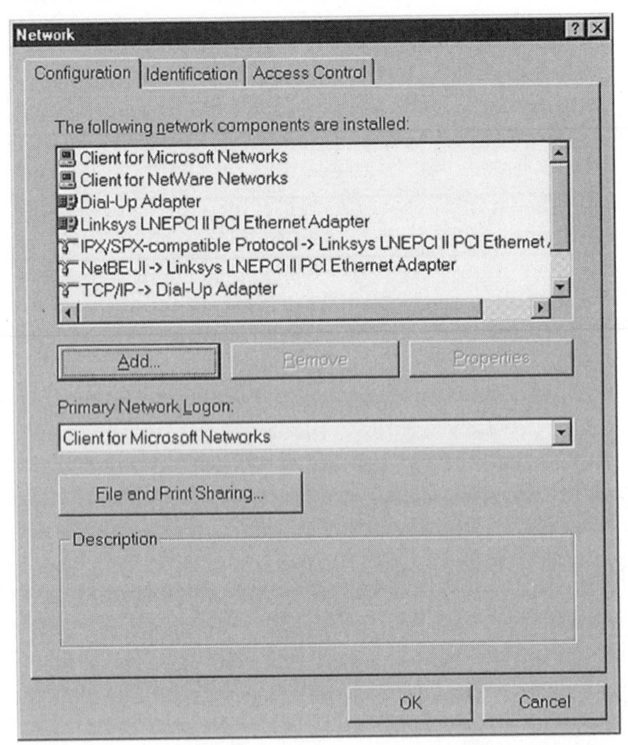

Figure H-5 By accessing the Windows Control Panel and double-clicking on the "Network" icon, this dialog box opens, where the user can configure the computer for the primary network logon, plus file and printer sharing.

To allow a printer to be shared, the user right-clicks on the printer icon in the Control Panel and selects "Sharing" from the pick list (Figure H-9). Next, the user clicks on the "Shared As" radio button and enters a unique name for the printer (Figure H-10). If desired, this resource can be given a password as well. When another computer tries to access the printer, the user will be prompted to enter a password. If a password is not necessary, the password field is left blank.

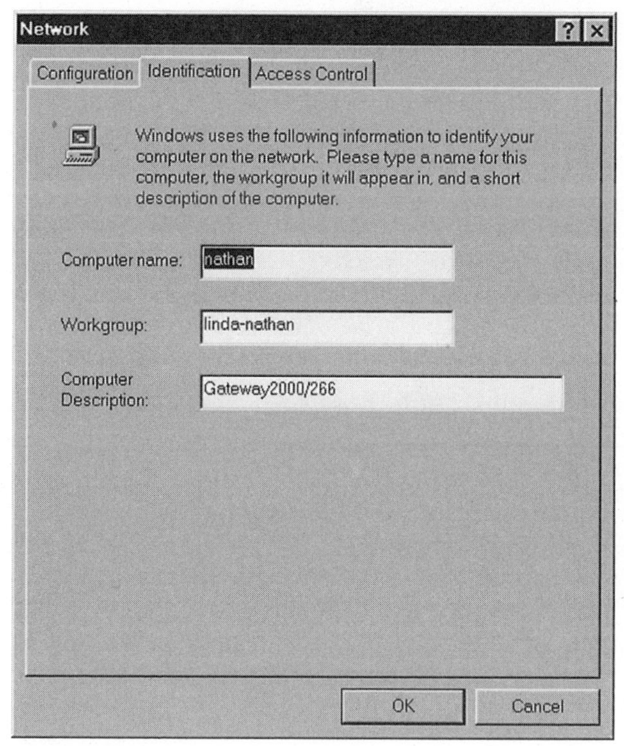

Figure H-6 A unique name for the computer and the workgroup to which it belongs and a brief description of the computer identify it to other users when they access Network Neighborhood to browse the network.

Another security option in the "Access Control" tab is user-level access, which is used to limit resource access by user name. This function eliminates the need to remember passwords for each shared resource. Each user simply logs onto the network with a unique name and password; the network administrator governs who can do what on the network. However, this requires the computers to be part of a larger network with a central server—one running Windows NT server, for example—that maintains the access-control

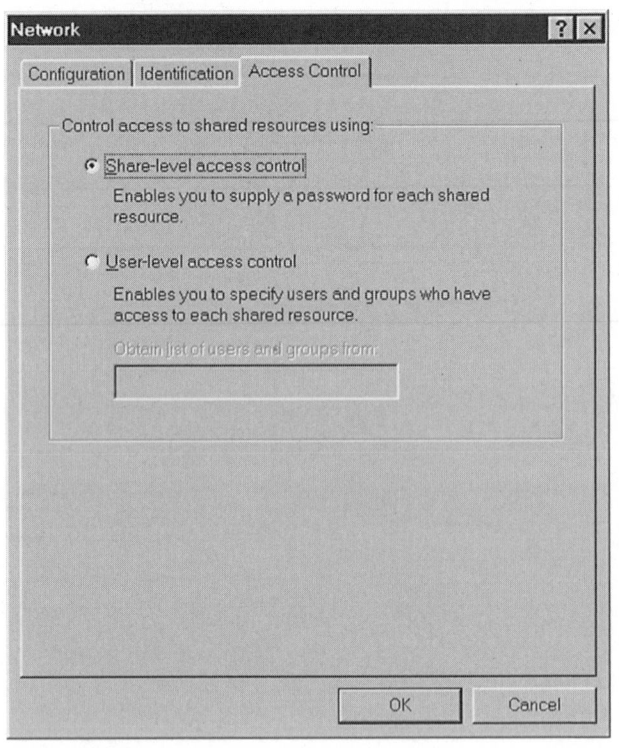

Figure H-7 Choosing share-level access allows the user to password-protect each shared resource.

list for the whole network. Since Windows 95/98/XP and Windows NT/2000 workstations support the same protocols, Windows 95/98/XP computers can participate in a Windows NT/2000 server domain.

Peer services can be combined with standard client-server networking. For example, if a Windows 98 computer is a member of a Windows NT network and has a color printer to share, the resource "owner" can share that printer with other computers on the network. The server's access-control list determines who is eligible to share resources.

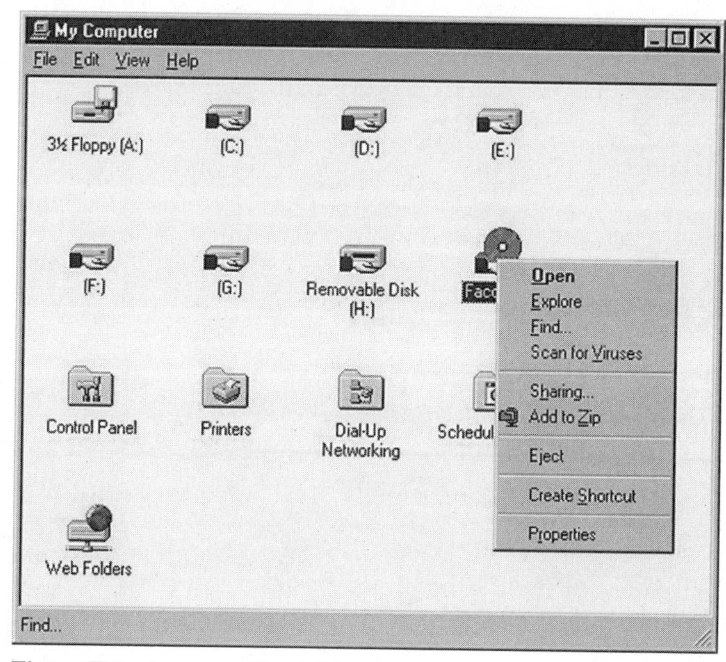

Figure H-8 Any type of hard drive or optical drive on any computer can be shared. Access privileges can be associated with each drive, such as Read-Only, Full, or Depends on Password.

Once the computers are configured properly and connected through the existing phone lines, the network is operational. Although designed for the home, this peer-to-peer network is an inexpensive way for small companies within a building to share resources among a small group of computers. This type of network provides many of the same functions as the traditional client-server network, including the ability to run network versions of popular software packages.

Standards

To standardize the products that interconnect computers over standard telephone wiring in the home, the vendor-oriented

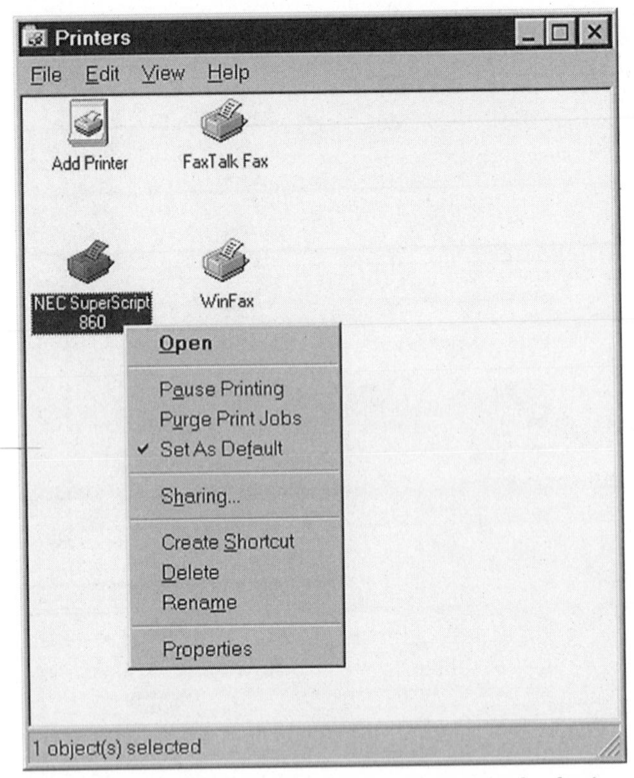

Figure H-9 Right-clicking on a "Printer" sets it up for sharing.

Home Phoneline Networking Alliance (HomePNA) was established in 1998. Products that adhere to the HomePNA standard permit the creation of simple, cost-effective home networks using existing phone wiring. The use of phone wiring for this purpose eliminates the need for cable installation and a hub yet allows shared Internet access with one ISP account without interfering with regular phone service. HomePNA Specification 2.0 delivers a 10-Mbps data rate for home phone-line networking while maintaining full backward compatibility and interoperability with the previous specification, which offered a data rate of 1 Mbps.

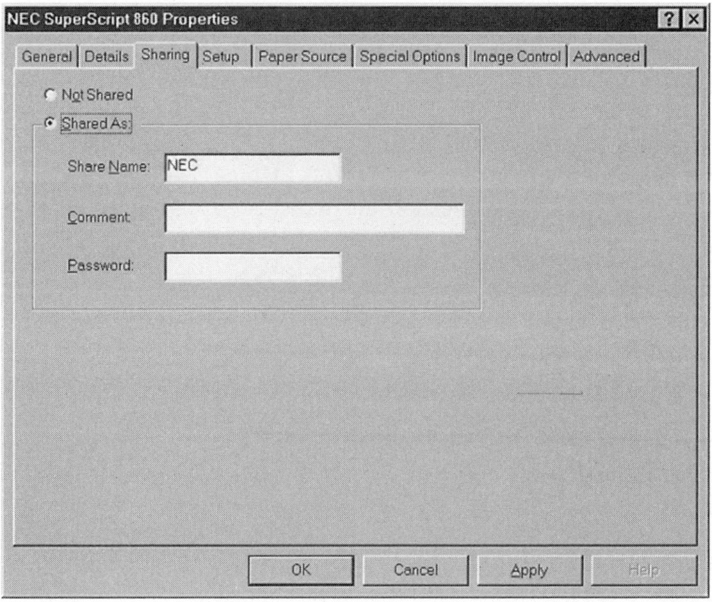

Figure H-10 A shared printer is named and may be password-protected if necessary.

Adding data to voice over existing phone wiring does not pose interference problems because different frequencies are used for each. Standard voice occupies the range from 20 Hz to 3.4 kHz in the United States (slightly higher internationally), while phone-line networking operates in a frequency range above 2 MHz. By comparison, DSL services such as ADSL occupy the frequency range from 25 kHz to 1.1 MHz. The frequencies used are far enough apart that the same wiring can support all three simultaneously.

Summary

As the number of PCs and peripherals in the home continues to increase, the need for a network to leverage these assets and provide shared access to the Internet becomes more apparent. PCs on home phone networks must run Windows 95, 98, or NT/2000 or some other software that supports file sharing. Once networked, the PCs can share a printer, as well as a modem dial-up connection for Internet or corporate network access. Users also can work high-speed DSL and cable modems into the mix. In fact, HomePNA lets the consumer choose the method of wide area network (WAN) access, which also can include Integrated Services Digital Network (ISDN) and wireless services.

See also

Cable Television Networks

Digital Subscriber Line Technologies

Ethernet

Hubs

HOME RADIO FREQUENCY NETWORKS

One method of implementing a wireless network in the home is to use products that adhere to the standards of the Home Radio Frequency Working Group. HomeRF is positioned as a global extension of Digitally Enhanced Cordless Telephony (DECT), the popular cordless phone standard that allows different brands to work together so that certified handsets from one vendor can communicate with base stations from another. DECT has been largely confined to Europe because its native 1.9-GHz frequency band requires a license elsewhere, but HomeRF extends DECT to other regions by using the license-free 2.4-GHz frequency band. It also adds functionality by blending several industry standards, including IEEE 802.11 Frequency Hopping for data and DECT for voice. This convergence makes HomeRF useful for broadband households.

As more PCs, peripherals, and intelligent devices are installed in the home, and as network connections proliferate, users are faced with new opportunities for accessing information as well as challenges for sharing resources. For example, users want to

- Access information delivered via the Internet from anywhere in the home.
- Share files between PCs and share access to peripherals no matter where they are located within the home.
- Control electrical systems and appliances whether in, around, or away from home.
- Effectively manage communications channels for phone, fax, and Internet usage.

Each of these capabilities requires a common connection between the various devices and networks found in the home. However, in order to truly be effective, any home network must meet certain criteria:

- It must not require additional home wiring. Most existing homes are not wired for networking, and retrofitting them would too labor-intensive and expensive. A wireless solution is a viable alternative.
- The wireless connections must be immune to interference, especially with the growing number of wireless devices and appliances emitting radio frequency (RF) noise in the home.
- The range of the wireless connection must be adequate to allow devices to communicate from anywhere within and around a typical family home.
- The network must be safe and protected from unwanted security breaches.
- It must be easy to install, configure, and operate for nontechnical users. Most home users do not have the expertise to handle complex network installation and configuration procedures.
- The entire system must be easily and spontaneously accessible—anytime and from anywhere in or even away from the home.

These issues have been addressed by a consortium of vendors called the Home Radio Frequency Working Group (HomeRFWG), which has developed a platform for a broad range of interoperable consumer devices. Its specification, called the Shared Wireless Access Protocol (SWAP), is an open standard that allows PCs, peripherals, cordless telephones, and other consumer electronic devices to communicate and interoperate with one another without the complexity and expense associated with installing new wires.

The SWAP is designed to carry both voice and data traffic and to interoperate with the Public Switched Telephone Network (PSTN) and the Internet. It operates in the 2.4-GHz ISM (industrial, scientific, medical) band and uses frequency-hopping spread-spectrum radio for security and reliability. The SWAP technology was derived from extensions of DECT and wireless LAN technologies to enable a new class of home cordless services. It supports both a Time Division Multiple Access (TDMA) service to provide delivery of interactive voice and other time-critical services and a Carrier Sense Multiple Access with Collision Avoidance

Table H-1 HomeRF Characteristics

Frequency-hopping network	50 hops/second
Frequency range	2.4-GHz ISM band
Transmission power	100 mW
Data rate	1.6 Mbps with HomeRF 1.0; 10 Mbps with HomeRF 2.0; 25 Mbps with HomeRF 3.0 (future)
Range	Covers up to 150 feet for typical home and yard
Total network devices	Up to 127
Voice connections	Up to 4 active handsets
Data security	Blowfish encryption algorithm (over 1 trillion codes)
Data compression	LZRW3-A algorithm
48-bit network ID	Enables concurrent operation of multiple co-located networks

SOURCE: HomeRF Working Group.

(CSMA/CA) service for delivery of high-speed packet data. Table H-1 summarizes the main characteristics of HomeRF.

Applications

The SWAP specification provides the basis for a broad range of new home networking applications, including

- Shared access to the Internet from anywhere in the home, allowing a user to browse the Web from a laptop on the deck or have stock quotes delivered to a PC in the den.
- Automatic intelligent routing of incoming telephone calls to one or more cordless handsets, fax machines, or voice mailboxes of individual family members.
- Cordless handset access to an integrated message system to review stored voice mail, faxes, and electronic mail.
- Personal intelligent agents running on the PC for each family member, accessed by speaking into cordless handsets. This new voice interface would allow users to access and control their PCs and all the resources on the home wireless network spontaneously, from anywhere within the home, using natural-language commands.
- Wireless LANs allowing users to share files and peripherals between one or more PCs, no matter where they are located within the home.
- Spontaneous control of security, electrical, heating, and air conditioning systems from anywhere in or around the home.
- Multiuser computer games playable in the same room or in multiple rooms throughout the home.

Network Topology

The SWAP system can operate either as an ad-hoc network or as a managed network under the control of a Connection

Point. In an ad-hoc network, where only data communication is supported, all stations are equal, and control of the network is distributed between the stations. For time-critical communications such as interactive voice, a Connection Point is required to coordinate the system. The Connection Point, which provides the gateway to the PSTN, can be connected to a PC via a standard interface such as the Universal Serial Bus (USB) that will enable enhanced voice and data services. The SWAP system also can use the Connection Point to support power management for prolonged battery life by scheduling device wakeup and polling.

The network can accommodate a maximum of 127 nodes. The nodes are of four basic types:

- Connection Point that supports voice and data services.
- Voice Terminal that only uses the TDMA service to communicate with a base station.
- Data Node that uses the CSMA/CA service to communicate with a base station and other data nodes.
- Integrated Node, which can use both TDMA and CSMA/CA services.

HomeRF uses intelligent hopping algorithms that detect wideband static interference from microwave ovens, cordless phones, baby monitors, and other wireless LAN systems. Once detected, the HomeRF hop set adapts so that no two consecutive hops occur within this interference range. This means that, with very high probability, a packet lost due to interference will get through when it retries on the next hop. While these algorithms benefit data applications, they are especially important for voice, which requires extremely low bit error rates and low latency.

Future Plans

Work has already begun on the future HomeRF 2.1 Specification, which will add features designed to reinforce its advantages for voice. Planned enhancements also will allow HomeRF to complement other wireless standards, including IEEE 802.11, also known as Wi-Fi.

HomeRF 2.0 already supports up to eight phone lines, eight registered handsets, and four active handsets with voice quality and range comparable to leading 2.4-GHz phone systems. With that many lines, each family member can have a personal phone number. HomeRF 2.1 plans to increase the number of active handsets with the same or better voice quality, thus supporting the needs of small businesses.

The 150-foot range of HomeRF already covers most homes and into the yard. HomeRF 2.1 will extend that range for larger homes and businesses by using wireless repeaters that are similar to enterprise access points but without the need to connect each one to Ethernet. HomeRF frequency-hopping technology also avoids the complexity of assigning RF channels to multiple access points (or repeaters) and offers easy and effective security and interference immunity. This is especially important because households and small businesses do not usually have network administrators.

To allow individuals to roam across very large homes and fairly large offices while talking on the phone and without loosing their voice connection, HomeRF 2.1 also will support voice roaming with soft handoff between repeaters.

HomeRF 2.0 supports Ethernet speeds of up to 10 Mbps with fallback speeds and backward-compatibility to earlier versions of HomeRF. Performance can be further enhanced to about 20 Mbps. The HomeRFWG is evaluating the need for such enhancements at 2.4 GHz in light of its planned support of 5 GHz.

A proposed change to Federal Communications Commission (FCC) Part 15 rules governing the 2.4-GHz ISM band will allow adaptive frequency hopping. While not legal today, these proposed techniques allow hoppers such as Bluetooth and HomeRF to recognize and avoid interference from static frequency technologies such as Wi-Fi. Since HomeRF already adjusts its hopping pattern based on interference to ensure that two consecutive hops do not land on interference, supporting this FCC proposal seems trivial.

The HomeRFWG believes in the peaceful coexistence of 2.4 and 5 GHz because each frequency band and technology has specific strengths that complement each other. Rather than draft a specification for 5 GHz, the group simply endorses IEEE 802.11a (also known as Wi-Fi5) for high-bandwidth applications such as high-definition video streaming and MPEG2 compression. It plans to write application briefs describing how to bridge between 2.4- and 5-GHz technologies, including how to handle differences in quality of service (QoS). This information, while written for IEEE 802.11a, also can apply to HiperLAN-2, IEEE 802.11h, and proprietary IEEE 802.11a extensions.

Many analysts expect IEEE 802.11a eventually to take over as the wireless standard for enterprise offices, gain needed QoS support from IEEE 802.11e, and start a slow migration into homes. It already supports 54 Mbps, and proprietary extensions increase performance to about 100 Mbps. But because of the higher frequencies used, IEEE 802.11a has disadvantages in cost, power consumption, range, and signal attenuation through materials. A combination of HomeRF and IEEE 802.11a combines the strengths of both technologies.

Summary

Home users have a need for a wireless network that is easy to use, cost-effective, spontaneously accessible, and can carry voice and data communications. Certified HomeRF products are available today from consumer brands such as Compaq, Intel, Motorola, Proxim, and Siemens through retail, online, and service provider channels. They come in a variety of form factors such as USB and PC card adapters, residential gateways, and a growing variety of devices that embed HomeRF.

See also

Bluetooth
Wireless Fidelity

HUBS

With today's networks becoming increasingly more complex, the conventional bus and ring LAN topologies have exhibited shortcomings, especially with regard to cable installation and maintenance. Furthermore, a fault anywhere in the cabling often brought down the entire network or a significant portion of it. This weakness was compounded by the inability of technicians to readily identify the point of failure from a central administration point, which tended to prolong network downtime. This situation led to the development of the wiring hub in the mid-1980s.

Hubs provide a central point at which all wires meet. They are at the center of the star configuration, with the wires (i.e., segments) radiating outward to connect the various network devices, which may include bridge/routers that connect to remote LANs via the WAN (Figure H-11). Wiring hubs physically convert the networks from a bus or ring topology to a star topology while logically maintaining their Ethernet or Token Ring characteristics. The advantage of this configuration is that the failure of one segment—which may be shared among several devices or dedicated to just one device—does not necessarily impact the performance of other segments.

Not only do hubs limit the impact of cabling faults to a particular segment, they also provide a centralized administration point for the entire network. If the wiring hub also employs some CPUs and management software to automate fault isolation, network reconfiguration, and statistics gathering operations, it is no longer just a hub but an "intelligent

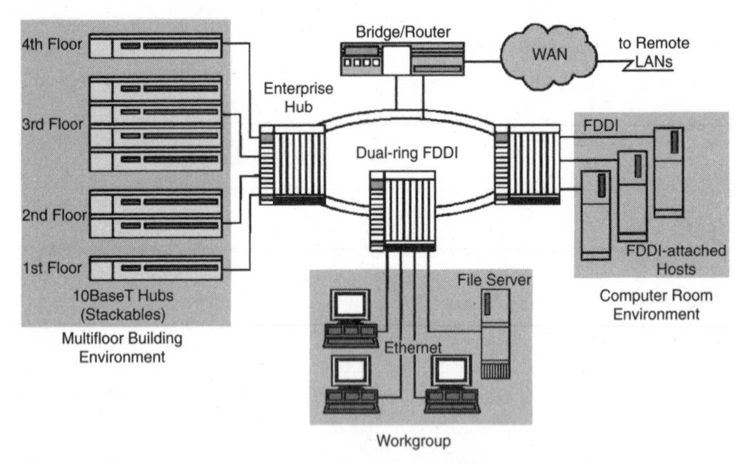

Figure H-11 Some connectivity options available through various types of hubs serving multifloor building, workgroup, and computer room environments.

hub," capable of solving a wide range of connectivity problems efficiently and economically.

Types of Hubs

High-end hubs are modular in design, allowing the addition of ports, network interfaces, and special features as they are needed by the organization. These enterprise-level hubs can support networks that combine different LAN topologies and media types in a single chassis. Ethernet, Token Ring, and Fiber Distributed Data Interface (FDDI) networks can coexist in a single hub. LAN segments using twisted-pair wiring, coaxial cable, and optical fiber also can be interconnected through the hub. In such cases, the hub is equipped with media conversion modules.

Other hubs are available in fixed configurations for departments or workgroups that do not anticipate future growth. A variation of the fixed-configuration hub is the stackable hub. A unique feature of stackable hubs is that they can be interconnected through a modular backplane. This offers managers the ability to economically expand workgroup and departmental networks as needed. Whereas the high-end modular systems are used to build large-scale enterprise networks, "stackables" are designed for small to medium-sized networks.

There are also workgroup hubs, which typically have eight ports or less. This type of hub not only provides connectivity among connected workstations, it also connects to a DSL or cable router to allow all the workstations to share the same Internet access connection. This type of hub is becoming popular for small businesses, small office–home office (SOHO) environments, and consumers who have multiple PCs in the home.

A relatively new category of hub is the "superhub." These are modular units that provide at least an uplink to a standalone Asynchronous Transfer Mode (ATM) switch, if not some level of integral ATM switching, in addition to 100-Mbps LAN support, and integrated LAN switching and routing. Fully populated superhubs support in excess of 500 ports of mixed-media, shared, and switched connectivity over a gigabit-per-second backplane in a software-manageable, fault-tolerant, hot-swappable modular chassis that can cost well over $100,000.

Hub Components

Enterprise-level intelligent hubs contain four basic components: chassis, backplane, plug-in modules, and a network management system.

Chassis The chassis is the hub's most visible component. It contains an integral power supply and/or primary controller unit and varies in the number of available module slots. The modules insert into the chassis and are connected by a series of buses, each of which may constitute a separate network or integrate with one or more backbone networks. The chassis holds the individual modules. In fitting into the chassis, each module is instantly connected to other modules via the hub's high-speed backplane.

Backplane The main artery of the hub is its backplane, a board that contains one or more buses that carry all communications between LAN segments. The hub's backplane is analogous to a PC bus through which various interface cards may be interconnected. The data path that carries traffic from card to card is often called a "channel"; unlike the PC, though, the hub's backplane typically consists of multiple physical or logical channels. Minimally, the hub accommodates one LAN segment for each channel on the backplane.

Segmenting the backplane in this way allows multiple independent LANs or LAN segments to coexist within the same chassis. There is usually a separate backplane channel to carry management information. The segmented backplane typically has dedicated channels for Ethernet, Token Ring, and FDDI. Some hubs employ a multiplexing technique across the backplane to divide the available bandwidth into multiple logical channels. Other hubs support load sharing that allows network modules to select the backplane channel that will transport the traffic. Still other hubs are designed to allow backplanes to be added or upgraded to accommodate network expansion and new technologies.

The potential bandwidth capacity of newer backplane designs supporting ATM switching is quite impressive, reaching well into the gigabit-per-second range—more than enough to accommodate several Ethernet, Token Ring, and FDDI networks simultaneously.

Modules The functionality of hubs is provided by individual modules, the types of which depend on the hub vendor. Typically, the vendor will provide multiuser Ethernet and Token Ring cards, LAN management, and LAN bridge and router cards. The use of bridge and router modules in hubs overcomes the distance limitations imposed by the LAN cabling and facilitates communication between LANs and WANs.

There are even plug-in modules for terminal servers, communications servers, file and application servers, and Systems Network Architecture (SNA) gateways. Hub vendors also offer a variety of WAN interfaces, including those for X.25, Frame Relay, ISDN, T-carrier, SMDS, and ATM. As many as 60 different types of modules may be available from a single hub vendor, many of them provided under third-party original equipment manufacturer (OEM), technology-swap, and other vendor-partnering arrangements.

Modules plug into vacant chassis slots. Depending on the vendor, the modules can plug into any vacant slot or slots specifically devoted to their function. Hubs supporting any-slot insertion automatically detect the type of module that is inserted into the chassis and establish the connections to other compatible modules. In addition, many vendors offer a "hot swap" capability that permits modules to be removed or inserted without powering down the hub.

Management System

Hubs occupy a strategic position on the network, providing the central point of connection for workstations, servers, hosts, bridges, and routers on the LAN and over the WAN. The hub's management system is used to view and control all devices connected to it, providing information that can greatly aid troubleshooting, fault isolation, and administration. The management tools typically fall into five categories: accounting management, configuration management, performance management, fault management, and security management.

Hub vendors typically provide proprietary management systems that offer value-added features that can make it easier to track down problem-causing workstations or servers. Most of these management systems support the Simple Network Management Protocol (SNMP), enabling them to be controlled and managed through an existing management platform such as IBM's SystemView for AIX and Hewlett-Packard's OpenView. Some hubs have Remote Monitoring (RMON) embedded in the hub, making possible more advanced network monitoring and analysis up to OSI Layer 7, the Application Layer.

Summary

Hubs are now the central point of control and management for the elements that make up departmental and enterprise networks. Hubs, which were developed to simplify the management of structured wiring as networks became bigger and more complex, allow the wiring infrastructure to expand in an orderly and cost-effective manner as the organization's computer systems grow and move and as interconnectivity requirements become more sophisticated.

See also

Bridges
LAN Switching
Routers

I

INCUMBENT LOCAL EXCHANGE CARRIERS

Incumbent Local Exchange Carriers (ILECs) is a term that refers to the 22 former Bell Operating Companies (BOCs) divested from AT&T in 1984, as well as Cincinnati Bell, Southern New England Telephone (SNET), and the larger independent telephone companies of GTE and United Telecommunications. In addition, some 1300 smaller telephone companies are also in operation, serving mostly rural areas. These, too, are considered incumbents, but the small markets they serve do not attract much competition.

After being spun off by AT&T in 1984, the BOCs were assigned to seven regional holding companies: Ameritech, Bell Atlantic, BellSouth, Nynex, Pacific Telesis, Southwestern Bell Communications (SBC), and US West. Over the years, some of these regional companies merged to the point that today only four are left. Bell Atlantic and Nynex were the first to merge in 1994. Bell Atlantic also completed a $53 billion merger with GTE in mid-1999 and changed its name to Verizon. SBC Communications merged with Pacific Telesis in 1997 and then Ameritech in 1999. It also acquired Southern New England Telephone (SNET).

Regulatory Approval

All the mergers passed regulatory approval at the state and national levels. The Federal Communications Commission (FCC) approves mergers with input from the Department of Justice (DoJ). In the case of the SBC-Ameritech merger, the FCC imposed 28 conditions on SBC in exchange for approving the transaction.

The approval package contained a sweeping array of conditions designed to make SBC-Ameritech's markets the most open in the nation, boosting local competition by providing competitors with the nation's steepest discounts for resold local service and full access to operating support systems (OSS).

It also required SBC to accelerate by 6 months its entry into new markets, forcing the company to compete in 30 new markets within 30 months after completion of the merger. The FCC's rationale was that increased competition in out-of-region territories would help offset reduced competition in the SBC-Ameritech service areas.

The conditions also required stringent performance monitoring, reporting, and enforcement provisions that could trigger more than $2 billion in fines if these goals were not met. Fortunately for SBC, the agreement required it to serve only three customers in each out-of-region market. According to SBC, it will not begin to seriously market its out-of-region services until it has obtained approval to offer long distance services in its 13 home states.

Summary

The monopoly status of the ILECs officially ended with passage of the Telecommunications Act of 1996. Not only can other types of carriers enter the market for local services in competition with them, but also their regional parent companies can compete in each other's territories. Through mergers, the reasoning went, the combined companies can enter out-of-region markets on a broad scale quickly and efficiently enough to become effective national competitors.

Unfortunately, this has not occurred on a significant scale. In fact, the lack of out-of-region competition among the "Baby Bells" means that consumers and businesses do not have as much choice in service providers, especially now that many Competitive Local Exchange Carriers (CLECs) are being hit hard by financial problems and the lack of venture capital. The ILECs are more concerned with being able to qualify for long-distance services in their own markets so that they can bundle local and long-distance services and Internet access—a package few, if any, competitors would be able to match.

See also

Competitive Local Exchange Carriers
Interexchange Carriers
Local Exchange Carriers

INFINIBAND

InfiniBand, short for "infinite bandwidth," is a bus technology that provides the basis for an input-output (I/O) fabric designed to increase the aggregate data rate between servers and storage devices. The point-to-point linking technology allows server vendors to replace outmoded system buses with InfiniBand to greatly multiply total I/O traffic compared with legacy system buses such as Peripheral Component Interconnect (PCI) and its successor PCI-X.

The current PCI bus standard supports up to 133 Mbps across the installed PCI slots, providing shared bandwidth of up to 566 Mbps, while PCI-X permits a maximum bandwidth of 1 Gbps. Fibre Channel offers bandwidth up to 2 Gbps and is used to build storage area networks (SANs). In contrast, InfiniBand utilizes a 2.5-Gbps wire speed connection with multiwire link widths. With a four-wire link width, Infiniband offers 10 Gbps of bandwidth. The InfiniBand specification supports both copper and fiber implementations.

The I/O fabric of the InfiniBand architecture takes on a role similar to that of the traditional mainframe-based channel architecture, which used point-to-point cabling to maximize overall I/O throughput by handling multiple I/O streams simultaneously. The move to InfiniBand means that I/O subsystems need no longer be the bottleneck to improving overall data throughput for server systems.

In addition to performance, InfiniBand promises other benefits such as lower latency, easier and faster sharing of data, built-in security and quality of service, and improved usability through a form factor that makes components much easier to add, remove, or upgrade than today's shared-bus I/O cards.

Distributed Computing Requirements

Today's computing model is becoming more distributed as companies reorganize to take advantage of the Internet for e-commerce and e-business. This computing environment is challenging the scalability, reliability, availability, and performance of servers. To meet these demands requires a balanced system architecture that provides higher performance in the memory, processor, and I/O subsystems. InfiniBand technology was designed to meet these requirements of the Internet economy in several ways:

- It assumes very fast processors capable of 1 GHz or more, which push more data than conventional system buses can handle. InfiniBand overcomes the technology mismatch in which the fast processor has no way to push multiplying data packets to nearby storage resources. This is one bottleneck InfiniBand seeks to unclog.
- It accommodates Internet data types, such as streaming video, multimedia, and high-resolution graphics, and will

drive more data traffic than traditional business-oriented data processing.

- InfiniBand can be used as an interconnect technology to link individual servers into clusters for purposes of high availability, scalability, and improved manageability.
- InfiniBand will improve link speeds between servers and storage, improving the performance of SANs.

Operation

InfiniBand technology works by connecting host-channel adapters to target-channel adapters. The host-channel adapters tend to be located near the servers' CPUs and memory, while the target-channel adapters tend to be located near the systems' storage and peripherals. A switch located between the two types of adapters directs data packets to the appropriate destination based on information that is bundled into the data packets themselves.

The connection between the host-channel and target-channel adapters is the InfiniBand switch, which allows the links to create a uniform fabric environment. One of the key features of this switch is that it allows data to be managed based on variables such as service-level agreements and a destination identifier. In addition, InfiniBand devices support both packet and connection protocols to provide a seamless transition between the system area network and external networks.

The InfiniBand specification is the culmination of the combined efforts of about 80 companies that belong to the InfiniBand Trade Association, led by industry giants Intel, Compaq, Dell, Hewlett-Packard, IBM, Microsoft, and Sun Microsystems.

Summary

InfiniBand will coexist with the wide variety of existing I/O standards that are already widely deployed in user sites.

Existing architectures include PCI, Ethernet, and Fibre Channel. Likewise, InfiniBand fabrics can be expected to coexist with newer I/O standards, including PCI-X, Gigabit Ethernet, and 10×Gigabit Ethernet. The key advantage of the InfiniBand architecture, however, is that it offers a new approach to I/O efficiency. Specifically, it replaces the traditional system bus with an I/O fabric that supports parallel data transfers along multiple I/O links. Furthermore, the InfiniBand architecture offloads CPU cycles for I/O processing, delivers faster memory pipes and higher aggregate data-transfer rates, and reduces management overhead for the server system.

See also

Fibre Channel

INFRARED NETWORKING

Infrared technology is used to implement wireless LANs as well as the wireless interface to connect laptops and other portable machines to the desktop computer equipped with an infrared transceiver. Infrared LANs are proprietary in nature, so users must rely on a single vendor for all the equipment. However, the infrared interface for connecting portable devices with the desktop computer is standardized by the Infrared Data Association (IrDA).

Infrared LANs

Infrared LANs typically use the wavelength band between 780 and 950 nanometers (nm). This is due primarily to the ready availability of inexpensive, reliable system components.

There are two categories of infrared systems that are commonly used for wireless LANs. One is directed infrared, which uses a very narrow laser beam to transmit data over 1 to 3 miles. This approach may be used for connecting LANs in different buildings. Although transmissions over laser beam are virtually immune to electromechanical interference and would be extremely difficult to intercept, such systems are not used widely because their performance can be impaired by atmospheric conditions, which can vary daily. Such effects as absorption, scattering, and shimmer can reduce the amount of light energy that is picked up by the receiver, causing the data to be lost or corrupted.

The other category is nondirected infrared, which uses a less focused approach. Instead of a narrow beam to convey the signal, the light energy is spread out and bounced off narrowly defined target areas or larger surfaces such as office walls and ceilings.

Nondirected infrared links may be further categorized as either line of sight or diffuse (Figure I-1). Line-of-sight links require a clear path between transmitter and receiver but generally offer higher performance.

The line-of-sight limitation may be overcome by incorporating a recovery mechanism in the infrared LAN that is managed and implemented by a separate device called a "multiple access unit" (MAU) to which the workstations are connected. When a line-of-sight signal between two stations

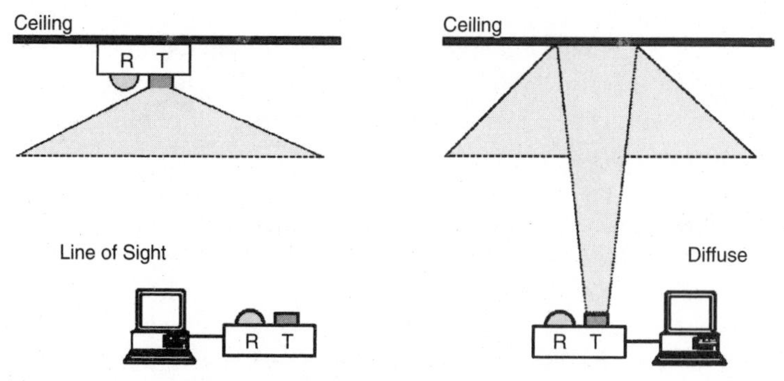

Figure I-1 Line of sight versus diffuse configurations for infrared links.

is temporarily blocked, the MAU's internal optical link control circuitry automatically changes the link's path to get around the obstruction. When the original path is cleared, the MAU restores the link over that path. No data are lost during this recovery process.

Diffuse links rely on light bounced off reflective surfaces. Because it is difficult to block all the light reflected from large surface areas, diffuse links are generally more robust than line-of-sight links. The disadvantage of diffused infrared is that a great deal of energy is lost, and consequently, the data rates and operating distances are much lower.

System Components

Light-emitting diodes (LEDs) or laser diodes (LDs) are used for transmitters. LEDs are less efficient than LDs, typically exhibiting only 10 to 20 percent electrooptical power conversion efficiency, while LDs offer an electrooptical conversion efficiency of 30 to 70 percent. However, LEDs are much less expensive than LDs, which is why most commercial systems use them.

Two types of low-capacitance silicon photodiodes are used for receivers: positive-intrinsic-negative (PIN) and avalanche. The simpler and less expensive PIN photodiode is typically used in receivers that operate in environments with bright illumination, whereas the more complex and more expensive avalanche photodiode is used in receivers that must operate in environments where background illumination is weak. The difference in the two types of photodiodes is their sensitivity.

The PIN photodiode produces an electric current in proportion to the amount of light energy projected onto it. Although the avalanche photodiode requires more complex receiver circuitry, it operates in much the same way as the PIN diode, except that when light is projected onto it, there is a slight amplification of the light energy. This makes it more appropriate for weakly illuminated environments. The avalanche photodiode also offers a faster response time than the PIN photodiode.

Operating Performance

Current applications of infrared technology yield performance that matches or exceeds the data rate of wire-based LANs: 10 Mbps for Ethernet and 16 Mbps for Token Ring. However, infrared technology has a much higher performance potential—transmission systems operating at 50 and 100 Mbps have already been demonstrated.

Because of its limited range and inability to penetrate walls, nondirected infrared can be easily secured against eavesdropping. Even signals that go out windows are useless to eavesdroppers because they do not travel far and may be distorted by impurities in the glass as well as by the glass's placement angle.

Infrared offers high immunity from electromagnetic interference, which makes it suitable for operation in harsh environments like factory floors. Because of its limited range and inability to penetrate walls, several infrared LANs may operate in different areas of the same building without interfering with each other. Since there is less chance of multipath fading (large fluctuations in received signal amplitude and phase), infrared links are highly robust.

Many indoor environments have incandescent or fluorescent lighting, which induces noise in infrared receivers. This is overcome by using directional infrared transceivers with special filters to reject background light.

Media Access Control

Infrared supports both contention-based and deterministic media access control techniques, making it suitable for Ethernet as well as Token Ring LANs.

To implement Ethernet's contention protocol, Carrier Sense Multiple Access (CSMA), each computer's infrared

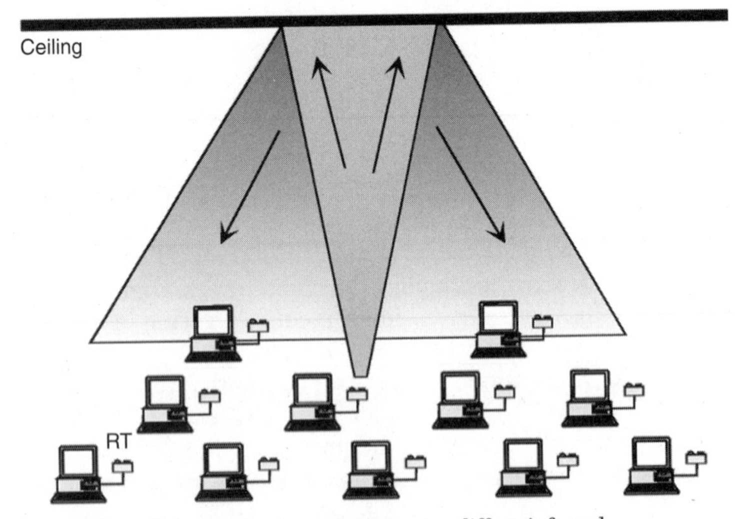

Figure I-2 Ethernet implementation using diffuse infrared.

transceiver is typically aimed at the ceiling. Light bounces off the reflector in all directions to let each user receive data from other users (Figure I-2). CSMA ensures that only one station can transmit data at a time. Only the station(s) to which packets are addressed can actually receive them.

Deterministic media access control relies on token passing to ensure that all stations in turn get an equal chance to transmit data. This technique is used in Token Ring LANs, where each station uses a pair of highly directive (line of sight) infrared transceivers. The outgoing transducer is pointed at the incoming transducer of a station down line, thus forming a closed ring with the wireless-infrared links among the computers (Figure I-3). With this configuration, much higher data rates can be achieved because of the gain associated with the directive infrared signals. This approach improves overall throughput, since fewer bit errors will occur, which minimizes the need for retransmissions.

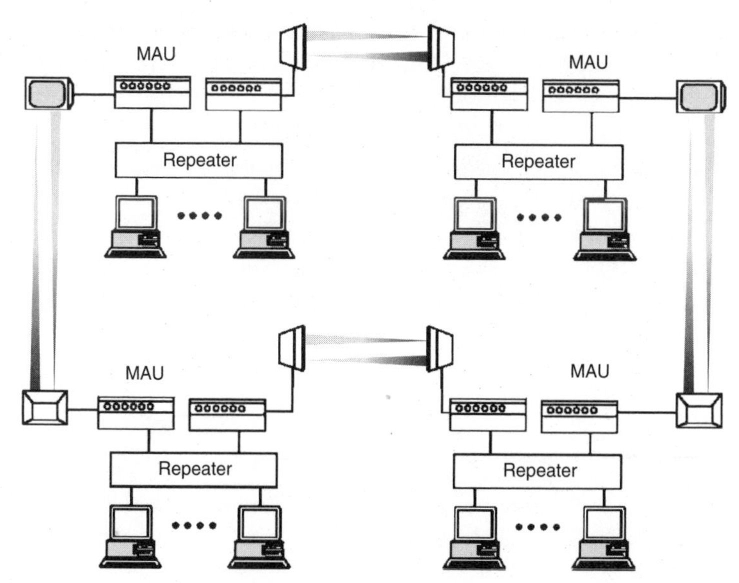

Figure I-3 Token Ring implementation using line-of-sight infrared.

Infrared Computer Connectivity

Most notebook computers and personal digital assistants (PDAs) have infrared ports. Every major mobile phone brand has at least one infrared-enabled handset, and even wristwatches are beginning to incorporate infrared data ports. Infrared products for computer connectivity conform to the standards developed by the Infrared Data Association (IrDA). The standard protocols include Serial Infrared (SIR) at 115 kbps, Fast Infrared (FIR) at 4 Mbps, and Very Fast Infrared (VFIR) at 16 Mbps. The complete IrDA protocol suite contains the following mandatory protocols and optional protocols.

Mandatory protocols:

- *Infrared Physical Layer* Specifies infrared transmitter and receiver optical link, modulation and demodulation schemes, and frame formats.

- *Infrared Link Access Protocol (IrLAP)* Has responsibility for link initiation, device address discovery, address conflict resolution, and connection startup. It also ensures reliable data delivery and provides disconnection services.
- *Infrared Link Management Protocol (IrLMP)* Allows multiple software applications to operate independently and concurrently sharing a single IrLAP session between a portable PC and network access device.
- *Information Access Service (IAS)* Used along with IrLMP and IrLAP, this protocol resolves queries and responses between a client and server to determine the services each device supports.

Optional protocols:

- *Infrared Transport Protocol (IrTTP) or Tiny TP* Has responsibility for data flow control and packet segmentation and reassembly.
- *IrLAN* Defines how a network connection is established over an IrDA link.
- *IrCOM* Provides COM (serial and parallel) port emulation for legacy COM applications, printing, and modem devices.
- *IrOBEX* Provides object exchange services similar to the Hyper Text Transfer Protocol (HTTP) used to move information around the Web.
- *IrDA Lite* Provides methods of reducing the size of IrDA code while maintaining compatibility with full implementations.
- *IrTran-P* Provides image exchange for digital image capture devices/cameras.
- *IrMC* Specifies how mobile telephony and communication devices can exchange information. This includes phonebook, calendar, and message data.

In addition, there is a protocol called IrDA Control that allows cordless peripherals such as keyboards, mice, game pads, joysticks, and pointing devices to interact with many types of intelligent host devices. Host devices include PCs, home appliances, game machines, and television/Web set-top boxes.

An extension called Very Fast Infrared (VFIR) provides a maximum transfer rate of 16 Mbps, a fourfold increase in speed from the previous maximum data rate of 4 Mbps. The extension brings end users faster throughput without an increase in cost and is backward-compatible with equipment employing the previous data rate. The higher speed is intended to address the new demands of transferring large image files between digital cameras, scanners, and PCs. Table I-1 summarizes the performance characteristics of the IrDA's infrared standard.

The IrDA has developed a "point and pay" global wireless point-of-sale (POS) payment standard for handheld devices called Infrared Financial Messaging (IrFM). In an electronic wallet application, consumers use their infrared-enabled PDAs to make purchases at the point of sale. Users "beam" their financial information to pay for a purchase and receive a digital receipt. The IrFM protocol defines payment usage models, profiles, architecture, and protocol layers to enable

TABLE I-1 Performance Characteristics of the Infrared Data Association's Infrared Standard

Feature/Function	Performance
Connection type	Infrared, narrow beam (30º angle or less)
Spectrum	Optical, 850 nanometers (nm)
Transmission power	100 milliwatts (mW)
Data rate	Up to 16 Mbps using Very Fast Infrared (VFIR)
Range	Up to 3 feet (1 meter)
Supported devices	Two
Data security	The short range and narrow angle of the infrared beam provide a simple form of security; otherwise, there are no security capabilities at the link level.
Addressing	Each device has a 32-bit physical ID that is used to establish a connection with another device.

hardware, software, and systems designers to develop IrFM-compliant products and ensure interoperability and compatibility globally. IrFM uses IrDA's Object Exchange (OBEX) protocol to facilitate interoperability between devices.

The Infrared Data Association has formed a Special Interest Group (SIG) to produce a standard for interappliance MP3 data exchange using infrared technology. The popularity of MP3-capable appliances beg for a standard connection between the MP3 players, computers, and the network, allowing consumers to easily move music from device to device without a cable or docking port. The handheld player should be able to transfer a song into a car stereo or home entertainment system. The MP3 SIG is identifying concerns specific to transferring MP3 data and building solutions into the protocol. Among the issues that must be addressed is how to identify copyrighted content and describe distribution restrictions to handle the MP3 content appropriately.

Summary

Infrared's primary impact will take the form of benefits for mobile professional users. It enables simple point-and-shoot connectivity to standard networks, which streamlines users' workflow and allows them to reap more of the productivity gains promised by portable computing. IrDA technology is supported in over 100 million electronic devices, including desktop, notebook, and palm PCs; printers; digital cameras; public phones/kiosks; cellular phones; pagers; PDAs; electronic books; electronic wallets; and other mobile devices.

When used on a LAN, infrared technology also confers substantial benefits to network administrators. Infrared is easy to install and configure, requires no maintenance, and imposes no remote-access tracking hassles. It does not disrupt other network operations, and it provides data security. And because it makes connectivity so easy, it encourages the use of high-productivity network and groupware applications on portables, thus helping administrators amortize the costs of these packages across a larger user base.

See also

Bluetooth
HomeRF
Wireless LANs

INSIDE CABLE WIRING

Inside cable wiring is located within a subscriber's home or apartment and is usually installed by a CATV operator or its contractor. It does not include such items as amplifiers, converters, decoder boxes, or remote control units. Extensions to the initially installed cable may be performed by the subscriber using materials purchased from retail sources and without the permission of the cable operator.

In accordance with FCC rules, after a subscriber voluntarily terminates cable service, the CATV operator may leave the home cabling in place or notify the consumer that it will remove the wiring unless the consumer purchases it from the cable operator on a per-foot replacement-cost basis. If the cable was transferred previously or sold to the subscriber, the subscriber owns it, and the cable operator cannot remove it or restrict its use, regardless of the reason for service termination. In either case, the cable operator cannot be held responsible for any signal leakage that occurs from the home cabling once the operator ceases providing service over that cable.

If the subscriber does not already own the cabling and declines to purchase it from the CATV operator, the operator may remove the home cabling within 30 days. The CATV operator must do so at no charge to the subscriber and pay the cost of any damage caused by its removal.

To leave the cabling inside and remove the cabling outside the subscriber's home, the CATV operator may, for single-unit dwellings, sever the cable approximately 12 inches outside the point where the cable wire enters the outside wall of the subscriber's home. For multiple-unit dwellings (MUDs), the cable operator may sever the cable approximately 12 inches outside the point where it enters the subscriber's individual dwelling unit, except in cases of loop-through or other similar series cable configurations not covered by the home cabling rules.

Where an operator fails to adhere to these procedures, it is deemed to have relinquished immediately any and all ownership interests in the home cabling and is not entitled to compensation for the cabling. Furthermore, the CATV operator may not subsequently attempt to remove the home cabling or restrict its use.

If the CATV operator informs the subscriber of his or her rights and the subscriber agrees to purchase the wiring, constructive ownership over the home cabling will transfer immediately to the subscriber, who may authorize a competing service provider to connect with and use the home cabling. If the subscriber declines to purchase the home cabling, the CATV operator has 7 business days to remove it or make no subsequent attempt to remove it or restrict its use.

If the CATV operator owns the inside cabling, it cannot charge its subscribers a separate cable maintenance fee because the lease rate for operator-owned cabling would already include a component for maintenance and repair. If the subscriber owns the inside cabling, the CATV operator cannot charge a lease rate but could offer a separate optional cable maintenance service and charge hourly rates for as-needed service.

Multiple-Dwelling Units

For MDUs, the cable demarcation point is set at (or about) 12 inches outside of where the cable enters the subscriber's individual dwelling unit. Generally, each subscriber in an MDU has a dedicated segment (often called a "home run") running to his or her premises from a common feeder or riser cable that serves as the source of video programming signals for the entire MDU. The riser cable typically runs vertically in a multistory building and connects to the dedicated segment at a tap or multitap, which extracts portions of the signal strength from the riser and distributes individual signals to subscribers.

Depending on the size of the building, the taps are usually located in a security box (often called a "lock box") or utility closet located on each floor or at a single point in the basement. Each time the riser cable encounters a tap, its signal strength decreases. In addition, the strength of a signal diminishes as the signal passes through the coaxial cable. As a result, cable wiring often requires periodic amplification within an MDU to maintain picture quality. Amplifiers are installed at periodic intervals along the riser based on the number of taps and the length of coaxial cable within the MDU.

Summary

The FCC allows subscribers to purchase the cable inside the premises up to the demarcation point. As with telephone wiring, a demarcation point generally is the point where a CATV operator's cabling ends and the customer-controlled cabling begins. From the customers' point of view, this point is significant because it defines the cabling that they may own or control for purposes of adding more televisions and/or set-top boxes. For the perspective of competition, the demarcation point is significant because it defines the point where an alternative service provider may attach its cabling to reach the customers.

See also

Building Local Exchange Carriers

INSIDE TELEPHONE WIRING

Part 68 of the FCC's rules governs the terms and conditions under which customer premises equipment (CPE) and wiring may be connected to the telephone network. Although not put into effect in until 1975, Part 68 traces its origins to the FCC's Carterfone decision of 1968.

The purpose of Part 68 is to ensure that terminal equipment and wiring can be connected to the Public Switched Telephone Network (PSTN) without causing harm to the network. Under Part 68, CPE and wiring restrictions should be no greater than necessary to ensure network protection. It is the responsibility of carriers to show that any particular Part 68 restriction is necessary.

In 1984, the FCC refined Part 68 to allow customers to connect one- and two-line business and residential telephone wiring to the network. The FCC established a demarcation point to mark the end of the carrier network and the beginning of customer-controlled wiring. The demarcation point would be located on the subscriber's side of the network.

The FCC also issued orders detariffing the installation and maintenance of telephone inside wiring. It first detariffed the installation of complex wiring. In 1986, the FCC extended detariffing to the installation of simple inside wiring and the maintenance of all inside wiring. The FCC allowed carriers to retain ownership of telephone inside wiring, but prohibited carriers from

- Using their ownership to restrict the removal, replacement, rearrangement, or maintenance of telephone inside wiring.
- Requiring customers to purchase telephone inside wiring.
- Imposing a charge for the use of such wiring.

In 1990, the FCC amended the definition of the demarcation point for both simple and complex wiring to ensure that the demarcation point would be near the point where the wiring entered the customer's premises. The revised definition required that the demarcation point generally be no further than 12 inches inside the customer's premises. For existing multiunit installations, the demarcation point is determined in accordance with the carrier's reasonable and nondiscriminatory standard operating practices. For new wiring installations in multiunit premises, including additions, modifications, and rearrangements of existing wiring, the carrier may establish a reasonable and nondiscriminatory practice of placing the demarcation point at the minimum point of entry. When the carrier does not have such a practice, the multiunit premises owner determines the location of the demarcation point or points. If there are multiple demarcation points for either existing or new multiunit installations, the demarcation point for any particular customer may not be further inside the customer's premises than 12 inches from the point at which the wiring enters the customer's premises.

In June 1997, the FCC clarified its position on inside telephone wiring, stating that the carrier standard operating practices for determining the demarcation point for multiunit installations are those practices that were in effect on August 13, 1990. Its subsequent rulings do not authorize changing the demarcation point for an existing building to the minimum point of entry. Reiterating that carriers may not require the customer or building owner to purchase or pay for the use of carrier-installed wiring that is now on the customer's side of the demarcation point, the FCC concluded that the carrier may not remove such wiring.

The FCC also amended its previous definition of the telephone demarcation point as follows:

- The demarcation point may be located within 12 inches of the point at which the wiring enters the customer's premises "or as near thereto as practicable."
- Only major additions or rearrangements of existing wiring can be treated as new installations.
- Multiunit building owners can restrict customer access to only that wiring located in the customers' individual units.

- Local telephone companies are required to provide building owners with all available information regarding carrier-installed wiring on the customer's side of the demarcation point to facilitate the servicing and maintenance of that wiring.

In addition, building owners have a right of access to wiring on their side of the demarcation point and have a responsibility to maintain it. Accordingly, the FCC has restricted the ability of carriers to interfere with customer access and maintenance activities. Carriers may not use claims of ownership as a basis for imposing restrictions on the building owner's removal, rearrangement, replacement, or maintenance of the wiring. Because there are already procedures under which carriers recover the costs of inside wiring that was originally installed or maintained under tariff, carriers are not entitled to additional compensation for wiring, nor can carriers require that the wiring be purchased or impose a charge for its use.

Finally, because carrier removal of installed wiring would prevent customer access and maintenance of that wiring and because the threat of removal could be used to force the customer to purchase the wiring, the FCC has ruled that carriers may not remove installed inside wiring.

Quality Standards

In January 2000, the FCC issued quality guidelines for newly installed inside telephone wiring. At a minimum, inside telephone wiring must be solid 24-gauge or larger twisted pairs that comply with the electrical specifications for Category 3, as defined in the American National Standards Institute (ANSI) EIA/TIA Building Wiring Standards.

Conductors must have insulation with a 1500-volt minimum breakdown rating. This rating is established by covering the jacket or sheath with at least 6 inches (15 centimeters) of conductive foil. A potential difference between the foil and all the individual conductors connected together must be established such that the potential difference is gradually increased over a 30-second time period to 1500 volts, 60 hertz, and then applied continuously for 60 seconds more. At no time during this 90-second time interval should the current between these points exceed 10 milliamperes peak.

All wire and connectors meeting these requirements must be marked, in a manner visible to the consumer, with the symbol "CAT 3" or a symbol consisting of a "C" with a "3" contained within the "C" character, at intervals not to exceed 12 inches along the length of the wire.

In establishing minimum standards for simple inside wiring that eventually connects to the PSTN, consumers and small businesses will benefit by having wiring that is capable of accommodating clear telecommunications and digital transmissions. Consumers and small businesses also will benefit from the decreased necessity for the expensive replacement of poor-quality simple inside wiring, as may be required to accommodate extra lines for additional telephones, personal computers, fax machines, and Integrated Services Digital Network (ISDN) or Digital Subscriber Line (DSL) services. The quality guidelines also will stem the increasing incidence of cross-talk and the risk of network harm associated with the installation of poor-quality inside wiring.

Summary

The FCC has consistently issued rulings aimed at encouraging competition in the inside wiring installation and maintenance markets, promoting new entry into those markets, and fostering the development of an unregulated, competitive telecommunications marketplace. This includes setting quality standards for inside telephone wiring that will enable consumers and small businesses to obtain access to broadband communications services such as DSL.

See also

Building Local Exchange Carriers
Inside Cable Wiring

INSTITUTE OF ELECTRICAL AND ELECTRONICS ENGINEERS

Local area network (LAN) specifications are developed within the Institute of Electrical and Electronics Engineers (IEEE), the world's largest technical professional society with over 350,000 members who conduct and participate in its activities through 1500 chapters in 150 countries. The IEEE and its predecessors—the American Institute of Electrical Engineers (AIEE) and the Institute of Radio Engineers (IRE)—date to 1884. In 1961, the AIEE and IRE merged to form the IEEE.

The technical objectives of the IEEE focus on advancing the theory and practice of electrical, electronics, and computer engineering and computer science—publishing nearly 25 percent of the world's technical papers in these fields. The IEEE sponsors technical conferences, symposia, and local meetings worldwide. It also provides educational programs for members.

Technical Societies

As of mid-2001, the IEEE had 40 technical societies that provide publications, conferences, and other benefits to members within specialized areas—from aerospace and electronic systems to vehicular technology. Each of these societies has technical committees that define and implement the technical directions of the society. For example, there are 21 technical committees within the Communications Society:

- Cable-Based Delivery and Access Systems
- Communications Software
- Communication Switching

- Communications Systems Integration and Modeling
- Communication Theory
- Computer Communications
- Enterprise Networking
- Gigabit Networking
- Information Infrastructure
- Interconnections in High-Speed Digital Systems
- Internet
- Multimedia Communications
- Network Operations and Management
- Optical Communications
- Personal Communications
- Quality Assurance Management
- Radio Communications
- Satellite and Space Communications
- Signal Processing and Communications Electronics
- Signal Processing for Storage
- Transmission and Access and Optical Systems

Standards Board

The IEEE Standards Board is responsible for all matters regarding standards in the fields of electrical engineering, electronics, radio, and the allied branches of engineering. There are currently 10 standing committees of the IEEE Standards Board:

Procedures Committee (ProCom) This committee is responsible for recommending to the IEEE Standards Board improvements and procedural changes to promote efficient discharge of responsibilities by the IEEE Standards Board, its committees, and other committees of the Institute engaged in standards activities.

New Standards Committee (NesCom) This committee is responsible for ensuring that proposed standards projects are within the scope and purpose of the IEEE, that standards projects are assigned to the proper society or other organizational body, and that interested parties are appropriately represented in the development of IEEE standards. This committee examines project authorization requests and makes recommendations to the IEEE Standards Board regarding their approval.

Standards Review Committee (RevCom) This committee is responsible for reviewing submittals for the approval of new and revised standards and for the reaffirmation or withdrawal of existing standards to ensure that the submittals represent a consensus of the parties having a significant interest in the covered subjects. The committee makes recommendations to the IEEE Standards Board regarding the approval of these submittals.

Awards and Recognition Committee (ArCom) This committee is responsible for the administration of all awards presented by the IEEE Standards Board. It acts on behalf of the Standards Board to approve nominations for IEEE Standards Awards. It also submits nominations for standards awards sponsored by other organizations.

New Opportunities in Standards Committee (NosCom) This committee is responsible for identifying and exploring avenues for enhancing IEEE leadership in areas of new technological growth and for recommending to the IEEE Standards Board actions to achieve this purpose.

Procedures Audit Committee (AudCom) This committee provides oversight of the standards development activities of the societies, their standards-developing entities, and the Standards Coordinating Committees (SCCs)[1] of the IEEE Standards Board.

---

[1] When a proposed standard does not fall into the subject area covered by one of the technical societies or a technical society cannot handle the workload, a Standards Coordinating Committee is established and coordinated by the IEEE Standards Board.

Seminars Committee (SemCom) This committee provides oversight for the operation of the seminar program by providing technical expertise and support. The committee reviews and proposes new seminars to ensure that the topics covered are appropriate.

International Committee (IntCom) This committee is responsible for coordinating IEEE standards activities with non-IEEE standards organizations. The committee also assists in the adoption by IEEE of non-IEEE standards when appropriate.

Administrative Committee (AdCom) This committee acts for the Standards Board between meetings and makes recommendations to the Standards Board for its disposition at regular meetings.

Patent Committee (PatCom) This committee provides oversight on the use of any patents and patent information in IEEE standards. PatCom also reviews any patent information submitted to the IEEE Standards Board to determine conformity with the patent procedures and guidelines.

Summary

The IEEE is not a member of the International Electrotechnical Commission (IEC) or the International Organization for Standardization (ISO), since only countries, not standards bodies, can have membership in the IEC and ISO. However, when IEEE working groups need global participation in their projects, they can go through any IEC or ISO member country to make submissions to their IEC or ISO technical committees.

See also

Open Systems Interconnection

INTEGRATED ACCESS DEVICES

Integrated access devices (IADs) support voice, data, and video services over the same $N \times$ T1/E1 access lines. They typically support Asynchronous Transfer Mode (ATM) technology (Figure I-4), which turns the different traffic types into fixed-length 53-byte cells for transmission through the carrier's ATM network. The consolidation of multiple traffic types over the same aggregate access facility eliminates the need for separate lines for each type of application and having to subscribe to separate services. Even though the IAD resides at the customer location, management of the device and the access links into the network is usually the responsibility of the service provider.

Carriers benefit from this arrangement as well. Today's IADs are feature-rich. They offer end-to-end support for ATM Adaptation Layer 2 (AAL2), the industry standard for

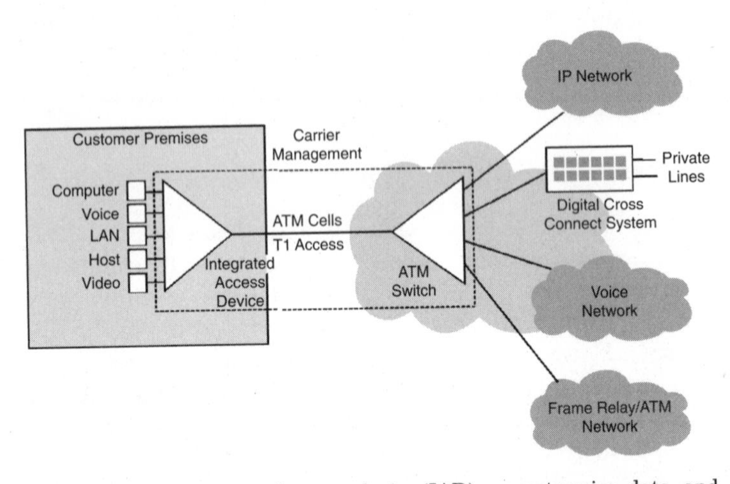

Figure I-4 An integrated access device (IAD) supports voice, data, and video services over the same access lines, which can be bonded together and used as a single higher-speed channel shared by multiple applications. Traffic is split out on the carrier side of the network and routed to the appropriate destinations.

transporting compressed voice over ATM. This enables service providers to deliver eight times the call traffic of traditional T1/E1 lines without degradation in the quality of voice. IADs allow carriers to offer not just bandwidth but also provide them a way to deliver integrated services to their customers with quality, reliability, and flexibility over that bandwidth. IADs also provide carriers with the means to support any service—ATM, Frame Relay, or Internet Protocol (IP)—over any channel within the same T1/E1 line simultaneously.

In addition, such products provide queuing algorithms that provide traffic prioritization capabilities that allow time-sensitive traffic like voice to use the access bandwidth ahead of routine data traffic, which can tolerate delay. With multiple classes of service available, carriers can guarantee differentiated levels of service for both data and voice traffic, enabling them to offer customers virtual private networks (VPNs) that support voice as well as data.

Quality of voice features such as silence suppression and echo cancellation, as well as support for fax/modem call detection and international code conversion, are also available. The common platform—customer premises and carrier point of presence (POP)—ensures interoperability and simplifies operations and management for carriers while providing cost savings to customers.

Modular routers can be turned into IADs very easily, starting with multiple T1/E1 ports and voice compression options, and be incrementally upgraded to support Inverse Multiplexing over ATM (IMA), DS3/E3, and OC-3 wide area network (WAN) connections. This scalability enables service providers to meet the demands of a growing enterprise customer with a single CPE solution.

Summary

IADs allow users to aggregate a variety of lower-speed services onto a high-speed ATM infrastructure. A Competitive

Local Exchange Carrier (CLEC), for example, can deploy IADs to provide integrated voice, data, and video services to business customers, letting them take advantage of applications necessary to succeed in a highly competitive marketplace. At the same time, IADs allow CLECs and other types of carriers to meet their business needs for cost containment in service provisioning.

See also

Asynchronous Transfer Mode

Quality of Service

INTEREXCHANGE CARRIERS

Interexchange Carriers (IXCs), otherwise known as "long-distance carriers," include the big three—AT&T, Worldcom, and Sprint. The ILECs have been limited since 1984 to providing local calling services within their own Local Access and Transport Areas (LATAs), except where they have received specific authorization by the FCC. Generally, long-distance calls between LATAs must be handed off to the IXCs who have established points of presence (POPs) within each LATA for the purpose of receiving and terminating interLATA traffic.

In addition to providing long-distance telephone service, the IXCs offer business services such as ISDN, Frame Relay, leased lines, and a variety of other digital services. Many IXCs are also Internet service providers (ISPs), which offer Internet access services, virtual private networks, electronic mail, Web hosting, and other Internet-related services.

Bypass

Traditionally limited to providing service between LATAs, the Telecommunications Act of 1996 allows IXCs to offer local exchange services in competition with the ILECs. But because the ILECs charge too much for local loop connections and services and do not deliver them in a consistently timely manner, the larger IXCs have implemented technologies that allow them to bypass the local exchange. Among the methods IXCs use to bypass the local exchange include CATV networks and broadband wireless technologies, such as Local Multipoint Distribution Service (LMDS) and Multichannel Multipoint Distribution Service (MMDS).

With regard to cable, AT&T, for example, has acquired the nation's two largest cable companies, TCI and MediaOne, to bring local telephone services to consumers, in addition to television programming and broadband Internet access. As these bundled services are introduced in each market, they are provided to consumers at an attractive price with the added convenience of a single monthly bill. Sprint uses MMDS to offer Internet access to consumers and businesses that are out of range for DSL services. XO Communications, a nationwide Integrated Communications Provider (ICP) uses LMDS to reach beyond its metropolitan fiber loops to reach buildings that are out of the central business districts.

Long-Distance Market

In January 2001, the FCC released the results of a study on the long-distance telecommunications industry. Among the findings from the report:

- In 1999, the long-distance market had more than $108 billion in revenues, compared to $105 billion in 1998. In 1999, long-distance carriers accounted for over $99 billion, and local telephone companies accounted for the remaining $9 billion.
- Interstate long-distance revenues increased by 12.8 percent in 1999 compared to 1.5 percent the year before.
- Since 1984, international revenues have grown more than fivefold from less than $4 billion in 1984 to over $20 billion

in 1999. The number of calls has increased from about half a billion in 1984 to almost 8 billion in 1999.

- In 1984, AT&T's market share was about 90 percent of the toll revenues reported by long-distance carriers. By 1999, AT&T's market share had declined to about 40 percent, WorldCom's share was 25 percent, Sprint's was 10 percent, and more than 700 other long-distance carriers had the remaining quarter of the market.
- According to a sampling of residential telephone bills, in 1999 the average household spent $64 monthly on telecommunications. Of this amount, $21 was for services provided by long-distance carriers, $34 for services by local exchange carriers, and the remainder for services by wireless carriers.
- According to the same sampling of residential telephone bills, 38 percent of toll calls in 1999 were interstate and accounted for 50 percent of toll minutes. Also, 33 percent of residential long-distance minutes were on weekdays, 30 percent on weekday evenings, and 37 percent on weekends.

Summary

Growing competition in long-distance services has eroded AT&T's market share from its former monopoly level to about 40 percent. With this competition has come increasing availability of low-cost calling plans for a broad range of consumers. As a result, average revenue per minute earned by carriers has been declining steadily for several years, while long-distance usage has increased substantially to make up for that revenue shortfall. As more ILECs get permission from the FCC to enter the in-region long-distance market, IXCs will come under increasing competitive pressure because the ILECs will be able to bundle local, long-distance, and Internet access into attractively priced service packages.

See also

Building Local Exchange Carriers
Incumbent Local Exchange Carriers

INVERSE MULTIPLEXERS

Inverse multiplexers allow users to put together increments of bandwidth and use it as a high-speed channel to support a given application. Originally, inverse multiplexing specifically addressed the bandwidth needs of videoconferencing, but the concept now applies to other applications and to providing scalable bandwidth for Internet access. Inverse multiplexers may be used to combine bandwidth on multiple dial-up or dedicated connections. The connections may be in the local loop or on the network (i.e., interoffice) side.

In the dial-up scenario, inverse multiplexing might come into play when the user wants to access the Internet at speeds greater than 56 kbps but does not have broadband services such as DSL or cable available. With the right software and a multiport modem, up to three dial-up connections can be established to the ISP to achieve a data rate of 168 kbps. In the case of dedicated connections, up to eight T1 lines can be bonded together by an inverse multiplexer to achieve up to 12 Mbps of access bandwidth to the Internet or to a carrier service such as ATM, which would be more economical that having to step up to a T3 line.

On the network side, inverse multiplexing could be used to dial up long-distance bandwidth within the PSTN in support of videoconferencing or document collaboration among distributed corporate locations. In this case, multiple 56/64-kbps channels would be dialed by the inverse multiplexer within the carrier's network. The company pays for the number of interexchange channels only when they are set up to handle the conference call. On completion of the call or as different locations drop off the conference, the channels are taken down and carrier billing stops. This method of access obviates the need for overprovisioning the corporate network to support infrequently used applications.

Implementation

Inverse multiplexing can be implemented in CPE or as a carrier-provided service. Either way, the advantages of inverse multiplexing include the immediate availability of extra bandwidth when needed, which eliminates of the need for standby leased lines that are billed to the user whether fully used or not. This adds up to significant cost savings for the organization.

In a typical videoconference application, for example, the inverse multiplexer accepts the data stream from the video codec and divides it among multiple 56/64- or 384-kbps channels that are dialed up as needed and aggregated to achieve what is, in effect, a higher-speed link. The inverse multiplexer synchronizes the information across the channels and transmits it via the PSTN to a similar device at the remote location. There the data are received as a single data stream (Figure I-5) and passed to another video codec.

Some inverse multiplexers can be configured to support multiple applications simultaneously. For example, an inverse multiplexer that can be used to link (1) multiple applications at a single site to the public network via a T1 or ISDN access facility, (2) a Private Branch Exchange (PBX) to a VPN, (3) a router to a Fractional T1 network, or (4) a video

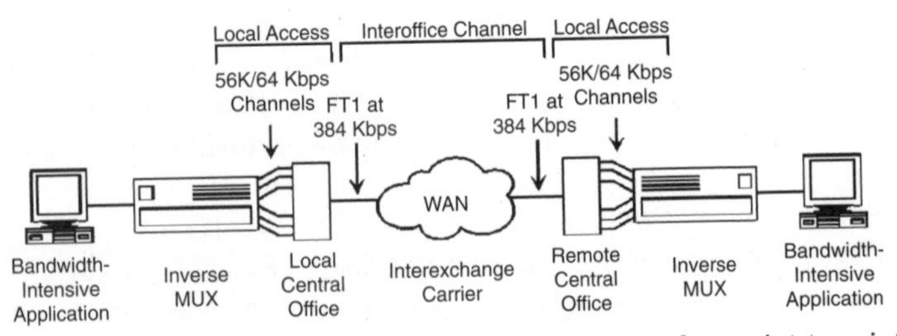

Figure I-5 A simple inverse multiplexer configuration for a point-to-point videoconference or image transfer.

codec to a switched digital service. This capability appeals to users who want to spread the cost of a T1 access line across multiple applications. Some products allow users to switch multiple applications on a call-by-call basis over different carriers' services simultaneously. While some inverse multiplexers interface only to switched services, others can access both switched and dedicated communications facilities.

Another capability of some inverse multiplexers is the transport of bandwidth-intensive data across multiple T1 circuits to achieve a Fractional T3 circuit. T3-level inverse multiplexers are intended for applications that require transport between the T1 and T3 rates of 1.544 and 44.736 Mbps. As many as eight T1 circuits can be aggregated to achieve the desired increment of bandwidth without the organization being forced to lease a more expensive T3 facility, much of which would go unused. If the organization needs more bandwidth than eight T1 circuits would provide, then it is more economical to lease a T3 line.

System Management

The system management interface usually consists of a microcomputer equipped with software that allows the network manager to define and monitor traffic flow, bandwidth requirements, access line quality, and various configuration parameters.

Through this interface, administrative functions are also performed, such as the creation of call profiles. A "call profile" is a file that contains the parameters of a particular data call so that a similar call can be quickly reestablished at another time simply by loading the call profile. Usually the call profile function includes a factory-loaded profile that acts as the template for creating and storing user-defined call profiles. Because each data call may involve as many as 25 separately configurable parameters, the use of call profiles can save a lot of time. Users typically load or edit a call profile using keyboard commands to the management software on the microcomputer.

Inverse multiplexer management interfaces often support remote devices. This capability allows a network administrator at a central location to configure, test, and otherwise manage other inverse multiplexers at remote locations in much the same way as is currently offered by the in-band management systems of some T1 multiplexers. This is accomplished by the management interface reserving a certain amount of the network bandwidth, usually not more than 2 percent, as a subchannel to implement remote management.

Most inverse multiplexers can be remotely monitored and controlled via Simple Network Management Protocol (SNMP). This is usually accomplished with SNMP agent software included with the product. The agent collects detailed error statistics, utilization ratios, and performance histories that can be retrieved for analysis.

Standards

The Bandwidth on Demand Interoperability Group (BONDING), formed in late 1991, defined interoperability standards for inverse multiplexers. BONDING is often used for non-Internet applications, such as videoconferencing. There is also a set of international standards for bandwidth-on-demand services called Global Bandwidth on Demand (GloBanD). The BONDING specification describes four modes of inverse multiplexer operation:

- *Mode 0* Enables inverse multiplexers to receive two 56-kbps calls from a video codec and initiate dual 56-kbps calls to support a videoconference.
- *Mode 1* Enables inverse multiplexers to spread a high-speed data stream over multiple switched 56/64-kbps circuits. Because this mode does not provide error checking, the inverse multiplexers operating in this mode have no way of knowing if one of the circuits in a multicircuit call has failed. In this case, it is up to the receiving node to

detect that it has not received the full amount of data and must request more bandwidth.

- *Mode 2* Adds error checking to each 56/64-kbps circuit by stealing 1.6 percent of the bandwidth from each circuit for the passage of information that detects circuit failures and reestablishes links.
- *Mode 3* Uses out-of-band signaling for error checking, which may be derived from a separate dial-up circuit or the unused bandwidth of an existing circuit.

When establishing calls, inverse multiplexers at both ends first determine whether they can interoperate using the vendor's proprietary protocol. If not, this means that the inverse multiplexers of different vendors are being used and that they should use the BONDING protocol to support the transmission.

Summary

The inverse multiplexer allows network managers to match bandwidth to the application. These devices (or a carrier-provided service) provide a degree of configuration flexibility that cannot be matched in efficiency or economy by any other technology. With inverse multiplexers, organizations no longer have to overprovision their networks to handle peak traffic or run occasional high-bandwidth applications. Instead, they can order bandwidth only when it is needed and, in the process, save on line costs. Another type of inverse multiplexer is the integrated access device (IAD), which supports multiple protocols, bandwidth contention among different applications, and quality of service (QoS) features to ensure optimal performance for all applications.

See also

Integrated Access Devices
Multiplexers

J

JITTER

While delay is the time it takes to get a unit of information from source to destination through a network, jitter is the variance of the delay. Both can have potentially disrupting effects on applications running over the network, particularly if they are time-sensitive. Examples of time-sensitive applications are telephone calls and videoconferences. Some data applications are also time-sensitive, such as pages and text chat sessions.

In the past, delay and jitter were not important aspects of computer networks. For example, it did not matter if file transfers or e-mail took half a second longer, independent of the total transfer time (delay). Similarly, it did not matter if—on a particular file transfer—70 percent of the data were sent during the first half of the transfer and 30 percent in the last half (jitter).

But jitter and delay matter when it comes to two-way or multiway conversations and conferences—they must have low delay and jitter to support the natural interaction among participants, since long pauses can be potentially disruptive to a conversation. Jitter and delay are especially important when the real-time application runs over a packet network like the Internet, where the individual packets take different paths to the destination and must be put back in the right order.

Multimedia applications that combine audio and video content are even more sensitive to delay and jitter. To prevent dropouts in an audio stream or jerkiness in video, jitter must be low. For one-way broadcasts, buffering can be used in the end stations to decrease the effect of jitter, but only at the cost of increased delay. While this delay is acceptable for one-way broadcasts, it is not acceptable for two-way conversation.

The electrical pulses, which are sent through a network as indications of 0s and 1s, are normally sent at very specific intervals of time. The repeaters, bridges, and switches on the network contain buffers to accept the signal from the input side and send it to the output side at a tightly controlled speed. This results in a clean output signal, which can be received by the next piece of equipment on the network.

However, every piece of equipment has a narrow tolerance within which it operates. If a signal passes through several pieces of equipment or cable segments, these tolerances can add up and cause the resulting signal to shift in phase compared with what was originally sent. This makes it difficult or impossible for the next device on the network to lock on to the signal, which causes errors (Figure J-1).

Summary

As clock speeds in computers and data rates in communications systems increase, timing budgets become tighter, and the need to measure and characterize jitter becomes more critical. Oscilloscopes offer various tools to measure jitter. On new networks with no performance history, a device called a "jitter generator" can be used to add controlled jitter to digital signals for the purpose of stressing the connections to Public Switched Telephone Network (PSTN), cellular, and PCS base stations, satellite modems, and microwave links. Not only does this ensure that the connections are error-free

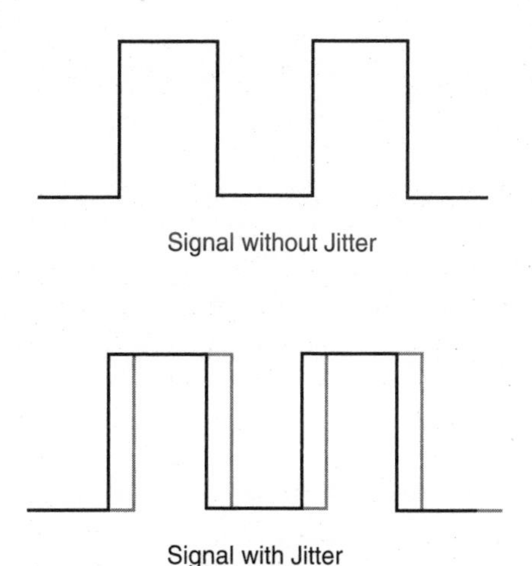

Figure J-1 A digital signal with no jitter (*top*) and the same signal with jitter (*bottom*).

before being turned over to user traffic, but once the jitter parameters are known, lower-cost oscilloscopes can be used for periodic quality checks.

See also

Latency

L

LAN ADMINISTRATION

The local area network (LAN) administrator's main focus is usually on keeping the network operating properly and making sure the needs of users are addressed in a timely manner, including hardware and software upgrades. To meet the needs of all users, the LAN administrator must have appropriate tools to accomplish a number of specific tasks. Many of these tasks can be automated to enable the LAN administrator to take care of multiple networks that may consist of hundreds of servers, desktop computers, and peripherals—the configurations of which may change on a daily basis to meet the varying needs of mobile professionals, telecommuters, workgroups, departments, or the organization as a whole. Many of these tools may come bundled with the LAN vendor's network management system. Some are bundled with help desk software. Others are available from third-party vendors as standalone products that can be launched from the network management system or help desk. All these different management and administration systems and tools can even share data via Application Programming Interfaces (APIs).

Whether bundled with other products or used separately, the right tools help the LAN administrator monitor, analyze, and adapt the LAN to changing organizational needs. The tools themselves are applications and utilities based on NetWare, Windows, or UNIX. In large, heterogeneous environments, the LAN administrator will have occasion to use tools that work with multiple network operating systems. With the right tools, the LAN administrator can access multiple functions and client operating systems through a consistent graphical user interface, which can greatly improve personal performance.

Console and Agents

The key concepts in LAN administration are the console and agents. The console is the workstation that is set up to view information collected by the agents. The agents are special programs that are designed to retrieve specific information from the network. An application agent, for example, works on each workstation to log application usage. Workstation users are not aware of the agent, and it has no effect on the performance of the workstation or the applications running on it. The collected information is organized into datasets and stored in a relational database where it can be retrieved for viewing on the LAN administrator's console. Information from multiple sets of data can be displayed in several ways—cells, charts, and text—and analyzed for such purposes as license management or inventory management and printed as a detail or summary report. The entire process is illustrated in Figure L-1.

A comprehensive tool set allows the LAN administrator to perform the following main functions:

- View and manipulate network data
- Automate file distribution
- Maintain hardware inventory
- Manage installed software, including application usage
- Receive notification of network events

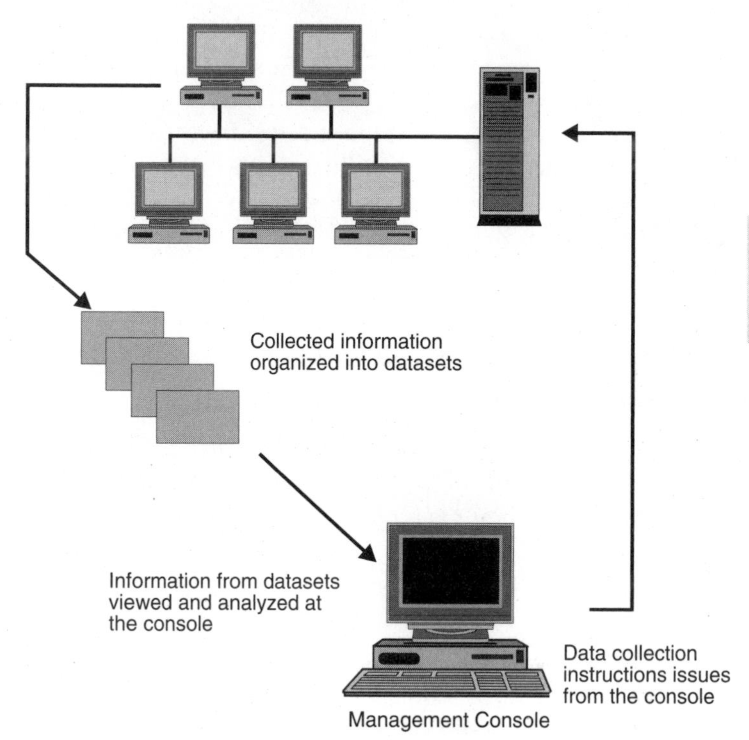

Figure L-1 Information flow between console and agent.

- Establish and manage network printer support
- Automate network processes, such as backup and virus detection
- Monitor disk and file usage
- Create task lists
- Work with text files
- Establish and maintain security
- Manage storage

All the agents that collect information in support of these functions are configured at the console using commands selected from the menu bar. Once configured, each type of agent can be assigned an icon that launches its associated viewer for displaying collected information.

With LANs increasingly being interconnected over wider geographic areas, network administrators can make use of agents to monitor wide area network (WAN) links as well. The agents play a role similar to the one monitors and protocol analyzers play in hardware. Although the agents collect the same information as the monitors, they also process the packets to provide detailed and high-level information regarding network traffic. In this way, they resemble protocol analyzers.

Hardware-based monitors and software-based agents can be used together distributed throughout a LAN as well as geographically dispersed via the WAN. Their packet capture with filtering and decoding capabilities allows early detection of suspect traffic patterns and identification of faulty network devices. Since agents use the network only when information is requested from the network management system, they do not burden the network with unnecessary overhead.

Intelligent Agents

A critical tool in the information technology (IT) department's arsenal of management tools is the "intelligent agent," which is an autonomous and adaptive software program that accomplishes its tasks by executing commands remotely. System administrators, network managers, and software developers can create and use intelligent agents to execute critical processes, including performance monitoring, fault detection and restoration, hardware and software asset management, virus protection, and information search and retrieval. One of the latest applications of intelligent agents is intrusion detection, in which the agent reports security breaches at a router or firewall and takes appropriate steps to prevent further attacks. With the agent concept enjoying increasing acceptance, vendors are offering integrated development environments for creating agents, agent managers for deploying and managing agents across a network, and sample intelligent agents that are ready to run and which can be customized for particular needs without requiring any programming skills.

What makes these agents so smart is the addition of programming code that tells them exactly what to do, how to do it, and when to do it. In essence, the intelligent agent plays the dual role of manager and agent. Under this scheme, polling is localized, events and alarms are collected and correlated, various tasks and trouble responses are automated, and only the most relevant information is forwarded to the central management station. In the process, network traffic is greatly reduced, as is the time for problem resolution.

The behavior of intelligent agents can be modified in two ways: templates and programs. The choice will depend on the level of an organization's in-house network and systems management expertise.

Some vendors, such as Hewlett-Packard, offer rules-based templates to modify the behavior of intelligent agents without the need for native-language programming. The role of the agent is defined in a template that tells the network management system what to do with the information collected by the agent. A network manager can bring up a representation of the template used for monitoring a particular application, for example, and edit the rules concerning responses to various alerts.

For instance, when a firewall issues an error message, under the rules described in its template, it sends all alerts to a particular system administrator. The network manager can change the rule so that an automated response is initiated instead, allowing agents to resolve problems and perform routine tasks (e.g., backups, batch jobs, file maintenance) locally. This prevents the system administrator from being overwhelmed by warning and informational messages so that he or she can focus only on potential service-disrupting conditions that cannot be resolved locally.

Agents can be built with Java and can be used to monitor and report on key performance metrics of systems, services, and applications. Since Java is a cross-platform development tool, agents built with Java can provide a single, unified management system to support any mix of Internet Protocol (IP)–based desktop, server, and network resources that also run Java—including hubs, switches, and routers. In addition to relieving the burden of front-line managers, who usually must cope with a collection of unrelated tools while demands on them are accelerating, the Java agents can self-populate through the network to add new resource support and functionality enhancements.

For programmers, many vendors offer tool kits that accelerate the development of the agent and manager components, which is normally a significant and time-consuming activity. Without a tool kit, each agent must be hand-coded, i.e., built from scratch—a process that can take days or weeks. The use of tool kits can reduce development time to only minutes, allowing developers to spend more time on the value-added components of their application, such as processing data gathered by the agent or communicating with, and controlling, external devices.

Agent Applications

Agent technology has been available for several years and still represents one of the fastest growing areas in network management—and for good reason. In a global economy that encourages the expansion of networks to reach new markets and discourages the addition of personnel to minimize operating costs, it simply makes sense to automate as many management tasks as possible through the use of intelligent agents. In recognition of these new business realities, the list of tasks that are being handled by agents is continually growing. Among the key capabilities are

- *Performance management* Network performance monitoring can help determine network service-level objectives by providing measurements to help managers understand typical network behavior and normal periods. The challenge is defining "typical" and "normal." Intelligent agents can help define the network's behavior and gather the information for documenting achieved performance levels. The following capabilities of intelligent agents are particularly useful for building a network performance profile:
- *Baselining and network trending* Identifies the true operating envelope of the network by defining typical and normal behavior. Baselining also provides long-term measurements to check service-level objectives and show out-of-norm conditions, which, if left unchecked, may have drastic consequences on the productivity of network users.
- *Application usage and analysis* Identifies the overall load of network traffic, what times of the day certain applications load the network, which applications are running between critical servers and clients, and what their load is throughout the day, week, and month. Application usage and analysis allows the network manager to discover important performance information on a real-time or historical basis.
- *Client-server performance analysis* Identifies which servers may be overutilized, which clients are hogging server resources, and what applications or protocols they are running. Such performance analyses help the network manager define and adhere to client-server performance objectives.
- *Internetwork perspective* Identifies traffic rates between subnets so the network manager can find out which nodes are using WAN links to communicate. This information can be used to define typical rates between interconnect devices. This perspective can show how certain applications

use the critical interconnect paths and define normal WAN use for applications.

- *Data correlation* Allows peak network usage intervals to be selected throughout the day to determine which nodes were contributing to the network load at that peak point in time. Traffic source and associated destinations can be determined with seven-layer protocol identification.

Applications Management There are client-side agents that continuously monitor the performance and availability of applications from the end user's perspective. A just-in-time applications performance management capability captures detailed diagnostic information at the precise moment when a problem or performance degradation occurs, pinpointing the source of the problem so it can be resolved immediately.

Such agents are installed on clients as well as application servers. They monitor every transaction that crosses the user desktop, traversing networks, application servers, and database servers. They monitor all distributed applications and environmental conditions in real time, comparing actual availability and performance with service-level thresholds.

This analysis enables network and application managers to understand the source of application response-time problems by breaking down response times into network, application, and server components. As a result, troubleshooting that sometimes takes weeks can be accomplished in a matter of minutes.

- *Fault Management* When faults on the network occur, it is imperative that problems be resolved quickly to decrease the negative impact on user productivity. Network managers must be able to respond quickly and have procedures in place to reestablish lost service and maintain beneficial service levels. The following capabilities of intelligent agents can be used to gather and sort the data needed to quickly identify the cause of faults and errors on the network:

- *Packet interrogation* Isolates the actual conversation that is causing the network problem, allowing the network manager to get to the heart of the problem quickly.
- *Data correlation* Since managers cannot always be on constant watch for network faults, it is imperative to have historical data available that provide views of key network metrics at the time of the fault. What was the overall error/packet rate and the types of errors that occurred? What applications were running at the time of the fault? Which servers were most active? Which clients were accessing these active servers, and which applications were they running? Data correlation can help to answer these questions.
- *Identification of top error generators* Identifies the network nodes that are generating the faults and contributing to problems such as bottlenecks caused by errors and network down time.
- *Immediate fault notification* With immediate notification of network faults, managers can instantly learn when a problem is occurring before users do. Proactive alarms help detect and solve the problem as it is happening.
- *Automated resolution procedures* The intelligent agents can be configured to automatically fix the problem when it occurs. The agent can even be programmed to automatically e-mail or notify help desk personnel with instructions on how to solve the problem, thus saving time and money.

Capacity Planning and Reporting Capacity planning and reporting services play a significant role in delivering sustainable network service levels to end users. They also provide documented proof to management and other organizations that pay for services to help ensure that network service levels are consistently achieved. Capacity planning and reporting allows for the collection and evaluation of information to make informed decisions about future network configurations, accommodating growth in client-server computing environments. The following capabilities of intelligent agents can be used to assist in managing network growth:

- *Baselining* Allows the network manager to determine the true operating envelope for the network, which can identify business cycle deviations.
- *Load balancing* Allows the network manager to compare internetwork service objectives from multiple sites at once to determine which subnets are over- or underutilized. It also helps the network manager discover which subnets can sustain increased growth and which require immediate attention.
- *Protocol/application distribution* Helps the network manager understand which applications have outgrown which domains or subnets. For example, these capabilities can find out if certain applications are continuously taking up more precious bandwidth and resources throughout the enterprise. With this kind of information, the network manager can better plan for the future.
- *Host load balancing* Allows the network manager to obtain a list of the top network-wide servers and clients using mission-critical applications. For example, the information collected from intelligent agents might reveal if specific servers always dominate precious LAN or WAN bandwidth or spot when a CPU is becoming overloaded. In either case, an agent on the LAN segment, WAN device, or host can initiate load balancing automatically when predefined performance thresholds are met. The information gathered by the agent can be used for resource planning.
- *Traffic profile optimization* Allows network managers to compare actual network configurations against proposed configurations. From the information gathered and reported by intelligent agents, traffic profiles can be developed that allow what-if scenarios to be put together and

tested before incurring the cost of physically redesigning the network. This takes the guesswork out of determining the best placement of client/server nodes and applications, for example.

Security Management A properly functioning and secure corporate network plays a key role in maintaining an organization's competitive advantage. Setting up security objectives related to network access must be considered before mission-critical applications are put in potentially compromising networked environments. Intelligent agents can help discover holes in network security by continuously monitoring network access with the following capabilities:

- *Monitor effects of firewall configurations* By monitoring postfirewall traffic, the network manager can determine if the firewall is functioning properly. For example, if the firewall was just programmed to disallow access of a specific protocol or external site, but the program's syntax was wrong, the intelligent agent will report it immediately.
- *Show access to and from secure subnets* By monitoring access from internal and external sites to secure data centers or subnets, the network manager can set up security service-level objectives and firewall configurations based on the findings. For example, the information reported by the intelligent agent can be used to determine whether external sites should have access to the company's database servers.
- *Trigger packet capture of network security signature* Intelligent agents can be set up to issue alarms and automatically capture packets upon the occurrence of external intrusions or unauthorized application access. This information can be used to track down the source of security breaches. Some intelligent agents even have the capability to initiate a trace procedure to discover a breach's point of origination.

- *Show access to secure servers and nodes with data correlation* This capability reveals which external or internal nodes are accessing potentially secure servers or nodes and identifies which applications they are running.
- *Show applications running on secure nets with application monitoring* This capability evaluates applications and protocol use on secure networks or traffic components to and from secure nodes.
- *Watch protocol and application use throughout the enterprise* This capability allows the network manager to select applications or protocols for monitoring by the intelligent agent so that the flow of information throughout the enterprise can be viewed. This information can identify who is browsing the Web, accessing database client-server applications, or using the email system, for example.

Summary

Whether bundled with other products or purchased separately, the right tools help the LAN administrator monitor, analyze, and adapt the LAN to the changing needs of users, workgroups, departments, and the organization by providing the means to centrally view and administer the network and automate many routine tasks. Such tools enable LAN administrators to be responsive to the fairly routine needs of users without becoming unduly burdened in the process and diverting attention from the main focus of keeping the network operating smoothly for the benefit of all.

See also

Asset Management
Electronic Software Distribution
Help Desks
LAN Restoration
Network Management Systems
Network Support

LAN RESTORATION

With data and applications residing in many more places and linked together via LANs, it becomes harder to protect vital information from loss, damage, or theft. It is also more difficult to monitor network performance, isolate problems, and implement corrective measures. The failure to give adequate attention to LAN restoration planning can result in poor performance, data loss, more frequent outages, and prolonged network downtime. Fortunately, a number of solutions are available to address these problems.

The LAN is a data-intensive environment requiring special precautions to safeguard one of the organization's most valuable assets—information. The procedural aspect of minimizing data loss entails the implementation of manual or automated methods for backing up all data on the LAN to avoid the tedious and costly process of recreating vast amounts of information. The equipment aspect of minimizing data loss entails the use of redundant circuitry, as well as components and subsystems that are activated automatically on the failure of various LAN devices to prevent data loss and maintain network availability.

In addition to the ability to respond to errors in transmissions by detection and correction, other important aspects of LAN operation are recovery and reconfiguration. Recovery deals with bringing the LAN back to a stable condition after an error, and reconfiguration is the mechanism by which the network is restored to its previous condition after a failure.

LAN reconfigurations involve mechanisms to restore service on loss of a link or network interface unit. To recover or reconfigure the network after failures or faults requires that the network possess mechanisms to detect that an error or fault has occurred and to determine how to minimize the effect on the system's performance. Generally, these mechanisms provide

- Performance monitoring
- Fault location

- Network management
- System availability management
- Configuration management

These mechanisms work in concert to detect and isolate errors, determine their effects on the system, and remedy these errors to bring the network to a stable state with minimal impact on network availability.

Reconfiguration is an error-management scheme used to bypass major failures of network components. This process entails detecting that an error condition has occurred that cannot be corrected by the usual means. Once it is determined that an error has occurred, its impact on the network is assessed so that an appropriate reconfiguration can be formulated and implemented. In this way, normal operations can continue under a new configuration.

Error detection is augmented by logging systems that keep track of failures over a period of time. This information is examined to determine trends that adversely affect network performance. This information, for example, might reveal that a particular component is continually causing errors to be inserted onto the network, or the monitoring system might detect that a component on the network has a higher than normal failure rate.

The configuration-assessment component of the reconfiguration system uses information about the current system configuration—including connectivity, component placement, paths, and traffic flows—and maps it against the failed component. This information is analyzed to indicate how that particular failure is affecting the system and to isolate the cause of the failure. Once this assessment has been performed, a solution can be worked out and implemented.

The solution may consist of reconfiguring most of the operational processes to avoid the source of the error. The solution determination component examines the configuration and the affected hardware or software components, determines how to move resources around to bring the network back to an operational state or indicates what must be eliminated because of the failure, and identifies network components that must be serviced.

The determination of the most effective course of action is based on the criticality of keeping certain functions of the network operating and maintaining the resources available to do this. In some environments, nothing can be done to restore service because of device limitations (e.g., lack of redundant subsystems) or the lack of spare bandwidth. In such cases, about all that can be done is to indicate to the servicing agent what must be corrected and keep users informed of the situation.

Once an alternate configuration has been determined, the reconfiguration system implements it. In most cases, this means rerouting transmissions, moving and restarting processes from failed devices, and reinitializing software that has failed because of some intermittent error condition. In some cases, nothing may need to be done except notify affected users that the failure is not severe enough to warrant system reconfiguration.

Geographically distributed LANs can be internetworked over the WAN using such devices as bridges and routers connected to leased lines and/or switched services. An advantage of using routers for this purpose is that they permit the building of large mesh networks. With mesh networks, the routers can steer traffic around points of congestion or failure and balance the traffic load across the remaining links. In addition, routers have flow control and more comprehensive error protection than bridges and can be equipped with protocols to apply quality-of-service parameters.to multiple applications.

Bridges are useful for partitioning sprawling LANs into discrete subnetworks that are easier to control and manage. Through the use of bridges, like devices, protocols, and transmission media can be grouped together into communities of interest. Such partitioning can yield many advantages, such as eliminating congestion and improving the response time of the entire network. Subnetworks are also useful for testing new applications before making them available over the enterprise network.

Server Restoration Capabilities

Sharing resources distributed over the LAN can better protect users against the loss of information and unnecessary downtime than a network with all its resources centralized at a single location. The vehicle for resource sharing is the server, which constitutes the heart of the LAN. The server gives the LAN its features, including those for security and data protection, as well as those for network management and resource accounting.

Because huge amounts of corporate data may be located at the server, the server must be able to implement recovery procedures in the event of a program, operating system, or hardware failure. For example, when a transaction terminates abnormally, the server must have the capability to detect an incomplete transaction so that the database is not left in an inconsistent state. The server's rollback facility is invoked automatically, which backs out of the partially updated database. The transaction can then be resubmitted by the program or user. A roll-forward facility recovers completed transactions and updates in the event of a disk failure by reading a transaction journal that contains a record of all updates.

The server is the most vulnerable resource on a LAN. When a server crashes, not only can it prevent the availability of crucial data, but it also can corrupt or destroy the data. For this reason, fault tolerance is becoming an increasingly important issue in the LAN world. As LANs take on crucial tasks that were once tasked to mainframes, restoration mechanisms that keep the server and its data alive through disk crashes and power outages are becoming a requirement. While there are some software-only and hardware-only solutions to choose from, the most effective way to achieve a high degree of fault tolerance is with a combination of both.

Hardware Solutions

Hardware in a fault-tolerant server must be duplicated so that there is an alternate hardware component that can carry on after a failure. Such redundancy extends to the server's CPUs, ports, network interfaces, memory, disks, tapes, and input-output (I/O) channels. This duplication is often implemented via a "hot standby" solution, where a complete duplicate system is used. The secondary system does nothing but monitor the tasks of the primary system, in most cases duplicating its processing. That way, when a component in the primary system fails, the secondary system is prepared to take over where the primary system left off.

There are obvious disadvantages to this method: Twice the amount of hardware must be purchased, with half of it remaining idle at any given time. The only way to cost-justify such a purchase may be to think of it as an insurance policy.

Another way to achieve fault tolerance is to have all hardware components function all the time, but with a load-balancing mechanism that reallocates the work to surviving components when a failure occurs. This arrangement requires a sophisticated operating system that continually monitors the system for errors and dynamically reconfigures the system upon sensing performance problems.

Software Solutions

For complete fault tolerance, software must react to hardware component failures. This allows the system to take advantage of duplicated hardware in order to keep the server running. It also ensures that the data are always available and uncorrupted.

The software monitors the system for failures and, in the event of failure, switches active primary components. The extent to which control is switched depends on whether the system is a hot-standby or a load-balancing system. With a hot-standby system, complete control is given to the standby system when a fatal error is detected, regardless of the component that failed.

With a load-balancing system, control is given to the backup just for the element that failed. Software designed for load-balancing fault-tolerant systems must be highly sophisticated, with the ability to orchestrate complex switching of control in the event of failure—with complete reliability. If the CPU fails, a secondary CPU gains control, but the primary I/O controller, disk controller, and disk drive remain in control. In the event of a CPU failure, the software must reallocate processes (such as the file server software) to active processors. When an I/O channel fails, the software must reroute disk access and communication I/O to an alternate channel. If the bus connecting the processors together fails, it must reroute inter-CPU communications to another bus. If a disk crashes, it must be able to switch to a backup disk drive.

Software has another function: to ensure that files remain available and uncorrupted during a failure and that the server can pick up where it left off upon system recovery. Several mechanisms are available to provide optimal data availability and integrity:

- Disk mirroring provides constant availability despite media failures. Disk mirroring allows all file updates to be written to two disks at the same time. If one disk (or the I/O channel to it) fails, the information can be read and written to the other disk.
- To maintain constant file availability, mirroring mechanisms provide the capability for tape backup of disk data to be made while updates continue. After a failure, users should have disk access during data rebuilds.
- Disk recovery after a failure requires the ability to bring the new secondary disk to the current state of the primary disk so that the data are once again protected. To maintain availability, disk recovery should be performable online, without taking the system down and depriving users of access.

Depending on the level of fault tolerance desired and the price the organization is willing to pay, the server may be configured in several ways: unmirrored, mirrored, or duplexed.

An unmirrored server configuration entails the use of one disk drive and one disk channel, which includes the controller, a power supply, and interface cabling. This is the basic configuration of most servers. The advantage is chiefly one of cost: The user pays only for one disk and disk channel. The disadvantage of this configuration is that a failure in either the drive or anywhere on the disk channel could cause temporary or permanent loss of the stored data.

The mirrored server configuration entails the use of two hard disks of similar size. There is also a single disk channel over which the two disks can be mirrored together. In this configuration, all data written to one disk are then automatically copied onto the other disk. If one of the disks fails, the other takes over, thus protecting the data and assuring all users of access to the data. The server's operating system issues an alarm notifying the network manager that one of the mirrored disks is in need of replacement.

The disadvantage of this configuration is that both disks use the same channel and controller. If a failure occurs on the channel or controller, both disks become inoperative. Because the same disk channel and controller are shared, the writes to the disks must be performed sequentially—that is, after the write is made to one disk, a write is made to the other disk. This can degrade overall server performance under heavy loads.

In disk duplexing, multiple disk drives are installed with separate disk channels for each set of drives. If a malfunction occurs anywhere along a disk channel, normal operation continues on the remaining channel and drives. Because each disk uses a separate disk channel, write operations are performed simultaneously, offering a performance advantage over servers using disk mirroring.

Disk duplexing also offers a performance advantage in read operations. Read requests are given to both drives. The drive that is closest to the information will respond and answer the request. The second request given to the other drive is canceled. In addition, the duplexed disks share multiple read requests for concurrent access.

The disadvantage of disk duplexing is the extra cost for multiple disks drives (also required for disk mirroring) as well as for the additional disk channels and controller hardware. However, the added cost for these components must be weighed against the replacement cost of lost information plus costs that accrue from the interruption of critical operations and lost business opportunities. Faced with these consequences, an organization might discover that the investment of a few hundred or even a few thousand dollars to safeguard valuable data is negligible.

One method of data protection is growing in popularity: redundant arrays of inexpensive disks (RAID). Instead of risking all its data on one high-capacity disk, the organization distributes the data across multiple smaller disks, offering protection from a crash that could wipe out all data on a single, shared disk.

Some applications require the deployment of hundreds of servers in a coordinated manner, which can be quite expensive using conventional chassis-based server configurations. The now-familiar process of rack-mounting servers has been the most common approach to assembling large numbers of servers. Newer "blade" technology extends the concept of rack mounting to allow hundreds of ultrathin servers to be vertically mounted into a single rack. The blades are comprehensive computing systems that include processor, memory, network connections, and associated electronics—all on a single motherboard called a "blade." Each blade can be configured to support a different operating system or application accessible from the LAN or WAN.

The blades slide into slots on a specially designed rack. While the server blade consists of essential processing and sometimes storage components, the rack unit provides for network and external storage connections, significantly reducing cabling and space requirements. Over 300 blades can fit into a standard sized rack. And if a blade fails, it can be easily replaced by simply pulling and replacing it.

Server blades have a lot of appeal for IT organizations looking for space-efficient solutions with higher levels of serviceability, scalability, and manageability. Installing, servicing, and removing blades is much easier than working with chassis-mounted servers. Shared power supply, cabling, fans, and storage reduce the number of redundant and "failable" components in the environment. Instead of equipping each server with redundant power supplies, for example, a single set of power supplies is shared by all the rack-mounted blades. Management tools allows IT managers to readily monitor, configure, and troubleshoot systems.

The ability to add and remove components quickly allows IT managers not only to deal with outages more efficiently than in a traditional network setup but also to adjust to fluctuations in traffic. For example, the management interface allows server blades to be set up to handle transaction software that is heavily used during the business day and then perform other tasks during other periods of the day. In addition, all blades in the system can be managed as a whole, or they can be assigned to different users, customers or partners and managed separately.

Automated Operations

With the right management tools, network backup can be automated under centralized control. Such tools are a virtual necessity for mixed-vendor environments. They can go a long way toward lowering operating and resource costs by reducing time spent on backup and recovery. Some tools even implement unattended network backup, eliminating operator intervention and further reducing costs.

These tools enhance media management by providing overwrite protection, log file analysis, media labeling, and the ability to recycle backup media. In addition, the journaling and scheduling capabilities of some tools relieve the operator of the time-consuming tasks of tracking, logging, and rescheduling network and system backups. Another useful feature of such tools is data compression, which reduces media costs by increasing media capacity. This feature also increases backup performance while reducing network traffic.

When these tools are integrated with high-level management platforms, such as Hewlett-Packard's OpenView, problems or errors that occur during automated network backup are reported to the OpenView management console. The console operator is notified of the problem or error via a color change of the respective backup application symbol on the OpenView map. By clicking on the symbol, the operator can directly access the network backup application to determine the cause of the problem or correct the error, enabling the backup operation to resume.

The trend in backup systems is toward increasing their levels of intelligence. Backup systems must ensure not only that files are backed up but also that they are easily located and restored. Systems intelligence has already progressed to the point where the user need not know the tape, the location on the tape, or even the name of a lost file in order to restore it.

Increasing levels of automation are facilitating the backup of very large networks. From expert software that determines what files to back up to automated tape changers that select, load, and unload tapes without operator intervention, backup is becoming as transparent and easily manageable as file sharing. Backup software is also supporting more types of applications, including online processing. As LAN backup systems move forward in intelligence, automation, application diversity, and media options, LANs can fulfill their potential as the primary means of corporate data management.

Off-Site Data Storage

Mission-critical data should be backed up daily or weekly and stored off-site. There are numerous services that provide off-site storage, often in combination with hierarchical storage management techniques. Carriers, computer vendors, and third-party service firms offer vault storage for secure off-site data storage of critical applications. Small companies need not employ such elaborate methods. They can back up their own data and have them delivered by overnight courier for storage at secure location or bring them to a bank safety deposit box. The typical bank vault can survive even a direct hit by a tornado.

In addition to backing up critical data, it is advisable to register all applications software with the manufacturer and keep the original program disks in a safe place at a different location. This minimizes the possibility of both copies being destroyed in the same catastrophe. Manuals and supplementary documentation also should be protected, as should the software licenses.

For companies that can afford it, a mirror site is a good alternative to a backup tape system. It contains copies of applications and data, perhaps located at the other end of a leased line that runs miles away from the main site, so that a natural or human-made disaster is unlikely to strike both sites simultaneously. If one site goes down, the other site takes over. Mirror sites are becoming more popular with high-volume e-commerce sites that cannot risk even a few minutes of downtime. The mirror site can be as large as a data center or merely a rack-mounted server at a carrier's collocation facility.

Both carrier-neutral and carrier-specific collocation facilities are commonplace. It is simply a secure environment that brings together the equipment and lines of multiple service providers and customers. Space is leased in the form of cages, cabinets, and racks into which customers place their own equipment. Customers pay a fixed monthly charge for the space. For small companies, this arrangement is often more economical than having to set up and maintain their own secure environment.

Some service providers offer site recovery options. This type of service is meant to deal with the loss of a primary data center that runs mission-critical applications. If an organization's data center suffers from a catastrophic fire or natural disaster, for example, traffic will be quickly rerouted to another comparably equipped site. When disaster strikes, the customer calls the carrier and requests activation of links to the alternate site, a process that may take about 2 hours to complete and which may entail the uploading of new routing tables to each router to reflect the changes. This service is far more economical than having to set up and maintain a live data center and supporting infrastructure.

Summary

A variety of LAN restoral and data protection methods can provide effective ways to meet the diverse requirements of today's information-intensive corporations that are both efficient and economical. Choosing wisely from among the available alternatives ensures that companies are not forced into making cost-benefit tradeoffs that jeopardize their information networks, productivity levels, and ultimately, their competitive positions.

See also

Hierarchical Storage Management
Network Backup
Network Restoral
Redundant Array of Inexpensive Disks

LAN TELEPHONY

LAN telephony integrates voice and data over the same medium, enabling automated call distribution, voice mail, and interactive voice response, as well as voice calls and teleconferencing between workstations on a LAN. The benefit of LAN-based telephony is that it can eliminate the costly, proprietary Private Branch Exchange (PBX) and replace it with a standards-based Ethernet/Internet Protocol (IP) solution. By carrying voice conversations in the form of IP packets, local calls can traverse the Ethernet LAN, while long-distance calls can go out to the wide area IP-based intranet. Through the use of IP/Public Switched Telephone Network (PSTN) gateways, calls can even reach conventional telephones off the IP network.

With LAN telephony, users working away from their offices—at home or in a hotel—can use a single phone line to carry both data and voice traffic. The user dials up to access the corporate intranet, which would be equipped and engineered to carry real-time voice traffic. Such a system provides an integrated directory view, enabling remote users to locate individuals within the corporation for voice- or e-mail connection in a unified way. Likewise, phone callers (internal or external to the corporation) can locate the mobile workers connected to any part of the intranet. Thus LAN telephony allows users to work seamlessly from any location.

By using the LAN-based conferencing standards, transparent connectivity of different terminal equipment can be achieved; the media used by any conference participant would be limited only by what is supported by his or her terminal equipment. Connectivity to room-based conference systems or analog telephones can be achieved by means of gateways, which would perform the required protocol and media translations.

IP PBX

The traditional PBX is a circuit switch that provides organizations with access to communications services and call handling features. It sets up a communication path between the calling and called party, supervises the circuit for various events (e.g., answer, busy, and disconnect), and tears down the path when it is no longer required. In many ways, the PBX mimics a telephone company's central office switch, except that it is smaller in scale and is privately owned or leased.

PBXs based on IP can transport intraoffice voice over an Ethernet LAN, a managed IP network, and via a gateway, the PSTN to reach off-net locations. Full-featured digital phone sets link directly to the Ethernet LAN via a 10BaseT interface without requiring connection to a desktop computer. Phone features can be configured using a Web browser. Existing analog devices, such as phones and fax machines, can be linked to the LAN via a gateway. In addition to IP nets, calls can be placed or received using T1, Integrated Services Digital Network (ISDN) Primary Rate Interface (PRI), or traditional analog telephone lines.

All the desktop devices have access to the calling features offered through the IP PBX management software running on a LAN server. The call-management software allows client devices on the network, such as phones and computers, to perform functions such as call hold, call transfer, call forward, call park, and calling party ID. In addition, advanced PBX functions, such as multiple lines per phone or multiple phones per line, can be offered through the management software. The software also offers directory services. Unified messaging capabilities allow voice-mail messages to be sent to an existing electronic mailbox, along with e-mail and faxes.

Standards

The building block of LAN telephony is the international H.323 Standard, which specifies the visual telephone system and equipment for packet-switched networks. H.323 is an umbrella standard that covers a number of audio and video encoding standards. Among these standards is H.225 for formatting voice into packets. H.225 is based on the Internet Engineering Task Force's (IETF) Real Time Protocol (RTP) specification and the H.245 protocol for capability exchange between workstations.

On the sending side, uncompressed audio/video information is passed to the encoders by the drivers and then given to the audio/video application program. For transmission, the information is passed to the terminal management application, which may be the same as the audio/video application; the media streams are carried over RTP/User Datagram Protocol (UDP), and call control is performed using H.225-H.245/Transmission Control Protocol (TCP).

Gateways provide the interoperability between H.323 and the PSTN as well as networks running other teleconferencing standards such as H.320 for ISDN, H.324 for voice, and H.310/H.321 for Asynchronous Transfer Mode (ATM). An example H.323 deployment scenario involves H.323 terminals interconnected in the same local area by a switched LAN. Gateways, routers, or integrated gateway/router devices provide access to remote sites. The gateways provide communication with H.320 and H.324 terminals remotely connected to the ISDN and PSTN, respectively. H.323-to-H.323 communication between two remote sites can be achieved using routers that directly carry IP traffic over the Point-to-Point Protocol (PPP) running on ISDN. For better channel efficiency, gateways can translate H.323 streams into H.320 to be carried over ISDN lines, and vice versa.

In addition to H.323, there is the Session Initiation Protocol (SIP) defined by the IETF. SIP offers mechanisms for call routing, call signaling, capabilities exchange, media control, and supplementary services. It is a newer protocol that offers scalability, flexibility, and ease of implementation in building complex systems, while H.323 is an older protocol valued for its manageability, reliability, and interoperability with the PSTN. Standards bodies are working on procedures to allow seamless internetworking between the two protocols.

Summary

A LAN-based PBX eliminates the need for IP telephony software to be loaded on each client PC. It also allows organizations to avoid having to set up and manage separate LAN and PBX infrastructures. A unified backbone to the desktop allows common delivery of voice and data for reduced wiring and maintenance costs. Using a switched 100-Mbps Ethernet, network engineers can design telephone networks with essentially unlimited capacity. When the need arises for more workstations (i.e., extensions), another Ethernet switch is added. Administering these systems is done locally through a Windows graphical user interface or remotely through a Web browser.

See also

Ethernet
Internet Telephony
LAN Switching
Transmission Control Protocol/Internet Protocol (TCP/IP)

LAN SWITCHING

When shared LANs become too slow, performance can be improved by creating segments linked together by bridges. Bridges keep local traffic on a particular LAN segment while allowing packets destined for other segments to pass through in a process called "filtering." But no matter how many segments are created, LAN performance tends to diminish, if only because more users are continually added to the network. The greater the number of workstations simultaneously accessing the LAN, the smaller each workstation's available bandwidth becomes.

With a switched LAN—Ethernet or Token Ring—each user can have access to the network at full native speed instead of having to share it with multiple users. Dedicated LAN links improve network performance for all users, allowing them to be more productive and making the network easier to manage. In some cases, switched LANs provide enough performance improvements to hold off purchases of more equipment.

LAN switching is implemented at Layer 2 of the Open Systems Interconnection (OSI) reference model in conjunction with an intelligent wiring hub or a dedicated LAN switch. Bandwidth can be controlled by restricting access to any logical segment to only authorized members of the workgroup (Figure L-2). Creation of these virtual workgroups also pro-

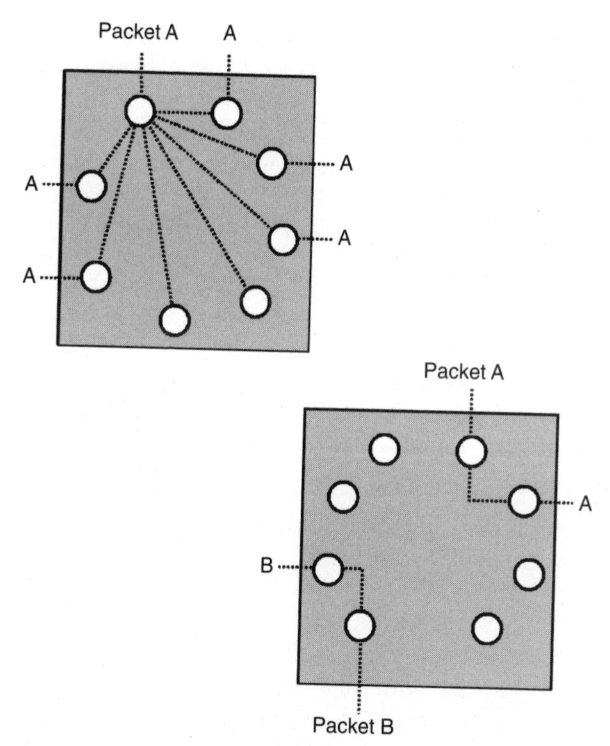

Figure L-2 Standard hub versus switching hub. A standard hub (*left*) broadcasts all packets to all ports, while a switching hub (*right*) supports virtual connections to limit traffic to only specific addressees. These types of configurations are also known as "virtual LANs."

vides security, since packets—broadcast and multicast—will be seen only by authorized users within that virtual workgroup. While users of one logical network cannot access another logical network—thus enforcing security—multiple virtual workgroups can still share centralized resources, such as e-mail servers. Highly secure logical segments can be established and modified from an Simple Network Management Protocol (SNMP)–based network management station and encompass any device on the network.

There are two basic types of switching hubs: port switching and segment switching. Port-switching hubs let administrators assign ports to segments via network-management software. In effect, these hubs act as software-controlled patch panels. Segment-switching hubs treat ports as separate segments and forward packets from port to port. How they accomplish this varies by vendor and has important implications depending on the application. What all segment-switching hubs have in common is that they can substantially increase available network bandwidth. Acting like high-speed multiport bridges, segment-switching Ethernet hubs, for example, offer 10/100 Mbps to each port.

Ethernet Switching

In the traditional Ethernet environment, stations contend with each other for access to the network, a process that is controlled by a statistical arbitration scheme. Each station "listens" to the network to determine if it is idle. On sensing that no traffic is currently on the network, the station is free to transmit. If the channel is already in use, the station backs off and tries again later. If multiple stations sense that the network is idle and send out packets simultaneously, a collision occurs that corrupts the data. When a collision is detected, the stations back off and try again at staggered intervals in a process called Carrier Sense Multiple Access with Collision Detection (CSMA/CD).

The problem with CSMA/CD is that collisions force retransmissions, which causes the network to slow down. In turn, this results in less bandwidth availability for all users. A related problem with Ethernet is that a station's packets are automatically broadcast to all other stations. This means that all other stations are aware of the packets, even though only one can actually read them. The broadcast nature of Ethernet increases the likelihood of collisions.

In Ethernet switching, the Media Access Control (MAC) Layer address determines which hub or switch port the packet will go to. Since no other ports are aware of the packet's existence, the stations do not have to be concerned about whether their packets will collide with data from other stations as they transmit toward the hub or switch. In Ethernet switching, a virtual connection is created between the sending and receiving ports (Figure L-3). This dedicated connection remains in place only long enough to pass packets between the sending and receiving stations.

LAN switches improve performance by creating isolated collision domains. By spreading users over several collision domains, packet collisions are minimized and performance improved. Many LAN switch installations assign just one user per port, which gives that user an effective bandwidth of 10/100 Mbps.

If a station has packets for a busy port, that station's port momentarily holds them in its buffer. When the busy port becomes free, a virtual connection is established, and the packets are released from the buffer and sent to the newly freed port. This mechanism works well unless the buffer gets filled, in which case packets are lost. To avoid this, some vendors offer a throttling capability. When a port's buffer begins to fill up, that port begins to send packets back to the workstation. This slows the workstation's transmission speed and evens out the "pressure" at the port. Because no packets are lost, many more packets make it through the hub than would otherwise be possible under heavy traffic conditions.

Some LAN switching products offer a choice of different packet switching modes: cut-through and store-and-forward. Cut-through speeds frame processing by beginning the transmission on the destination LAN before the entire frame

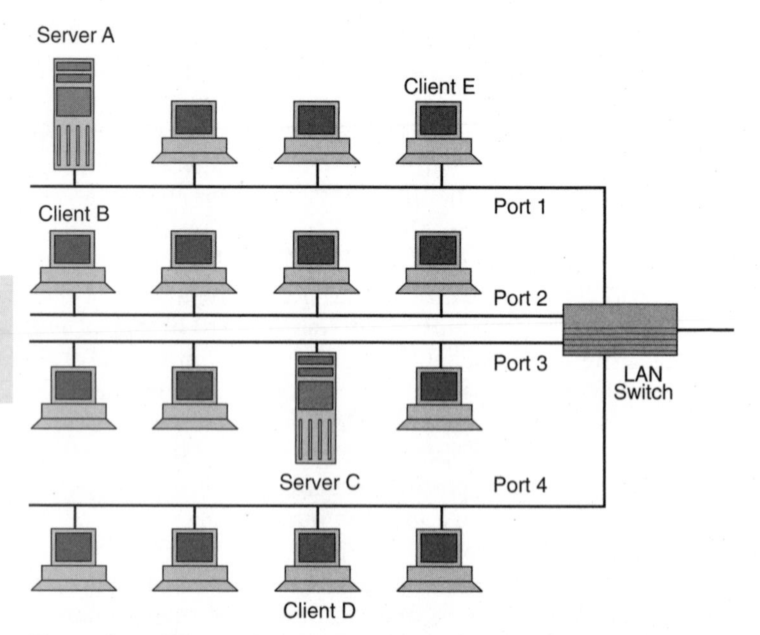

Figure L-3 If server A on port 1 needs to transmit to client B on port 2, the LAN switch forwards Ethernet frames from port 1 to port 2, thus sparing port 3 and port 4 from frames destined for client B. If server C needs to send data to client D at the same time that server A sends data to client B, it can do so because the LAN switch can forward frames from port 3 to port 4 at the same time it is forwarding frames from port 1 to port 2. If server A needs to send data to client E, and they are both on the same segment, the LAN switch does not need to be involved in frame forwarding.

arrives in the input buffer. Store-and-forward mode works like a traditional LAN bridge: Frame forwarding begins only after the entire frame arrives in the input buffer. Store-and-forward mode results in more of a performance hit but ensures error-free delivery. Some vendors support automatic switching between cut-through and store-and-forward modes. With this technique, called "error-free cut-through switching," the switch changes from cut-through to store-and-forward mode if the percentage of bad packets flowing through the switch exceeds a predefined threshold.

Some LAN switching devices support only one address per port, while others support 1500 or more. Some devices are capable of dynamically learning port addresses and allowing or disallowing new port addresses. Disallowing new port addresses enhances hub security: In ignoring new port addresses, the corresponding port is disabled, preventing unauthorized access.

Some vendors offer full-duplex Ethernet connections, providing each user with 10/100 Mbps of dedicated bandwidth (send and receive) over unshielded twisted-pair wiring. With half-duplex signaling and collision detection disabled, one pair of wires can transmit at the full speed, while the other pair receives at the full speed. This creates a collision-free connection that can ease bottlenecks between similarly equipped switching hubs or servers.

Token Ring Switching

LAN switching is also available for the Token Ring environment. Token Ring switching is a technology for dedicating 4 or 16 Mbps of bandwidth to each user on a LAN. Software-controlled switching of individual ports into any one of a number of Token Rings eliminates the need to patch network cables physically to change LAN configurations and provides centralized, per-port control over Token Ring LAN configurations.

Like switched Ethernet LANs, switched Token Ring LANs can be configured to operate in full duplex to send and receive data simultaneously. The total per-port capacity of each switch module would then be 32 Mbps.

Associated with Token Ring switching is the capability to automatically protect the ring from potential disruptions. Although some vendors offer the ability to mix 4 and 16 Mbps on the same module, the ports can detect and deny entry to any device attempting to connect to a port with a transmission frequency different from the port's predefined configuration. This prevents a device configured for 16-Mbps transmission from accessing a port configured for 4-Mbps transmission, for example.

Today's Token Ring switches feature High-Speed Token Ring (HSTR) uplinks, providing 100-Mbps connectivity, which significantly reduces bottlenecks between the switch and enabled servers. HSTR is a cost-effective solution for upgrading Token Ring networks while preserving all the essential characteristics that make Token Ring so effective, including source routing and large frame size support. HSTR frames can be both smaller and larger (up to 18 kB) than Fast Ethernet frames, enabling better utilization of the network's bandwidth while lowering CPU utilization, particularly for servers.

Summary

Enterprise-level LAN switches are available that provide seamless switching between other high-speed LAN technologies including Fiber Distributed Data Interface (FDDI), Fast Ethernet (100Base-T), and 100VG-AnyLAN while providing a safe migration from these and existing Ethernet and Token Ring LANs to ATM. There are even gigabit switches for supporting large departments in enterprise LANs. Combined with LAN bandwidth management intelligence, LAN switches provide users with the ability to build and manage reliable high-performance switched intranetworks.

See also

Ethernet
Hubs
Token Ring

LASER TRANSMISSION

A relatively new category of wireless communication uses laser technology to interconnect LANs in different buildings. Sometimes called "free space optics," laser systems operate in the near-infrared region of the light spectrum. Utilizing coherent laser light, these wireless line-of-sight links are used in campus environments and urban areas where the installation of cable is impractical and the performance of leased lines is too slow. Laser links between sites can be operated at the full LAN channel speed. And unlike microwave transmission, laser transmission does not require an Federal Communications Commission (FCC) license, and data traveling by laser beam cannot be intercepted.

Performance Impairments

The lasers at each location are aligned with a simple bar graph and tone lock procedure. Fiberoptic repeaters are used to connect the LANs to the laser units. Alternatively, a bridge equipped with a fiberoptic-to-AUI transceiver may be used. Connections to and from the laser are made using standard fiberoptic cable, protecting data from radio-frequency interference (RFI) and electromagnetic interference (EMI). Monitors can be attached to the laser units to provide operational status, such as signal strength, and to implement local and remote loop-back diagnostics.

The reason that laser products are not used very often for business applications is that transmission is affected by atmospheric conditions that produce such effects as absorption, scattering, and shimmer. All three can reduce the amount of light energy that is picked up by the receiver and corrupt the data being sent.

Absorption refers to the ability of various frequencies to pass through the air. Absorption is determined largely by the water vapor and carbon dioxide content of the air, along the transmission path, which, in turn, depends on humidity and altitude. The gases that form in the atmosphere have many resonant bands that allow specific frequencies of light to pass. These transmission windows occur at various wavelengths, such as the visible light range. Another window occurs at the near-infrared wavelength of approximately 820 nanometers (nm). Laser products tuned to this window are not greatly affected by absorption.

Scattering has a much greater effect on laser transmission than absorption. The atmospheric scattering of light is a function of its wavelength and the number and size of scattering particles in the air. The optical visibility along the transmission path is directly related to the number and size of these particles. Fog and smog are the main conditions that tend to limit visibility for optical-infrared transmission, followed by snow and rain.

Shimmer is caused by localized differences in the air's index of refraction. This is caused by a combination of factors, including time of day (daytime heat), terrain, cloud cover, wind, and the height of the optical path above the source of shimmer. These conditions cause fluctuations in the received signal level by directing some of the light out of its intended path. Beam fluctuations may degrade system performance by producing short-term signal amplitudes that approach threshold values. Signal fades below these threshold values result in error bursts.

Vendors have taken steps to mitigate the effects of absorption, scattering, and shimmer. For example, such techniques as frequency modulation (FM) in the transmitter and an automatic gain control (AGC) in the receiver can minimize the effects of shimmer. Also, selecting an optical path several meters above heat sources can greatly reduce the effects of shimmer. However, all these distorting conditions can vary greatly within a short time span or persist for long periods, requiring onsite expertise to constantly fine-tune the system.

Many businesses simply cannot risk frequent or extended periods of downtime while the necessary compensating adjustments are being made. As if all this were not enough, there are other potential problems to contend with, such as thermal window coatings and the laser beam's angle of incidence, both of which can disrupt transmission.

These problems are being overcome with newer lasers that operate in the 1550-nanometer wavelength. A 1550-nanometer delivery system is powerful enough to go through windows, can deliver signals under the fog blanket, and is safe enough that it does not blind the casual viewer who happens to look into the beam. Up to 1 Gbps of bandwidth is available with these systems—the equivalent bandwidth capacity of 660 T1 lines (Figure L-4).

Figure L-4 Terabeam Magna, a free-space optics system from TeraBeam Corp.

There is also a distance limitation associated with laser. The link generally cannot exceed 1.5 kilometers, and 1 kilometer is preferred. With 1550-nanometer systems, the practical distance of the link is only 500 meters.

Summary

Despite its limitations, laser, or free-space optics, can provide a valuable last link between the fiber network and the end user—including as a backup to more conventional methods, including fiber. Free space optics, unlike other transmission technologies, is not tied to standards or standards development. Vendors simply attach their equipment into existing fiber-based networks and then use any laser transmission methods they like. This encourages innovation, differentiation, and speed of deployment.

See also

Infrared Networking

LATENCY

Latency is the amount of delay that affects all types of communications links. Delay on telecommunications networks is usually measured in milliseconds (ms), or thousandths of a second.

A rule of thumb used by the telephone industry is that the round-trip delay for a telephone call should be less than 100 milliseconds. When delay exceeds 100 milliseconds, participants perceive a slight pause in the conversation and use it as the opportunity to begin speaking. But by the time their words arrive at the other end, the other speaker has already begun the next sentence and feels that he or she is being interrupted. When telephone calls go over satellite links, the round-trip delay is typically about 250 milliseconds, and conversations become full of awkward pauses and accidental interruptions.

Latency affects the performance of applications on data networks as well. On the Internet, for example, excessive delay can cause packets to arrive at their destination out of order, especially during busy hours. The reason packets may arrive out of sequence is that they can take different routes on the network. The packets are held in a buffer at the receiving device until all packets arrive and are put in the right order. While this does not affect e-mail and file transfers, which are not real-time applications, excess latency does affect the performance of multimedia and real-time applications.

If the packets containing voice or video do not arrive within a reasonable time, they are dropped. When packets containing voice are dropped, a condition known as "clipping" occurs, which is the cutting off of the first or final syllables in a conversation. Dropped packets of video cause the image to be jerky. Excessive latency also causes the voice and video components in a videoconference to arrive out of synchronization with each other, causing the video component to run slower than the voice component. For example, a person's lips will not match what he or she is really saying.

The effects of latency can be overcome by assigning an appropriate quality of service (QoS) to each application and prioritizing the traffic for transport through the network. QoS parameters can be programmed into the operating systems of routers, switches, and integrated access devices (IADs). When traffic is set to go onto the network, prioritization ensures that mission-critical applications obtain the bandwidth before routine applications.

Summary

The availability of policy-based network management tools from a variety of vendors has made it easier for large enterprises to implement QoS policies and traffic prioritization schemes with enough granularity to ensure that all applications are served in an appropriate manner without the company having to constantly shell out for more bandwidth and associated resources in a futile effort to stay ahead of the performance curve.

See also

Jitter

Ping

LOCAL EXCHANGE CARRIERS

Local Exchange Carriers (LECs) provide residential, business, and interexchange access services. In addition to centrex, many of the larger LECs are developing and/or offering value-added services such as voice and data messaging via cellular and Personal Communications Services (PCS) networks. The LECs, which are commonly called “telephone companies” (or telcos), include the 22 former Bell Operating Companies (BOCs) divested from AT&T in 1984, as well as Cincinnati Bell, Southern New England Telephone (SNET), and the telephone companies of GTE and Sprint. These companies are now referred to as ILECs—Incumbent Local Exchange Carriers—to distinguish them from competitors in the local market and the hundreds of smaller telephone companies serving largely rural areas.

In addition to providing local phone service and providing Interexchange Carriers (IXCs) with access to the local loop, the LECs provide billing services. Phone bills that come from a LEC actually can represent charges from a number of services and providers. A telephone bill can comprise many basic elements, including charges for local message units (MSUs), special service offerings, directory service, 911 emergency service, cellular calls, and Internet access. These and other charges are identified in a consolidated monthly invoice. In addition, charges for long-distance calls carried by IXCs and cellular service providers may appear on the monthly invoice as well.

Another type of LEC is the Competitive Local Exchange Carrier (CLEC). This type of service provider offers business and residential users lines and services on a resale basis or from its own facilities-based network, enabling customers to save money on their communications bill. Regional teleports, metropolitan fiber carriers, and CATV operators are among the types of companies that are now involved in providing competitive local exchange services. Typically, these alternative access carriers offer service in major markets, where traffic volumes are greatest and, consequently, users are hardest hit with high local service charges.

There are also Data Local Exchange Carriers (DLECs), which specialize in providing data services such as Digital Subscriber Line (DSL) for high-speed Internet access. The DLEC usually provisions its service over the same line that provides telephone service to the subscriber. Since the voice and data use different frequencies, the subscriber can talk on the phone and surf the Web at the same time. A DSL access multiplexer in the central office combines all the user traffic onto a fiber link that is ultimately connected to the Internet backbone via a network access point (NAP). DSL services are offered by ILECs and CLECs as well, sometimes in partnership with the DLECs.

Building Local Exchange Carriers (BLECs) specialize in setting up integrated voice and data services in office buildings. Typically, the BLEC targets buildings with 10 or more tenants and over 100,000 square feet of office space. It decides on what buildings to approach based on tenant profile and anticipated demand, the economic opportunity in the building, and access to broadband circuits. The BLEC selects the buildings and portfolio owners in the target markets it desires to secure rights in and negotiates an arrangement with the real estate owner that will benefit both parties. The real estate owners may be paid either a fixed rental fee per month or 5 percent of the revenue generated in the building. The typical lease or license agreement with an owner is for a term of 10 or more years. A typical network costs between $175,000 and $200,000 per building, with the BLEC picking up the entire amount.

Summary

With passage of the Telecommunications Act of 1996, new entities are allowed into the market for local telephone service, including cable operators, electric utilities, Internet service providers (ISPs), and entertainment companies. Until late

1999, the ILECs were restricted to providing local service within their assigned serving areas, called LATAs—Local Access and Transport Areas. The first ILEC to obtain FCC permission to offer long-distance service in its own territory was Bell Atlantic, which changed its name to Verizon. Ultimately, all the ILECs will receive permission to offer long-distance service in their respective territories. The objective is to offer customers bundled services consisting of local, long-distance, and Internet access at very attractive prices and, in the process, limit competition from carriers that are not in a position to offer such bundles.

See also

Building Local Exchange Carriers
Competitive Local Exchange Carriers
Incumbent Local Exchange Carriers
Interexchange Carriers

LOCAL MULTIPOINT DISTRIBUTION SERVICE

Local Multipoint Distribution Service (LMDS) is a two-way millimeter microwave technology that operates in the 27- to 31-GHz range. This broadband service allows communications providers to offer a variety of high-bandwidth services to homes and businesses, including broadband Internet access. LMDS offers greater bandwidth capabilities than a predecessor technology called Multichannel Multipoint Distribution Service (MMDS) but has a maximum range of only 7.5 miles from the carrier's hub to the customer premises. This range can be extended, however, through the use of optical fiber links.

Applications

LMDS provides enormous bandwidth—enough to support 16,000 voice conversations plus 200 channels of television programming. Figure L-5 contrasts LMDS with the bandwidth available over other wireless services.

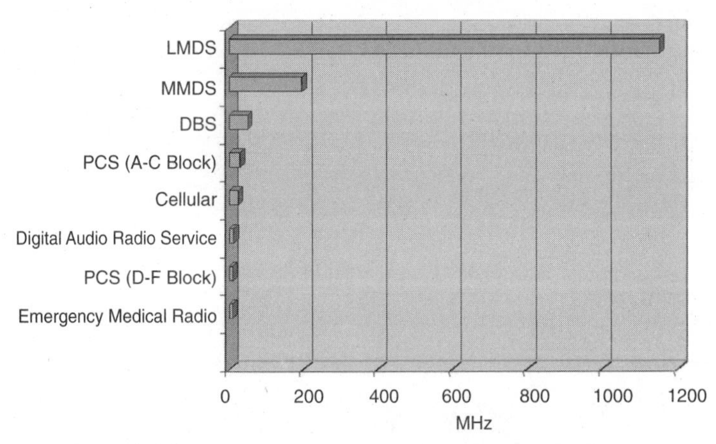

Figure L-5 Local Multipoint Distribution Service (LMDS) operates in the 27- to 31-GHz range and offers 1150 MHz of bandwidth capacity, which is over two times more than all other auctioned spectrum combined.

CLECs can deploy LMDS to completely bypass the local loops of the ILECs, eliminating access charges and avoiding service-provisioning delays. Since the service entails setting up equipment between the provider's hub location and customer buildings for the microwave link, LMDS costs far less to deploy than installing new fiber. This allows CLECs to very economically bring customer traffic onto their existing metropolitan fiber networks and, from there, to a national backbone network.

The strategy among many CLECs is to offer LMDS to owners of multitenant office buildings and then install cable to each tenant who subscribes to the service. The cabling goes to an on-premises switch, which is run to the antenna on the building's roof. That antenna is aimed at the service provider's antenna at its hub location. The line-of-sight wireless link between the two antennas offers a broadband "pipe" for multiple voice, data, and video applications. Subscribers can use LMDS for a variety of high-bandwidth applications, including television broadcast, videoconferencing, LAN interconnection, broadband Internet access, and telemedicine.

Operation

LMDS operation requires a clear line of sight between the carrier's hub station antenna and the antenna at each customer location. The maximum range between the two is 7.5 miles. However, LMDS is also capable of operating without having a direct line of sight with the receiver. This feature, highly desirable in built-up urban areas, may be achieved by bouncing signals off buildings so that they get around obstructions. At the receiving location, the data packets arriving at different times are held in queue for resequencing before they are passed to the application. This scheme does not work well for voice, however, because the delay resulting from queuing and resequencing disrupts two-way conversation.

At the carrier's hub location there is a roof-mounted multisectored antenna (Figure L-6). Each sector of the antenna receives/transmits signals between itself and a specific customer location. This antenna is very small, some measuring only 12 inches in diameter. The hub antenna brings the multiplexed traffic down to an indoor switch (Figure L-7), which processes the data into 53-byte ATM "cells" for transmission over the carrier's fiber network. These individually addressed cells are converted back to their native format before going off the carrier's network to their proper destinations—the Internet, PSTN, or the customer's remote location.

At each customer's location there is a rooftop antenna that sends/receives multiplexed traffic. This traffic passes through an indoor network interface unit (NIU), which provides the gateway between the radio-frequency (RF) components and the in-building equipment, such as a LAN hub, PBX, or videoconferencing system. The NIU includes an up/down converter that changes the frequency of the microwave signals to a lower intermediate frequency (IF)

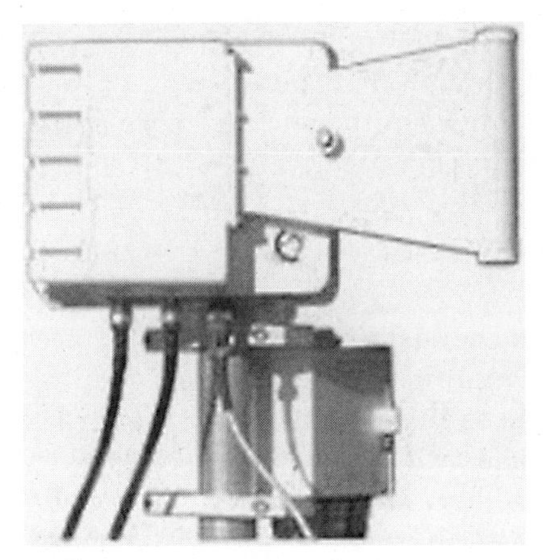

Figure L-6 A multisectored antenna at the carrier's hub location transmits/receives traffic between the antennas at each customer location.

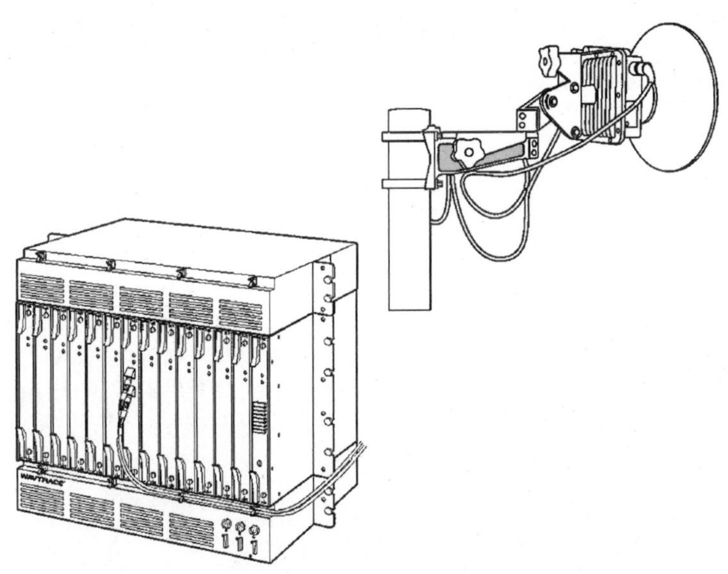

Figure L-7 A microwave transceiver (*top right*) handles multiple point-to-point downstream and upstream channels to customers. The transceiver is connected via coaxial cables to an indoor switch (*bottom left*) that provides the connectivity to the carrier's fiber network. The traffic is conveyed over the fiber network in the form of 53-byte ATM cells. *(Source: Wavtrace, Inc.)*

that the electronics in the office equipment can more easily (and inexpensively) manipulate.

Spectrum Auctions

In May 1999, the FCC held the last auction for LMDS spectrum. Over 100 companies qualified for the auctions, bidding against each other for licenses in select basic trading areas (BTAs).[1] The FCC auctioned two types of licenses in each market: An "A-block" license permits the holder to provision 1150 MHz of spectrum for distribution among its customers, while a "B-block" license permits the holder to provision 150 MHz. Most of the A-block licenses in the largest BTAs were won by major CLECs, while the B-block licenses were taken by smaller companies, ISPs, universities, and government agencies.[2] The licenses are granted for a 10-year period, after which the FCC can take them back if the holder does not have service up and running.

Potential Problems

A potential problem for LMDS users is that the signals can be disrupted by heavy rainfall and dense fog—even foliage can block a signal. In metropolitan areas where new construction is a fact of life, a line-of-sight transmission path can disappear virtually overnight. For these reasons, many information technology (IT) executives are leery of trusting mission-critical applications to this wireless technology.

[1] Basic trading area (BTA) is a term used in the geographic definition of economic activity, based on data compiled by Rand McNally. Most large cities are metropolitan trading areas (MTAs), and most of the larger U.S. towns are classed as BTAs. These are not the same as local access and transport areas (LATAs), which have defined the local service boundaries of the former Bell Operating Companies (BOCs) since their divestiture from AT&T in 1984.

[2] The ILECs, such as the regional BOCs, were forbidden to enter the LMDS market for 3 years. In 2002, they may use LMDS, among other technologies, to bypass each other's local loops to extend services to target markets.

Service providers downplay this situation by claiming that LMDS is just one local access option and that fiber links are the way to go for mission-critical applications. In fact, some LMDS providers offer fiber as a backup in case the microwave links experience interference.

There is controversy in the industry about the economics of point-to-multipoint architecture of LMDS, with some experts claiming that the business model of going after low-usage customers is fundamentally flawed and will never justify the service provider's cost of equipment, installation, and provisioning. With an overabundance of fiber in the ground and metropolitan area Gigabit Ethernet services coming online at a competitive price, the time for LMDS may have come and gone. In addition, newer wireless technologies like free-air laser hold a significant speed advantage over LMDS, as does submillimeter transmission in the 60- and 95-GHz bands.

Another problem that has beset LMDS is that the major license holders have gotten caught up in financial problems, some declaring Chapter 11 bankruptcy. These carriers built their networks quickly, incurring massive debt, without lining up customers fast enough. This strategy worked well as long as the capital markets were willing to continue funding these companies. But once the capital markets dried up in 2000, so did the wireless providers' coffers and their immediate prospects. The uncertain future of these financially strapped carriers has discouraged many companies from even trying LMDS.

Summary

Fiberoptics is the primary transmission medium for broadband connectivity today. However, of the estimated 4.6 million commercial buildings in the United States, 99 percent are not served by fiber. Businesses are at a competitive disadvantage in today's information-intensive world unless they have access to broadband access services, including high-speed Internet access. These businesses, including many data-intensive high-technology companies, can be served adequately with LMDS. Despite the financial problems of LMDS providers, the technology has the potential to become a significant portion of the global access market, which will include a mix of many technologies, including DSL, cable modems, broadband satellite, and fiberoptic systems.

See also

Asynchronous Transfer Mode
Cable Television Networks
Digital Subscriber Line Technologies
Multichannel Multipoint Distribution Service
T1 Lines

M

MEDIA CONVERTERS

As the term implies, a media converter makes the conversion from one network medium type—defined by cable and connector types and bandwidth—to another. By performing this transition, the media converter makes it possible for organizations to extend legacy networks with the latest technology instead of being tied to what the network was started with or—even worse—tearing it out and starting over. Other types of media converters allow standard Category 5 local area network (LAN) cabling to be tied into a fiberoptic backbone.

Functionally, a media converter is two transceivers or media attachment units (MAUs) that can pass data to and from each other and a power supply. Each of the MAUs has a different industry-standard connector to join the different media—one medium goes in; the other comes out. The connectors themselves comply with IEEE specifications and utilize standard data encoding rules and link tests. Media converters support connections to and from switches, hubs, routers, and even direct to servers.

Benefits and Applications

Media converters can be used virtually anywhere in the network—from the server to the workstation (Figure M-1). They can be used to enhance the flexibility of the network by facilitating upgrades to the network to better, faster, more secure technology—as with fiber cabling—without requiring a full network retrofit. Legacy copper cabling can be left in place, while the fiber can be used for additions and extensions to the network.

Conversion devices also provide a means to extend the network. Using a media converter to integrate optical fiber allows the network to support the longer cable distances that are available through the use of fiber. In standard Ethernet and Fast Ethernet networks, fiber specifications prescribe a maximum distance of 2000 meters versus the twisted-pair

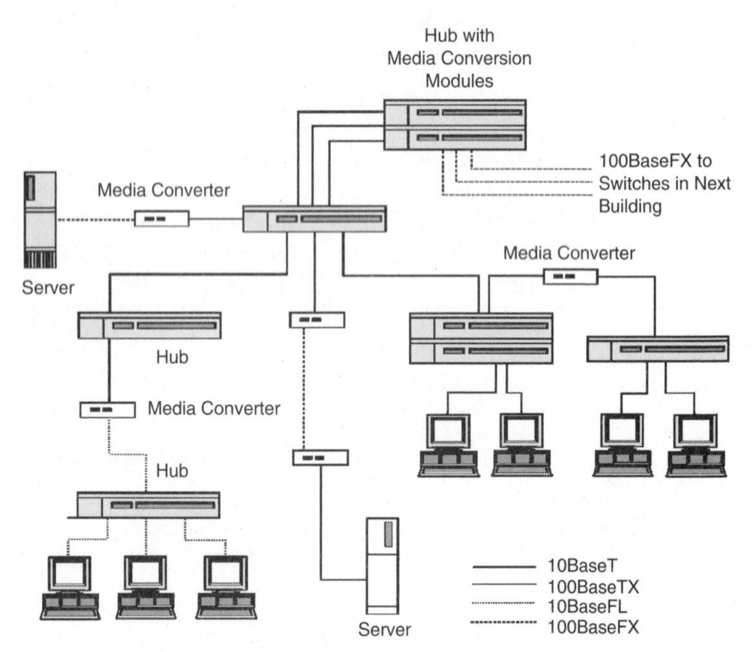

Figure M-1 The use of media converters throughout a network.

wiring limit of only 110 meters. Gigabit Ethernet, a fiber-only specification, supports single-mode fiber in addition to the multimode fiber supported by its other Ethernet cousins. The same 2000-meter limit applies to multimode fiber, but single-mode fiber is supported for Gigabit Ethernet to a distance of 3000 meters.

The use of media converters also makes it easier to add new devices to the network—including the newer high-bandwidth switches and hubs—regardless of connector restrictions. Switches solve many of the problems that are common to larger networks, but the majority of Ethernet and Fast Ethernet switches on the market today are equipped with twisted-pair connectors. Where the entire network is built of twisted-pair wiring, the switches are easy to integrate—just plug one in. However, for many new installations, network managers are looking to optical fiber for their cabling infrastructure because of its security features, bandwidth capacity, and ability to span longer distances. For older installations that use BNC connectors, the same incompatibility arises. A simple fiber-to-twisted-pair media converter or a BNC-to-twisted-pair connector can make these devices work together on the same network.

The most commonly used media converters support twisted-pair-to-fiber connections. Standard fiber connectors typically are classified as either ST (simple-twist) or SC (subscriber-channel). The media converters used most widely today provide quick, reliable, cost-effective connections between

- 10-Mbps twisted-pair cable segments or devices and 10-Mbps fiberoptic, single-mode, or multimode (10BaseT to 10BaseFL) segments or devices.
- 10-Mbps 10BaseT segments or devices and 10-Mbps Ethernet coaxial cable (10BaseT to 10Base2) segments or devices.
- 100-Mbps twisted-pair cable segments or devices and 100-Mbps fiberoptic, single-mode, or multimode (100BaseTX to 100BaseFX) segments or devices.

Newer fiber connectors are available in smaller form factors. These new fiber connectors are the MT-RJ, the VF-45, and the LC. These connectors are being put to use on various types of network hardware, including the latest hubs and switches. Because of their smaller size, these connectors enable more ports to be placed in a given device. A stackable 12-port Ethernet hub, for example, can now accommodate 24 ports—without increasing the size of the hub. This higher port density results in lower network costs. Media converters are available for all three of the new small-form-factor connector types, providing organizations with even more flexibility in designing and expanding their networks.

Summary

Media converters can be inserted almost anywhere in the network. The option of mounting media converters in a rack-mount chassis is useful where multiple converters are in use or where they are anticipated in the future. The ability of media converters to mix media and speeds provides organizations with more flexibility in designing or extending their networks, as well as integrating legacy devices with today's advanced systems. This flexibility, in turn, enables organizations to achieve network performance goals while containing costs.

See also

Transceivers

MULTIPLEXERS

Multiplexers first appeared in the mid-1980s when private networks became popular among larger corporations. These devices combine voice, data, and video traffic from various sources such as Private Branch Exchanges (PBXs), LANs and videoconferencing systems so that they can be transmitted over a single higher-speed digital link—such as a T1 or T3 line. At the other end, another multiplexer separates the individual lower-speed channels, sending the traffic to the appropriate terminals. Multiplexers enable businesses to reduce telecommunications costs by making the most efficient use of the leased line's available bandwidth. Since the line is billed at a flat monthly rate, there is ample incentive to load it with as much traffic as possible. Using different levels of voice and/or data compression, the channel capacity of the leased line can be doubled or quadrupled easily to save even more money.

There are several types of multiplexing in common use today. On private leased lines, the dominant technologies are Time Division Multiplexing (TDM) and Statistical Time Division Multiplexing (STDM). Each lends itself to particular types of applications. TDMs are used when most of the applications must run in real time, including voice, videoconferencing, and multimedia. However, when an input device has nothing to send, its assigned channel is wasted. This inefficiency is often justified by the need to have bandwidth immediately available to support real-time applications so that delay does not become a problem.

With STDM, if a device has nothing to send, the channel it would have used is dynamically reassigned to a device that does have something to send. If all channels are busy, input devices wait in queue until a channel becomes available. This type of device is used in situations where efficient bandwidth usage is valued and the applications are not bothered by delay.

Although used mostly on private networks, both types of multiplexers can interface with the Public Switched Telephone Network (PSTN) as well. For example, if a private T1 line degrades or fails, the multiplexer can be configured to automatically switch traffic to an Integrated Services Digital Network (ISDN) link until the private line is restored to service.

Time Division Multiplexing

With TDM, each input device is assigned its own time slot, or channel, into which data or digitized voice is placed for transport over a high-speed link. When a T1 line is used, there are 24 channels, each of which operates at 64 kbps. Multiplexers can also bond multiple 64-kbps channels to support higher-bandwidth applications. The digital link carries the channels from the transmitting multiplexer to the receiving multiplexer, where they are separated out and sent onto assigned output devices. As noted, if an input device has nothing to send, the assigned channel is left empty, and that increment of bandwidth goes unused.

The TDM manages access to the high-speed line and cyclically scans (or polls) the terminal lines, extracts bits or characters, and interleaves them into the assigned time slots (e.g., frames) for output to the high-speed line. The multiplexer includes channel cards for each low-speed channel and its associated device, a scanner/distributor, and common equipment to handle various processing functions. The low-speed channel cards handle the data and control signals for the terminal devices. They also provide storage capacity through registers that provide bit or character buffering for placing or receiving data from the time slots in the high-speed data stream.

The TDM's scanner/distributor scans and integrates information received from the low-speed devices into the message frame for transmission over the high-speed line and also distributes data received from the high-speed line to the appropriate terminals at the other end.

The common equipment provides the logical functions used to multiplex and demultiplex incoming and outgoing signals. It contains the necessary logic to communicate with both the low-speed devices and the high-speed device. It also generates data, control, and clock signals that ensure that the time slots are perfectly synchronized at both ends of the link.

Since digital facilities are used on the network side, the TDM must have a channel service unit/digital service unit

(CSU/DSU). This device comes in the form of a plug-in module and is a required network interface for carrier-provided digital facilities. The CSU is positioned at the front end of a circuit to equalize the received signal, filter both the transmitted and received waveforms, and interact with the carrier's test facilities. The DSU element transforms the encoded waveform from alternate mark inversion (AMI) to a standard business equipment interface, such as RS-232 or V.35. It also performs data regeneration, control signaling, synchronous sampling, and timing.

Operation TDM technology supports asynchronous, synchronous, and isochronous data transmission. Asynchronous data transfer requires the framing of each character by a start bit and a stop bit. This allows the originating terminal to control the timing of each transmitted character. Synchronous data transfer timing is controlled by the multiplexer. Terminals send synchronous blocks of data framed by characters. Bits within a block are synchronized to clock signals generated by the TDM. Synchronous terminals operate at higher speeds than their asynchronous counterparts, but both are multiplexed in a similar manner.

Isochronous transmission supports multimedia applications where voice and other data must arrive together. In this type of transmission, the individual terminals generate their own clock signals, with all clocks running at the same nominal rate. Isochronous multiplexers provide some buffering and rate adjustment to compensate for slight variations among the clock rates.

A TDM samples data from each terminal input channel and integrates it into a message frame for transmission over the high-speed line. Message frames consist of time slots, and each time-slot position is allocated to a specific terminal. Interleaving is the technique that multiplexers use to format data from multiple devices for aggregate transmission over the link (Figure M-2).

Most of the market leaders offer multiplexers that will interface with public networks via the byte-interleaving

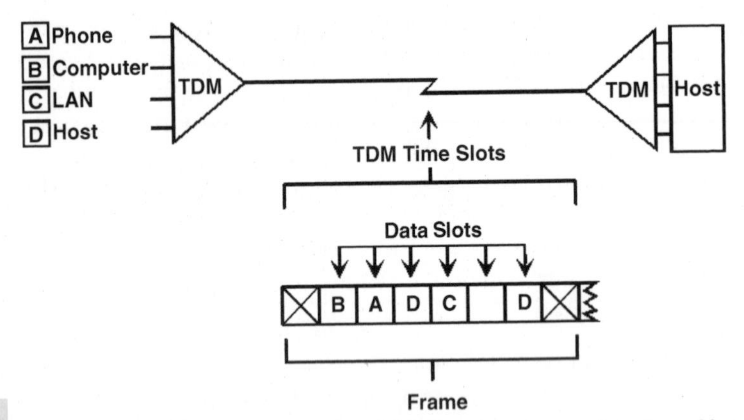

Figure M-2 Data channels from multiple input sources are interleaved by the Time Division Multiplexer for transport over the high-speed link. Note the empty time slot. If a device has nothing to send, this amount of bandwidth goes unused.

technique. Some vendors support both bit and byte interleaving, enabling their products to readily interconnect with both private facilities for maximum efficiency (i.e., bit) and public switched services for increased connectivity (i.e., byte). This configuration flexibility enables companies to take advantage of both environments according to shifting economics or application needs.

Features High-end T1 multiplexers provide numerous standard and optional features, including

- *Drop-and-insert* With this feature, a multiplexer is able to accept a high-speed composite data stream from another multiplexer, demultiplex (remove) a portion of the data stream, modify it (e.g., add additional data frames), and transmit the altered high-speed data stream to a third multiplexer. A complementary feature, called "bypass," allows a multiplexer to pass-through a high-speed data stream without modification.

- *Subrate data multiplexing* Provides programmable data rates at 56 kbps and below for both synchronous and asynchronous data. Subrate channels, together with normal channels (e.g., DS0), are carried over the same physical T1 facility. To maintain transparency, the subrate multiplexing technique accommodates independent clocking of the transmitted and received data.
- *ISDN* Via plug-in cards, many T1 vendors support primary rate ISDN (23B + D). In addition to the twenty-three 64-kbps B (bearer) channels and one 64-kbps D channel for out-of-band signaling, there are two high-capacity ISDN channels that can be supported by the multiplexer: the 384-kbps H0 channel and the 1.536-Mbps H11 channel. These channels are best suited for interconnecting LANs, high-speed data applications, videoconferencing, and backing up private network links in case of failure.
- *LAN adapters* Optional plug-in LAN adapters are available with most multiplexers. A 10BaseT module, for example, allows Ethernet traffic to be combined with voice, synchronous data, and video traffic over a fractional T1 or full T1 link.
- *Bridges and routers* To interconnect LANs, multiplexers can accommodate plug-in bridge/router modules. A variety of protocols are supported, including Transmission Control Protocol/Internet Protocol (TCP/IP) and Novell's IPX protocol.
- *Frame Relay* Frame Relay interfaces give users added flexibility in tying branch office and workgroup LANs to the backbone network. With Frame Relay, users can create a virtual packet network that can be overlaid onto a high-end T1 or statistical multiplexer network. The subnetwork can expand and contract to use available network resources. Some vendors even support voice traffic over the Frame Relay network. A separate voice compression module is used to digitize analog voice and put it into packet format for priority transmission.

- *ATM* Low-speed Asynchronous Transfer Mode (ATM) interfaces for T1 multiplexers provide constant-bite-rate (CBR) capability, enabling the ATM interface to handle delay-sensitive applications such as voice and videoconferencing. When the videoconference is over, the CBR capability can be disabled and the bandwidth made available to other data applications.
- *SMDS* Some T1 multiplexers can be equipped with the distributed queue dual bus (DQDB) subscriber-network interface (SNI) and data exchange interface (DXI) for SMDS. In addition, the ports can be individually configured to convert SMDS to Frame Relay or ATM or to convert Frame Relay to ATM.
- *Testing* TDMs support local loopback tests on both the high- and low-speed ports. They also support remote loopback tests for the low-speed ports at the other end of the link. System diagnostic tests can be performed when the loopback tests are combined with character generation and error detection capabilities. Some multiplexers automatically "busy out" individual remote low-speed ports when a failure is detected on a low-speed modem or computer port.

System Management With support for the Simple Network Management Protocol (SNMP), the network manager can configure, monitor, and control T1 multiplexers from the same console that controls the LAN devices. A management information base (MIB) for the T1 multiplexer gives administrators at a remote site the same configuration flexibility they would have if they were at the device's control panel. A special MIB gives network administrators access to, and control of, every configurable element of the device. SNMP GET messages let a network administrator receive status information from a device, while TRAP messages report alarm conditions and SET messages are used to reconfigure network devices.

A number of advanced system management features are available with T1 multiplexers. To help network managers effectively control system resources, some multiplexer management systems maintain a detailed physical inventory of every card in the network. When new cards are plugged into a node, the card ID information is automatically reported to the network management system, ensuring that equipment inventory is always current. This information includes card slot, card type, serial number, hardware revision level, and firmware revision level.

The network management systems of some multiplexers provide centralized order entry and order tracking. When a request is received for network service, for example, it is necessary to generate a network order for more bandwidth, which can be put into service immediately or at a specific time.

The alarm filter allows network operators to monitor alarms in real time and set filters to determine what information should be displayed at the network management system console. Most network management systems provide a visual and audible indication of most recent alarm summary information and track and update alarm conditions automatically in the database. A level of severity can be assigned to each alarm type: critical, major, minor, alert, and ignore.

The network management database contains an event log. This information is date/time-stamped and describes how the event was created. For example, if a technician pulls a card out of a multiplexer, a state change (event) is registered. Another state change (event) is registered when a new card is inserted. Other types of events include

- Operator actions
- Alarms
- Status changes
- Order installation and removal
- System errors
- Capacity

The event log provides a chronological history of all activity within the network, giving the network manager a single source of information to track and analyze network performance.

Security A multilevel password capability allows unique, individual access to the network. On an individual password basis, the system administrator can restrict access of any menu, submenu, or operating screen in the network management system. For example, a network supervisor may have access to all screens and functions, while an order entry clerk may have access only to the order management area to add or view orders, not to modify or delete them.

Redundancy The redundant components of a multiplexer can include CPUs, buses, trunk cards, power supplies and fans, and network management systems. If the primary CPU-bus pair fails, for example, the secondary CPU-bus pair will take over nodal processing and clocking functions. When the switch takes place, the network management system is notified, and an alarm is presented to the operator. Each CPU maintains a mirror image of information contained in the redundant CPU so that information is not lost in the event of switchovers. Switchover from an active to a redundant CPU will not corrupt active circuits. The same applies to other redundant system components. Management Reports Both standard and user-customized reports are available from most network management systems. Standard reports can be modified to facilitate network maintenance, capacity planning, inventory tracking, cost allocation, and vendor relations. In addition, ad-hoc reports can be created from the management system's database using Structured Query Language (SQL).

Statistical Time Division Multiplexers

STDM is a more efficient multiplexing technique in that input-output (I/O) devices are not assigned their own channels. If an input device has nothing to send at a particular time, another input device can use the channel. This uses the available bandwidth more efficiently. An STDM can be purchased as a stand-alone device, or it can be an add-on feature to a TDM, providing service over one or more assigned channels.

Operation STDM operation is similar to that of TDMs; the high-speed side appears very much like a TDM high-speed side, while the low-speed side is quite different. The STDM allocates high-speed channel capacity based on demand from the devices connected to the low-speed side. This allocation by demand (or contention) provides more efficient use of the available capacity on the high-speed line. In the variable-allocation scheme of an STDM frame, the time slots do not occur in a fixed sequence.

An STDM increases high-speed line usage by supporting input channels whose combined data rates would exceed the maximum rate supported by the high-speed port. When any given channel is idle (not sending or receiving data), input from another active channel is used in the time slot instead. The STDM has the option of turning off the flow of data from a sender if there is insufficient line capacity and then turning the flow back on when the capacity becomes available.

Features TDMs and STDMs share many of the same operational and management features. One is data compression. Like TDMs, STDMs support techniques for compressing data so that they can actually transmit fewer bits per character. Data compression shrinks the time slot for the STDM and allows it to transmit more time slots per frame.

While TDMs detect and flag errors, STDMs are able to correct them. The sending STDM stores each transmitted data frame and waits for the receiving STDM or computer to acknowledge receipt of the frame. A positive acknowledgment (ACK) or negative acknowledgment (NAK) is returned. If an ACK is received, the STDM discards the stored frame and continues sending the next frame. If a NAK is returned, the STDM retransmits the questionable frame and any subsequent frames. The process is repeated until the problematic frame is accepted or a frame retransmission counter reaches a predetermined number of attempts and activates an alarm.

For applications such as asynchronous data transmission, where error detection is not performed as part of the protocol, STDM error detection and correction is a valuable feature. However, for protocols such as IBM's binary synchronous control (BSC), which contains its own error-detection algorithm, STDM-performed error control adds additional delays and redundancy that may not be appropriate for the application.

Several throughput enhancements are available for STDMs, which can be added later when needs change. These features include

- Per-channel compression, in which each channel has its own compression table.
- Fast-packet technology, which increases throughput by sending part of a frame before the entire frame is built.
- Data prioritization, which inserts shorter interactive frames between larger variable run-length frames.
- Run-length compression, which removes redundant characters from the transmission to improve performance.

Summary

Despite the continuous price reductions on leased lines since the 1980s, businesses are always looking to cut the cost of telecommunications. One of the most effective ways to do this is through the deployment of multiplexers that increase the bandwidth utilization of leased lines. Carriers bill for these lines at a flat monthly rate, regardless of the amount of traffic they carry. Not only can the business save money by using multiplexers, the cost of the devices themselves can be recovered in a matter of a few months out of the money saved. Using different levels of voice and/or data compression, the channel capacity of a leased line can be easily doubled or quadrupled to save even more money, enabling the business to recover the cost of the equipment even faster.

See also

Inverse Multiplexers

MULTICHANNEL MULTIPOINT DISTRIBUTION SERVICE

Multichannel Multipoint Distribution Service (MMDS) is a microwave technology that traces its origins to 1972 when it was introduced to provide an analog service called Multipoint Distribution Service (MDS). For many years, MMDS was used for one-way broadcast of television programming, but in early 1999, the Federal Communications Commission (FCC) opened up this spectrum to allow for two-way transmissions, making it useful for delivering telecommunications services, including high-speed Internet access to homes and businesses.

This technology, which has now been updated to digital, operates in the 2- to 3-GHz range, enabling large amounts of data to be carried over the air from the operator's antenna towers to small receiving dishes installed at each customer location. The useful signal range of MMDS is about 30 miles, which beats Local Multipoint Distribution Service (LMDS) at 7.5 miles and Digital Subscriber Line (DSL) at 18,000 feet. Furthermore, MMDS is easier and less costly to install than cable service.

Operation

With MMDS, a complete package of TV programs can be transmitted to homes and businesses. Since MMDS operates within the frequency range of 2 to 3 GHz, which is much lower than LMDS at 28 to 31 GHz, it can support only up to 24 stations. However, operating at a lower frequency range means that the signals are not as susceptible to interference as those using LMDS technology.

Most of the time the operator receives TV programming via a satellite downlink. Large satellite antennas installed at the head end collect these signals and feed them into encoders that compress and encrypt the programming. The encoded video and audio signals are modulated, via amplitude modulation (AM) and frequency modulation (FM), respectively, to an intermediate-frequency (IF) signal. These IF signals are up-converted to MMDS frequencies and then amplified and combined for delivery to a coaxial cable that is connected to the transmitting antenna. The antenna can have an omnidirectional or sectional pattern.

The small antennas at each subscriber location receive the signals and pass them via a cable to a set-top box connected to the television. If the service also supports high-speed Internet access, a cable also goes to a special modem connected to the subscriber's PC. MMDS sends data as fast as 10 Mbps downstream (toward the computer). Typically, service providers offer downstream rates of 512 kbps to 2.0 Mbps, with burst rates up to 5 Mbps whenever spare bandwidth becomes available.

Originally, there was a line-of-sight limitation with MMDS technology, but this has been overcome with a complementary technology called Vector Orthogonal Frequency Division Multiplexing (VOFDM). Because MMDS does not require an unobstructed line of sight between antennas, signals bouncing off objects en route to their destination require a mechanism for being reassembled in their proper order at the receiving site. VOFDM handles this function by leveraging multipath signals, which normally degrade transmissions. It does this by combining multiple signals at the receiving end to enhance or recreate the transmitted signals. This increases the overall wireless system performance, link quality, and availability. It also increases service providers' market coverage through non-line-of-sight transmission.

Channel Derivation

MMDS equipment can be categorized into two types based on the duplexing technology used: Frequency Division Duplexing (FDD) or Time Division Duplexing (TDD). Systems based on FDD are a good solution for voice and bidirectional data because forward and reverse use separate and equally large frequency bands. However, the fixed nature of this scheme limits overall efficiency when used for Internet access. This is so because Internet traffic tends to be "bursty" and asymmetric. Instead of preassigning bandwidth with FDD, Internet traffic is best supported by a more flexible bandwidth allocation scheme.

This is where TDD comes in; it is more efficient because each radio channel is divided into multiple time slots through Time Division Multiple Access (TDMA) technology, which enables multiple channels to be supported. Because TDD has flexible time slot allocations, it is better suited for data delivery—specifically, Internet traffic. TDD enables service providers to vary uplink and downlink ratios as they add customers and services. Many more users can be supported by the allocation of bandwidth on a nonpredefined basis.

Summary

MMDS is being used to fill the gaps in market segments where cable modems and DSL cannot be deployed because of distance limitations and cost concerns. Like these technologies, MMDS provides data services and enhanced video services such as video on demand, as well as Internet access. MMDS will be another access method to complement a carrier's existing cable and DSL infrastructure, or it can be used alone for direct competition. With VOFDM technology, MMDS is becoming a workable option that can be deployed cost-effectively to reach urban businesses that do have line-of-sight access and in suburban and rural markets for small businesses and telecommuters.

See also

Cable Television Networks
Digital Subscriber Line Technologies
Local Multipoint Distribution Service
Microwave Communications

MULTIPROTOCOL LABEL SWITCHING

With the explosive growth of the Internet in recent years, there is growing dissatisfaction with its performance. New techniques are available to improve performance, such as Multiprotocol Label Switching (MPLS), which delivers quality-of-service (QoS) and security capabilities over IP networks, including virtual private networks (VPNs) used for interconnecting LANs.

MPLS attaches tags, or labels, to IP packets as they leave the edge router and enter the MPLS-based network. The labels eliminate the need for intermediate router nodes to look deeply into each packet's IP header to make forwarding and class-of-service handling decisions. The result is that packet streams can pass through an MPLS-based wide area network (WAN) infrastructure very fast, and time-sensitive traffic can get the priority treatment it requires.

The same labels that distinguish IP packet streams for appropriate class-of-service handling also provide secure isolation of these packets from other traffic over the same physical links. Since MPLS labeling hides the real IP address and other aspects of the packet stream, it provides data protection at least as secure as other Layer 2 technologies, including Frame Relay and ATM.

Operation

To enhance the performance of IP networks, the various routes are assigned labels. Each node maintains a table of label-to-route bindings. At the node, a Label Switch Router (LSR) tracks incoming and outgoing labels for all routes it can reach, and it swaps an incoming label with an outgoing label as it forwards packet information (Figure M-3). Since MPLS routers do not need to read as far into a packet as a traditional router does and perform a complex route lookup based on destination IP address, packets are forwarded much faster, which improves the performance of the entire IP network.

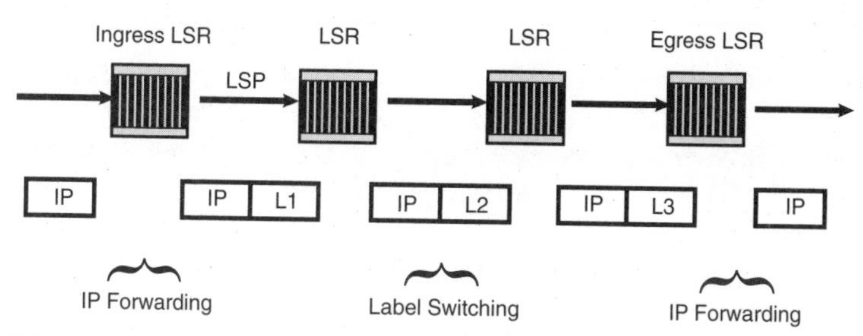

Figure M-3 A label-switched route (LSR) is defined by fixed-length tags appended to the data packets. At each hop, the LSR strips off the existing label and appends a new label, which tells the next hop how to forward the packet. These labels enable the data packets to be forwarded through the network without the intermediate routers having to perform a complex route lookup based on destination IP address.

Although MPLS routers forward packets on a hop-by-hop basis, just like traditional routers, they operate more efficiently. As a packet arrives on an MPLS node, its label is compared to the label information base (LIB), which contains a table that is used to add a label to a packet while determining the outgoing interface to which the data will be sent. After consulting the LIB, the MPLS node forwards the packet toward its destination over a Label-Switched Path

(LSP). The LIB can simplify forwarding and increase scalability by tying many incoming labels to the same outgoing label, achieving even greater levels of efficiency in routing. The LSPs can be used to provide QoS guarantees, define and enforce service-level agreements, and establish private user groups for VPNs.

MPLS provides a flexible scheme in that the labels could be used to manually define routes for load sharing or to establish a secure path. A multilevel system of labels can be used to indicate route information within a routing domain (interior routing) and across domains (exterior routing). This decoupling of interior and exterior routing means MPLS routers in the middle of a routing domain would need to track less routing information. That, in turn, helps the technology scale to handle large IP networks.

MPLS could provide a similar benefit to corporations that have large ATM-based backbones with routers as edge devices. Normally, as such networks grow and more routers are added, each router may need additional memory to keep up with the increasing size of the routing tables. MPLS alleviates this problem by having the ATM switches use the same routing protocols as routers. In this way, the routers on the edge of the backbone and the ATM-based label switches in the core would maintain summarized routing information and only need to know how to get to their nearest neighbor—not to all peers on the network.

MPLS also offers benefits to Internet service providers and carriers. It allows Layer 2 switches to participate in Layer 3 routing. This increases network scalability because it reduces the number of routing peers that each edge router must deal with. It also enables new traffic tuning mechanisms in router-based networks by integrating virtual circuit capabilities available previously only in Layer 2 fabrics. With label switching, packet flows can be directed across the router network along predetermined paths, similar to virtual circuits, rather than along the hop-by-hop routes of normal routed networks. This enables routers to perform advanced traffic management tasks, such as load balancing, in the same manner as ATM or frame relay switches.

Finally, MPLS can be applied not only to the IP networks but also to any other network-layer protocol. This is so because tag switching is independent of the routing protocols employed. While the Internet runs on IP, a lot of campus backbone traffic is transported on protocols such as IPX, making a pure IP solution inadequate for many organizations.

Summary

MPLS came about as a result of Cisco's Tag Switching concept, which was given over to the Internet Engineering Task Force (IETF) for further development and standardization. In 1996, the framework document published by the IETF presented MPLS as a label-switching architecture suitable for any protocol. The label process takes place without referencing the content of the data packet, eliminating the need for protocol-specific handling. By having the data-handling layer of MPLS separate from the control layer, multiple control layers—one for each protocol—could be supported. The IETF, however, has focused on MPLS as a means of improving IP networking, where the commercial opportunity is greatest. MPLS many encourage more service providers to migrate core infrastructures from ATM to IP. Now that MPLS provides IP with high speed, QoS, and security, there may be less reason for service providers to build an ATM infrastructure, which provides these advantages but at a much higher cost than IP.

See also

Asynchronous Transfer Mode

Quality of Service

Routers

Transmission Control Protocol/Internet Protocol (TCP/IP)

N

NETWORK BACKUP

Network backup is the capability to protect information on various local area network (LAN) servers from loss by storing it on appropriate media—hard disk, tape, or optical disc. If a disk in the server crashes or a virus wipes it clean, the data can be restored from the backup copy. The currency of the data that are reloaded to the server depends on the frequency of network backup—daily, weekly, monthly. The more an organization depends on information for its core functions, the greater the frequency of backup to protect this vital asset.

Network backup is not a simple matter for most businesses. One reason is that it is difficult to find a backup system capable of supporting different network operating systems and data, especially if midrange systems and mainframes are involved. Protecting mission-critical data stored on LANs requires backup procedures that are well defined and rigorous. These procedures include backing up data in a proper rotation, using proper media, and testing the data to ensure that they can be restored easily and quickly in an emergency. Enterprise-wide backups are especially problematic. This is so because typically there are multiple servers and operating systems, as well as isolated workstations that often hold mission-critical data. Moreover, the network and client/server environments have special backup needs: Back up too often and throughput suffers; back up too infrequently and data can be lost.

Backup Procedures

Deciding which files to back up can be more complicated than picking the right storage medium. The most thorough backup is a full backup in which every file on every server is copied to one or more tapes or disks. However, the size of most databases makes this impractical to do more than once a month.

Incremental backups copy only files that have changed since the last backup. Although this is faster, it requires careful management because each tape may contain different files. To restore a system made with incremental backups requires all the incremental backups (in the correct order) made since the last full backup.

Differential backups split the difference between full and incremental techniques. Like an incremental backup, a differential backup requires a tape with the full set of files. However, each differential tape contains all the files that have changed since the last full backup, so restoration requires just that full set and the most recent differential.

Scheduling and Automation

The scheduling of backups is determined by several factors, including the criticality of business applications, network availability, and legal requirements. Network backup software with calendar-based planning features allows the system administrator to do such things as schedule the weekly archiving of all files on LAN-attached workstations. The backup can be scheduled for nonbusiness hours both to avoid disrupting user applications and to avoid network congestion.

Some scheduling tools allow the system administrator to set precise parameters with regard to network backups. For example, the backup can target only files that have not been accessed in the past 60 days with the objective of freeing at least 100 MB of disk space on a particular server. When the backup is complete, a report is generated listing the files that have been archived to tape, along with their file size and date of last access. The total number of bytes is also provided, allowing the system administrator to confirm that at least 100 MB of storage has been freed on the server.

Event-based scheduling allows the system administrator to run predefined workloads when dynamic events occur in the system, such as the close of a specific file or the start or termination of a job. With regard to network backups, the administrator can decide what events to monitor and what the automated response will be to those events. For example, the administrator can decide to archive all files in a directory after the last print job or do a database update after closing a particular spreadsheet.

Although most network backup programs can grab files from individual workstations on the LAN, there may be thousands of users with similar or identical system configurations. Instead of backing up 1000 copies of Windows, for example, the network backup program can be directed to copy only each user's system configuration files. That way, if a workstation experiences a disk crash, a new copy of Windows can be downloaded from the server, along with the user's applications, data, and configuration files.

With the right management tools, network backup can be automated under centralized control. Such tools can go a long way toward lowering operating and resource costs by reducing time spent on backup and recovery. These tools enhance media management by providing overwrite protection, log file analysis, media labeling, and the ability to recycle backup media. In addition, the journaling and scheduling capabilities of some tools relieve the operator of the time-consuming tasks of tracking, logging, and rescheduling network and system backups.

Another useful feature of such tools is data compression, which reduces media costs by increasing media capacity. This automated feature also increases backup performance while reducing network traffic.

When these tools are integrated with high-level management platforms—such as Hewlett-Packard's OpenView or IBM's NetView—or operating systems—such as Sun's Solaris—problems or errors that occur during automated network backup are reported to the central management console. The console operator is notified of any problem or error via a color change of the respective backup application symbol on the network map. By clicking on the symbol, the operator can directly access the network backup application to determine the cause of the problem or correct the error to resume the backup operation.

Capabilities and Features

Depending on the particular operating environment, some of the key areas to consider when evaluating network backup software include

- Storage capacity and data transfer rate of the backup system.
- A fast-start capability that allows full network backups to be performed immediately and fine-tuning of the backup parameters later.
- The ability to back up the NetWare bindery (if applicable), security information, and file and directory attributes.
- Support for multiple file systems including NetWare, the Apple File Protocol (AFP), OS/2 High Performance File System, Sun Microsystems' Network File System (NFS), and OSI File Transfer, Access, and Management (FTAM).
- Ability to monitor, back up, and log the activities of multiple file servers simultaneously.
- Tape labeling, rotation, and script file schemes for automating the backup and recovery process.

- Reporting and audit log capability.
- Fast-search capability, which allows a system administrator easily and quickly locate and retrieve files.
- File archiving and grooming methods, which allow automatic file and directory storage, including the ability to delete data that have not been accessed for a specified period of time.
- Integrated network virus protection.
- Security features that limit access to backups to only authorized users.

Another key feature of network backup software is the availability of agents that enable such programs to bypass operating system constraints to store files that are still open, even if the application is accessing or updating them while the backup is in progress. This capability eliminates incomplete backups that typically result when files are not closed. It is of particular value to organizations that need around-the-clock access to information while performing complete backups.

In evaluating software for LAN backups in the mainframe environment, some of the key areas to consider include

- Whether all or most of the platforms at the server level are supported, such as LAN Manager, NetWare, and UNIX.
- Whether all or most workstation platforms are supported, such as DOS/Windows, OS/2, UNIX, and Macintosh.
- Whether non-LAN-connected PCs are supported, such as those with 3270 emulation cards with direct connections to controllers or front-end processors (FEPs).
- Whether other wide area network (WAN) connections are supported or just Transmision Control Protocol/Internet Protocol (TCP/IP).
- Whether users are able to set windows of availability to force backups and recoveries to take place during nonpeak hours.

- Whether the product supports options for restores to be performed by the database administrator, the LAN administrator, or individual workstation users.
- Whether the product supports both a command-line interface for expert use and a graphical user interface for end-user access.
- Whether the product supports automatic archiving of files that have not been accessed for a specified period, thus freeing up server or workstation disk space.
- Whether the product supports the skipping of redundant files in the backup process.
- Whether the product supports other features such as heterogeneous file transfers, remote command execution, and job submission from PC to host.

Summary

As LANs continue to carry increasing volumes of critical data in varying file formats, vendors continue to push the limits of backup technology. On the software side, the trend is toward increasing levels of intelligence. Not only must backup systems ensure that files are backed up, but they also must ensure that they are easily located and restored. Systems intelligence has already progressed to the point where the user need not know the tape, the location on the tape, or even the exact name of a lost file in order to restore it.

See also

Hierarchical Storage Management

Storage Area Networks

NETWORK COMPUTING

The concept of network computing originated with Oracle Corp. in 1995 when the company articulated its vision of a minimally equipped computer that would depend mostly on the LAN for its applications—specifically, local servers. With applications deployed, managed, supported, and executed on servers, organizations need not go through the greater expense of equipping every desktop with its own resources for independent operation. Instead, desktops could be equipped with cheaper, application-specific thin clients.

This type of server-based computing model is especially useful in that it allows enterprises to overcome the critical application deployment challenges of management, access, performance, and security. As a result, organizations can more quickly realize value from the applications and data required to run their businesses, receive the greatest return on computing investment, and accommodate both current and future enterprise computing needs economically.

When Oracle first introduced the concept of network computing, the company's motives were questioned. Critics charged that Oracle was merely trying to increase corporate dependence on servers so that it could boost sagging sales. However, the thin-client architecture offered compelling benefits. Today, there are competing thin-client architectures, each attracting third-party developers to build hardware components, systems, applications, and management tools that plug into the overall framework. In this regard, Oracle's nearest competitor is Citrix Systems, which offers its Independent Computing Architecture (ICA). Through its Citrix Business Alliance program, third-party vendors work with Citrix to develop complementary products and markets for the company's WinFrame and MetaFrame thin-client/server software.

In November 1998, Oracle officially abandoned its original vision of network computing in favor of Internet computing. Instead of requiring a larger number of smaller databases placed on every LAN, as called for in the network computing model, the Internet computing model relies on a smaller number of larger databases to which users connect over the Internet to access data and applications via thin clients.

The cornerstone of the revised vision is Oracle8i, billed as the "world's only Internet database," that runs in conjunction with prepackaged server software on industry-standard hardware. Oracle8i supports both interpreted and compiled Java. The platform consolidates not only data but also Java objects and Windows files through its Internet File System (IFS). Users can drag and drop application files into IFS and search on the fields just as they would search and query database data.

The value of Internet computing is its ability to help small, medium-sized, and large businesses lower computing costs without the complexities of general-purpose operating systems. Ostensibly, Oracle8i simplifies a company's systems by consolidating business data onto large servers for easy management, global Internet access, and higher-quality business information. Oracle also offers subscription-based remote support services to help customers overcome labor costs.

Oracle estimates that the Oracle8i platform could deliver a 10 to 1 cost savings over client/server computing. The company is so convinced of the superiority of this new approach to network computing that it no longer offers traditional client/server versions of its products. According to Oracle, client/server distributes complexity and takes a tremendous amount of work to back up and maintain all the data and applications on users' desktops. The Internet computing model combines the best of the mainframe and client/server worlds by centralizing backups and giving users the benefit of an intuitive graphical interface.

Summary

There is now widespread recognition that thin clients and PCs are not mutually exclusive and that both are valued for certain tasks. Furthermore, it is fairly easy to integrate the two environments under central management, thereby realizing a significant reduction on total cost of ownership (TCO) for both. Consequently, there is not going to be the wholesale replacement of PCs with thin clients, as many vendors originally predicted. But that is not the end of the story. A new model of computing is emerging—namely, Oracle's "Internet computing"—that seeks to leverage existing investments in IP infrastructure.

See also

Client/Server Networks

Peer-to-Peer Networks

Thin-Client Architecture

NETWORK DESIGN TOOLS

A typical corporate network consists of different kinds of transmission facilities, equipment, LAN technologies, and protocols—all cobbled together to meet the differing needs of workgroups, departments, branch offices, divisions, subsidiaries, and increasingly, strategic partners, suppliers, and customers. Building such networks presents special design challenges that require comprehensive design tools.

Fortunately, a number of automated design tools have become available in recent years. With built-in intelligence, these tools take an active part in the design process, from building a computerized model of the network, validating its design, and gauging its performance to quantifying equipment requirements and exploring reliability and security issues before the purchase and installation of any network component. Even faulty equipment configurations, design flaws, and standards violations are identified in the design process.

Data Acquisition

The design process usually starts by opening a blank drawing window from within the design tool into which various vendor-specific devices—workstations, servers, hubs, and routers—can be dragged from a product library and dropped into place (Figure N-1). The devices are further defined by type of components, software, and protocols as appropriate. By drawing lines, the devices are linked to form a network, with each link assigned physical and logical attributes. Rapid prototyping is aided by the ability to copy objects—devices, LAN segments, network nodes, and subnets—from one drawing to the next, editing as necessary, until the entire network is built. Along the way, various simulations can be run to test virtually any aspect of the design.

The autodiscovery capabilities found in such management platforms as Hewlett-Packard's OpenView and IBM's NetView—which automatically detect various network elements and represent them with icons on a topology map—are often useful in accumulating the raw data for network

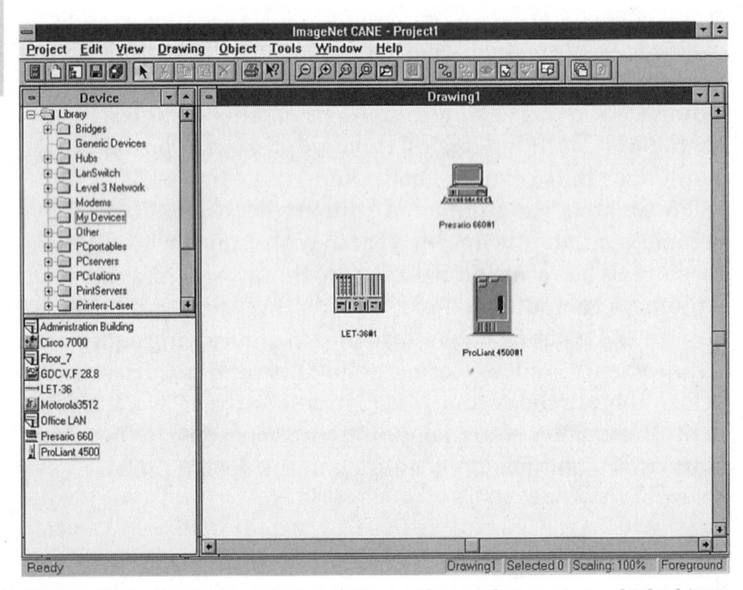

Figure N-1 Typically, a design tool provides workspace into which objects are dragged and dropped from device libraries to start the network design process from scratch. *(Source: Netformx.)*

design. Some stand-alone design tools allow designers to import this data from network management systems, which eases the task of initial data compilation. Although these network management systems offer some useful design capabilities, they are not as feature-rich as high-end stand-alone tools that are also able to incorporate a broader range of network technologies and equipment makes and models.

Designing a large, complex network requires a multifaceted tool—ideally, one that is graphical, object-oriented, and interactive. It should support the entire network life cycle, starting with the definition of end-user requirements and conceptual design and going to the very detailed vendor-specific configuration of network devices, the protocols they use, and the various links between them. At each phase in the design process, the tool should be able to test different design alternatives in terms of cost, performance, and validity. When the design checks out, the tool generates network diagrams and a bill of materials—all this before a single equipment vendor or carrier sales rep is contacted or the Request for Proposal (RFP) is written.

With the right tools, modules, and device libraries, every conceivable type of network can be designed, including legacy networks such as Systems Network Architecture (SNA) and DECnet, voice networks including Integrated Services Digital Network (ISDN), as well as T1, X.25, Asynchronous Transfer Mode (ATM), and TCP/IP nets. Some tools even take into account the use of satellite, microwave, and other wireless technologies.

The designer can take a top-down or bottom-up approach to building the network. In the former, the designer starts by sketching out the overall network; subsequent drawings add increasing levels of detail until every aspect of the network is eventually fleshed out. The bottom-up approach might start with a LAN on a specific floor of a specific building, with subsequent drawings linked to create the overall network structure.

As the drawing window is populated, devices can be further defined by type of component such as chassis, interface cards, and daughter boards. Even the operating system can be specified. Attributes can be added to each device taken from the library—to specify a device's protocol functionality, for example. Once the devices have been configured, a simulation profile is assigned to each device that specifies its traffic characteristics for purposes of simulating the network's load and capacity.

With each device's configuration defined, lines are drawn between them to form the network. With some design tools, the links can be validated against common protocols and network functions. This prevents NetWare clients from being connected to other clients instead of servers, for example. Such online analysis also can alert the designer to undefined links, unconnected devices, insufficient available ports in a device, and incorrect addresses in IP networks. Some tools are even able to report violations of network integrity and proper network design practices.

Network Simulation

Once the initial network design is completed, it can be tested for proper operation by running a simulation that describes how the actual network devices behave under various real-world conditions. The simulator generates network events over time based on the type of device and traffic pattern recorded in the simulation profile. This enables the designer to test the network's capacity under various what-if scenarios and fine-tune the network for optimal cost and performance. Simulators can be purchased as stand-alone programs or may be part of the design tool itself.

Some tools are more adept at designing WANs, particularly those that are based on Time Division Multiplexers (TDMs). With a TDM component taken from the device library, for example, a designer can build an entire T1 network within specified parameters and constraints. The designer can strive for the lowest transmission cost that supports all traffic, for instance, or strive for line redundancy between all TDM nodes. By mixing and matching different operating characteristics of various TDM components, overall design objectives can be addressed, simulated, and fine-tuned. Some tools come with a tariff database to price transmission links and determine the most economical network design.

Such tools also may address clocking in the network design. Clocks are used in TDM networks to regulate the flow of data transmitted between nodes. All clocks on the network therefore must be synchronized to ensure the uninterrupted flow of data from one node to another. The design tool automatically generates a network topology synchronization scheme, taking into account any user-defined criteria, to ensure that there are no embedded clock loops.

Bill of Materials

Once the design is validated, the network design tool generates a bill of materials that includes order codes, prices, and discounts. This report can be exported to any Microsoft Windows application, such as Word or Excel, for inclusion in the proposal for top management review or an RFP issued to vendors and carriers who will build the network. Through the tool's capability to render multiple device views, network planners can choose either a standard schematic or an actual as-built rendering of the cards and the slot assignments of the various devices. Some tools also generate Web-enabled output, which allows far-flung colleagues to discuss and annotate the proposal over the Internet—even allowing each person to drill down and extract appropriate information from the network device library.

Summary

Today's networks are more complex by orders of magnitude than networks envisioned only a few years ago. New Internet services, new technologies, and new trends toward

VPNs and voice-data convergence, plus the sheer number of new equipment offerings, have made reliance on traditional manual solutions to network engineering problems simply unworkable. Intelligent design tools with built-in error-detection, simulation, and analysis capabilities and plug-in modules for ancillary functionality are now available. They do not require managers and planners to be intimately familiar with every aspect of their networks. The essential information can be retrieved on a moment's notice—often with point-and-click ease—analyzed, queried, manipulated, and reanalyzed if necessary, with the results displayed in easy-to-understand graphical form or exported to other applications for further manipulation and study.

See also

Network Drawing Tools
Network Management Systems

NETWORK DIRECTORY SERVICES

Network directory services provide an easy way for users to access resources and find other people on the network. Applications such as e-mail, facsimile, personal information managers (PIMs), personnel management systems, messaging products, and numerous others all come with directories to facilitate user communication. When a company or individual needs a new application, it probably will need a directory to manage users, user groups, routing, security, and other information. With each type of product having its own directory structure, the only way to keep the content consistent is to manually enter the same information into each one. This consumes valuable time, increases the chance of error, and interferes with productivity.

Maintaining directory information for multiple applications is a costly and burdensome chore for most administrators and organizations. The Lightweight Directory Access Protocol (LDAP) is intended to provide a common method of accessing server directories, and it enables directories to be extended across intranets and the Internet, allowing them to be accessed by e-mail applications and Web browsers.

LDAP is based on the standards contained within the international X.500 standard but is significantly simpler. And unlike X.500, LDAP supports TCP/IP, which is necessary for any type of Internet access. Because it is a simpler version of X.500, LDAP is sometimes referred to as "X.500-lite."

To enable the LDAP to run directly over the TCP/IP stack, it had to shed many of X.500's overhead functions. However, LDAP makes up for this loss of power in the following ways:

- Whereas X.500 requires special network access software, LDAP was designed to run over TCP, making it ideal for Internet and intranet applications.
- LDAP has simpler functions, making it easier and less expensive for vendors to implement.
- LDAP encodes its protocol elements in a less complex way than X.500, thereby streamlining coding/decoding of requests.
- LDAP servers return only results or errors, which lightens their processing burden.
- LDAP servers take responsibility for "referrals" by handing off the request to the appropriate network resource. X.500 returns this information to the client, which must then issue a new search request.

Although LDAP enjoys widespread industry support, there is incompatibility among LDAP-compliant applications because the standard does not specify a consistent naming scheme for accessing directories by such fields as name, address, phone number, and e-mail address. Thus vendors have been using different ways for storing and maintaining this information. This problem has been addressed by the Lightweight Internet Person Schema (LIPS).

Lightweight Internet Person Schema

LIPS is designed to ensure easier implementation of LDAP through the definition of common terms for attribute names and content. For example, a messaging client may want to browse an LDAP directory to retrieve a name and phone number. Without LIPS, one server could define "phone number" as a field called PHONE with a length of 10 characters, and another vendor could define the field as BUS_PHNE with a length of 20 characters to accommodate international numbers. LIPS solves this problem by defining the field name, size, and acceptable characters (syntax) for 37 common attributes.

This is not intended to be an exhaustive list of attributes; in fact, most directories have far more than 37 fields. LIPS presents a baseline schema containing only the minimum number of common fields that loosely define an individual.

By adhering to these standardized attributes, client software vendors can build server-independent products using the LDAP standard. To be fully compliant, a vendor must expose all the LIPS attributes with the given field names and minimum sizes (larger values are allowed). However, there is no requirement that the attributes contain any data.

LIPS is not designed to be a server-to-server synchronization solution; it only defines how the data are presented to a client. There is no facility for initiating a server-to-server connection and replicating information, nor is that planned in the future. However, products that use LDAP and LIPS can be used to perform server-to-server directory synchronization.

XML for Directory Access

An eXtensible Markup Language (XML)–based standard for directory access is available that defines how applications running on the Web or mobile devices can access a directory without needing a special client, as required with LDAP.

This means that a cell phone or personal digital assistant (PDA) can use XML to access a directory instead of requiring it to have a bulkier LDAP implementation on that client.

Directory Services Markup Language (DSML) provides a standard way for a client application to read, query, update, and search a directory. DSML also simplifies application creation because developers can write exclusively in XML without having to know LDAP.

DSML eliminates the special-purpose client code. Although it standardizes basic directory functions such as query and update, it does not address user identification and authorization or chaining, the act of stringing directories together. The specification supports referrals, however, which lets one directory refer queries to another directory. The advantage of wrapping XML around LDAP is that vendors and companies do not have to reinvent their current LDAP products. Every directory vendor supports LDAP today, and DSML merely adds a more efficient way to deliver queries to their directories. This results in a broader reach of directory services to a new level of client applications.

Summary

LDAP offers a method of accessing directories, making it possible for almost any application running on virtually any computer platform to obtain directory information, such as e-mail addresses and public keys. Because LDAP is an open protocol, applications do not have to be tailored to the specific type of server hosting the directory. DSML improves directory services in that it does not require the client device, which may be memory-constrained, to run an LDAP client. For these devices, DSML provides a more efficient way to deliver queries to directories on the network.

See also

Electronic Mail

NETWORK DRAWING TOOLS

Network drawing tools are applications that facilitate the design and documentation of large networks. Network administrators faced with managing detailed and often large quantities of information on local and worldwide corporate networks require tools that can accurately depict these complex infrastructures. While the automatic discovery capabilities of high-end network management systems can help in this regard, they are not very useful for documenting the equipment at the level of detail that is now required by network planners.

A number of drawing tools have become available that can aid the network design process. Such tools provide the five major features considered critical to network planners:

- An easy-to-use drawing engine for general graphics.
- An extensive library of predrawn images representing vendor-specific equipment.
- A drill-down capability, which allows multiple drawings to be linked to show various views of the network.
- A database capability to assign descriptive data to the device images.
- A high degree of embedded intelligence that makes images easy to create and update.

Most network drawing tools are Windows-based and employ the drag-and-drop technique to move images of network equipment from a device library to a blank workspace. Many also allow network designs to be published on the corporate intranet or the public Internet, allowing any authorized user to view them with a Web browser. Some drawing tools can automatically discover devices on an existing network to ease the task of drawing and documenting the network.

Device Library

A device library holds images of such things as modems, telephones, hubs, Public Branch Exchanges (PBXs), and channel service units/data service unit (CSUs/DSUs) from different manufacturers. Representations of LANs and WANs, databases, buildings and rooms, satellite dishes, microwave towers, and a variety of line connectors are included. There are also shapes that represent such generic accessories as power supplies, PCs, towers, monitors, keyboards, and switches. There are even shapes for racks, shelves, patch panels, and cable runs (Figure N-2).

Typically, an annual subscription provides unlimited access to the hundreds of new network devices, adapters, and

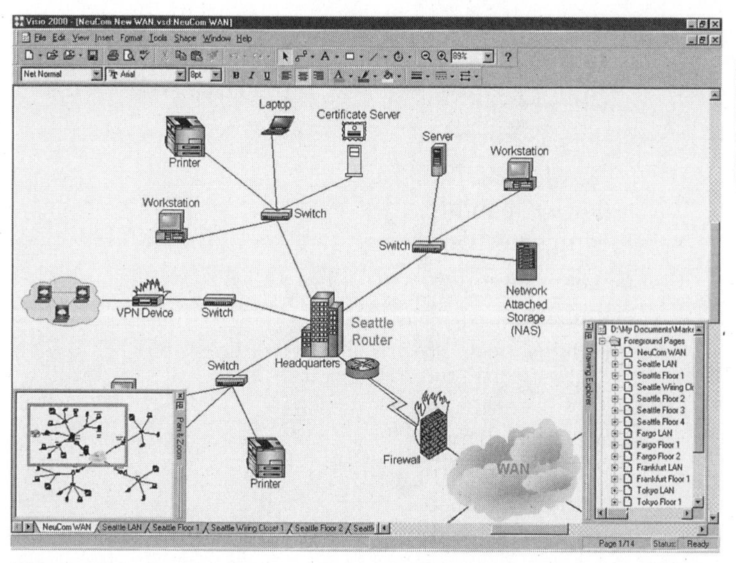

Figure N-2 From a library of network shapes, items are dragged and dropped into place as needed to design a new node or build a whole network. (*Source: Visio Corp., a Division of Microsoft Corp.*[1])

[1] As of January 2000, Visio Corp. became the Visio Division operating within Microsoft's Business Productivity Group.

accessories added to the device library. Depending on the drawing tool vendor, new objects may even be downloadable from the company's Web site.

While many drawing tools offer thousands of exact-replica hardware device images from hundreds of network equipment manufacturers, some tools have embedded intelligence into the shapes, which enables components such as network cards to snap into equipment racks and remain in place even when the rack is moved.

In addition, each shape can be annotated with product-specific attributes, including vendor, product name, part number, and description (Figure N-3). This permits users to

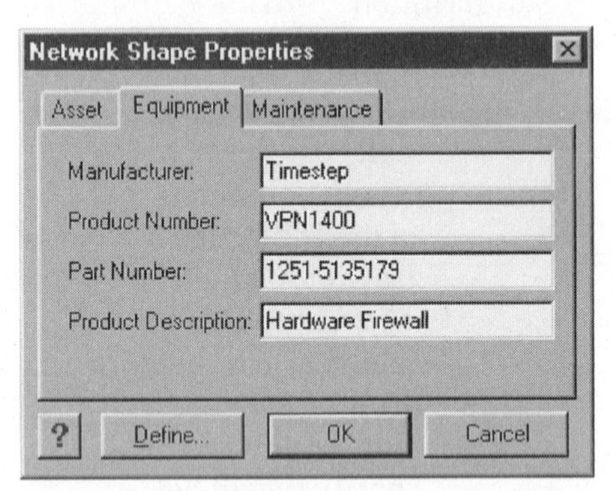

Figure N-3 Details about network equipment can be stored using custom property fields. Device-specific data for each network shape keeps track of asset, equipment, and manufacturer records that can be accessed from within network diagrams. *(Source: Visio Corp., a Division of Microsoft Corp.)*

generate detailed inventory reports for network asset management.

The shapes are even programmable so that they can behave like the objects they represent. This reduces the need for manual adjustments while drawing and ensures the accuracy of the final diagram. For example, the shape representing an equipment rack from a specific vendor can be programmed to know its dimensions. When the user populates the drawing with multiple instances of this shape, it could issue an alert if there is a discrepancy between the space available on the floor plan and the space requirements of the equipment racks.

Each shape also can be embedded with detailed information. For example, the user can associate a spreadsheet with any network element—to provide cost information on a new switch node or LAN segment, for example—along with a bar chart to perhaps illustrate the cost data by system component. The spreadsheet data can be manipulated until costs fit within budgetary parameters. The changes will be reflected in the bar chart the next time it is opened.

The Drawing Process

To start a network diagram, typically the user opens the template for the manufacturer whose equipment will be placed in the diagram. This causes a drawing page to appear that contains rules and grid. The drawing page itself can be sized to show the entire network or just a portion of it.

Various other systems and components can be added to the diagram using the drag-and-drop technique. The user has the option of having (or not having) the shapes snap into place within the drawing space so that they will be precisely positioned on grid lines. Once placed in the drawing space, the shapes can be moved, resized, flipped, rotated, and glued together. Expansion modules, for example, can be dragged onto the chassis so that the modules' end points glue to the connection points on the chassis expansion slots. This allows the chassis and modules to be moved anywhere in the diagram as a single unit. Via the cut-and-paste method, the user can add as many copies of the component as desired to quickly populate the network drawing.

To show the connections between various systems and components, the user can choose shapes that represent different types of networks, including LANs, X.25, satellite, microwave, and radio. Alternatively, the user can choose to connect the shapes with simple lines that can have square or curved corners.

Each network equipment shape has properties associated with it. Custom properties can be assigned to shapes for use in tracking equipment and generating reports, such as inventories. Text can be added to any network system or component, including a Lotus Notes field, specifying font, size, color, style, spacing, indent, and alignment. Text blocks can be moved and resized. Some tools even include a spell checker and a search-and-replace tool. The user can add words that are not in the standard dictionary that comes with the program. The user can specify a search of the entire drawing, a particular page, or selected text only.

AutoCAD files and clip art can be added to network drawings. The common file formats usually supported for importing graphics from other applications, including Encapsulated PostScript (.EPS), Joint Photographic Experts Group (.JPG), Tag Image File Format (.TIF), and ZSoft PC PaintBrush Bitmap (.PCX).

The various shapes used in a network drawing can be kept organized using layers. A "layer" is a named category of shapes. For example, the user can assign walls, wiring, and equipment racks to different layers in a space plan. This allows the user to

- Show, hide, or lock shapes on specific layers so that they can be edited without affecting shapes on other layers.
- Select and print shapes on the basis of their layer assignments.
- Temporarily change the display color of all shapes on a layer to make them easier to identify.
- Assign a shape to more than one layer, as well as assign the member shapes of a group to different layers.

The user also can group shapes into customizable stencils. If the same equipment is used at each node in a network, for example, the user can create a stencil containing all the devices. All the graphics and text associated with each device will be preserved in the newly created stencil. This saves time in drawing large-scale networks, especially those based on equipment from a variety of manufacturers.

At any step in the design process the user can share the results with other network planners by sending copies via e-mail. The diagram is converted to an image file, which is displayed as an icon in the message box, and sent as an attachment. When opened by the recipient, the attachment with all embedded information is displayed. The document can then be edited by creating a separate layer for review comments, each of which is done in a different color. The use of separate layers and colors protects the original drawing and makes comments easier to view and understand.

Some network drawing tools provide a utility that converts network designs and device details into a series of hyperlinked HyperText Markup Language (HTML) documents that can be accessed over the Web. These documents show device configurations, port usage, and even device photographs. Users can activate the links to navigate from device to device to trace connectivity and review device configurations (Figure N-4). In addition to supporting fault identification, the hyperlinked documents aid in planning design changes.

There are several ways that the network diagrams can be protected against inadvertent changes, especially if they are shared via e-mail or posted on the Web:

- The shapes can be locked to prevent them from being modified in specific ways.
- The attributes of a drawing file (styles, for example) can be protected against modification.
- The file can be saved as read-only so that it cannot be modified in any way.
- The shapes on specific layers can be protected against modification.

Users can password-protect their work to prevent attributes of a drawing file from being changed. For example, a

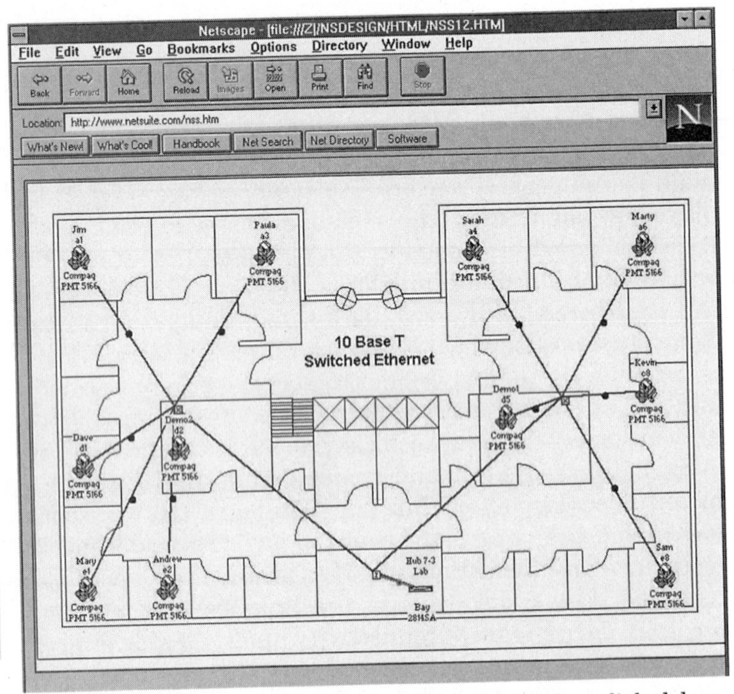

Figure N-4 This floor plan of a 10BaseT network is a hyperlinked drawing rendered by Netscape Navigator.

background containing standard shapes or settings can be password-protected. Users also can set a password for a drawing's styles, shapes, backgrounds, or masters. A password-protected item can be edited only if the correct password is entered.

Embedded Intelligence

Some products are so intelligent that they can no longer be considered merely drawing tools. Visio 2000 Enterprise Edition, for example, supports switched WANs through its AutoDiscovery feature. This technology includes support for Layer 2 (switched data link), Layer 3 (IP network), and Frame Relay network environments. AutoLayout technology makes it simple for users to automatically generate network diagrams of the discovered devices—including detailed mappings.

In addition to allowing information technology (IT) specialists to create conceptual, logical, and physical views of their information systems, Enterprise Edition owners can purchase Visio's add-on solution for monitoring network performance. Working with Enterprise Edition's AutoDiscovery, Real-Time Statistics documents the behavior of a network environment, capturing real-time data from any SNMP-manageable device on LANs and WANs. In being able to monitor the network's performance, managers have the information they need to redistribute network traffic and prevent overloads. Real-Time Statistics then turns this performance data into graphs that can be printed or exported for analysis.

Enterprise Edition enables developers to visualize and quickly start software development projects. They can visualize the design architecture of existing systems by reverse engineering source code from Microsoft Visual Studio. They also can decrease development time by generating fully customizable code skeletons for Visual Basic, C++, and Java from Unified Modeling Language (UML)[2] class diagrams.

Summary

Unlike traditional computer-aided design (CAD) programs, today's drawing tools are specifically designed for network and IT planners. They can improve communications and productivity with their easy-to-use and easy-to-learn graphics capabilities that offer seamless integration with other applications on the Windows desktop. Their graphical representation of complex projects also enables more people to understand and participate in the planning process. Despite their origins as simple drawing tools, this new generation of

[2] Pioneered by Rational Software Corp. and officially adopted as a standard by the Object Management Group (OMG), the Unified Modeling Language (UML) is an industry-standard language for specifying, visualizing, constructing, and documenting the elements of software systems. UML simplifies the complex process of software design, making a "blueprint" for construction.

tools provides a high degree of intelligence, programmability, and Web awareness that makes them well suited for the demanding needs of planners.

See also

Asset Management
Network Design Tools
Network Management Systems

NETWORK INTEGRATION

Distinct from systems integration, which focuses on getting different computer systems to communicate with each other, network integration is concerned with getting diverse, far-flung LANs and host systems interconnected over the WAN. Typically, network integration requires that attention be given to a plethora of different physical interfaces, protocols, frame sizes, and data transmission rates. Companies also face a bewildering array of carrier facilities and services to extend the reach of information systems globally over high-speed WANs. To get everything working properly may require a careful hardware selection, software customization, and applications tweaking.

A network integrator brings objectivity to the task of tying together diverse products and systems to form a seamless, unified network. To do this, the network integrator draws on its expertise in information systems (IS), office automation, LAN administration, telephony, data communications, and network management systems. Added value is provided through strong business planning, needs analysis, and project management skills, as well as accumulated experience in meeting customer requirements in a variety of industry segments and operating environments.

A qualified network integrator will have in place a stable support infrastructure capable of handling a high degree of ambiguity and complexity, as well as any technical challenge that may step in the way of the integration effort. In addition to financial stability, this support infrastructure includes staff representing a variety of technical and management disciplines and strategic relationships with specialized companies such as equipment providers, cable installers, and software firms.

Integration Services

There are a number of discrete services that are provided by network integration firms, including

- *Consulting* Includes needs analysis, business planning, systems/network architecture, technology assessment, feasibility studies, RFP development, vendor evaluation and product selection, quality assurance, security auditing, disaster recovery planning, and project management.
- *Design and development* Involves such activities as network design, facilities engineering, equipment installation and customization, and acceptance testing.
- *Systems implementation* Entails procurement, documentation, configuration management, contract management, and program management.
- *Facilities management* Provides operations, technical support, hot-line services, move and change management, and trouble ticket administration.
- *Network management* Includes network optimization, remote monitoring and diagnostics, network restoration, technician dispatch, and carrier and vendor relations.
- *Technology migration* Includes technology assessments, financial justification, project planning, pilot studies, project rollout and support, and training.
- *Contingency planning* Includes risk assessment and development of countermeasures to minimize damage if a disaster occurs that disrupts computer and telecommunications systems and services.

- *Life-cycle services* Identifies the system development and technical support resources available during the network's life span, detailing the different product stages starting from the first commercial shipment to eventual product retirement.

Network integration may be performed by in-house technical staff or through an outsourcing arrangement with a computer company, local- or interexchange carrier, management consulting firm, traditional IS-oriented systems integrator, or an interconnect vendor.

Each type of firm has specific strengths and weaknesses. The wrong selection can delay the implementation of new information systems and LANs, disrupt network expansion plans, and impede applications development—any of which can inflate operating costs over the long term and have adverse competitive impacts.

It is therefore advisable to choose an integrator whose products and services are particularly pivotal to the application. For example, if the network integration application is such that the computer requirements are extremely well defined and no significant computer changes are expected but a range of new communications services might be involved, a carrier would be a better choice of integrator than a computer vendor. On the other hand, if the project is narrow in scope and the needs are well understood, in-house staff might be able to handle the integration project, with as-needed assistance from a computer firm or carrier.

Summary

While some companies have the expertise required to design and install complex networks, others are turning to network integrators to oversee the process. The evaluation of various integration firms should reveal a well-organized and staffed infrastructure that is enthusiastic about helping to reach the customer's networking objectives. This includes having the methodologies already in place, the planning tools already available, and the required expertise already on staff. Beyond that, the integrator should be able to show that its resources have been deployed successfully in previous projects of a similar nature and scope.

See also

Downsizing

NETWORK INTERFACE CARDS

A network interface card (NIC) is an adapter that plugs into a computer, enabling it to connect to a LAN for the purpose of communicating with other computers and devices. The NICs are network-specific—there are adapters for Ethernet, Token Ring, Fiber Distributed Data Interface (FDDI), ATM, and other types of networks. NICs are also medium-specific—there are adapters for shielded and unshielded twisted-pair wiring, thick and thin coaxial cabling, and single-mode and multimode optical fiber. NICs are also bus-specific—there are adapters for the Industry Standard Architecture (ISA), Extension to Industry Standard Architecture (EISA), Micro-Channel Architecture (MCA), and Peripheral Component Interconnect (PCI/PCI-X) architectures. NICs are also available in the PC card form factor for connecting mobile notebook users to the LAN. Major vendors offer software that adds management capabilities of their NICs.

MAC Addresses

Devices on conventional LANs such as Ethernet and Token Ring use Media Access Control (MAC) addresses. These are the 6-byte hardware-level addresses of the NICs that provide workstations and other devices with the means to interconnect with each other through a hub or switch. An example of a MAC address is

00 00 0C 00 00 01

The first 3 bytes contain a manufacturer code (the one above is for Cisco Systems), and the last 3 bytes contain a unique station ID, both of which are burned into the NIC's firmware. Manufacturer IDs are assigned by the Institute of Electrical and Electronics Engineers (IEEE).

These addresses provide the means to implement virtual LANs (VLANs). This technology lets the same server link—via one NIC or a team of NICs—carry traffic for up to 64 logical subgroups created via software. This capability provides additional bandwidth management and security features and helps reduce the administrative overhead required to manage workstation moves and changes.

Client NICs

Client NICs provide the means to connect desktop computers, printers, and other devices to the LAN. Today's Ethernet and Token Ring NICs have an autosensing capability that allows the NIC, when connected to a switch or hub port, to automatically sense and connect at the highest network speed. NICs are available at different speeds. With Ethernet, for example, there are 10BaseT, 100BaseT, and 1000BaseT cards. Some cards support 10/100BaseT or combine all three Ethernet speeds onto the same card. By simultaneously performing multiple processing tasks, some NICs provide the fastest data transfer speeds available for the PCI bus.

NICs that feature 32-bit multimaster concurrency technology permit the card to communicate directly with the computer's CPU, bypassing sluggish interrupts and I/O channels. NICs that feature an onboard boot ROM socket allow for remote workstation boot-up from the server. Light-emitting diodes (LEDs) on the card report link status, packet activity, transmission speed, and transmission mode (half or full duplex).

Many NICs are optimized to work in specific operating environments. For example, NICs for Windows environments—specifically PCs running Windows 95, 98, or NT/2000—and are Plug and Play–compliant. The installation software allows connection of the PC to Novell NetWare networks as well. Windows-based diagnostics and configuration utilities facilitate installation and troubleshooting. Different NICs are available for the Macintosh and some UNIX environments.

Server NICs

Server NICs include the functions of client NICs but have additional functionality and provide higher bandwidth. For example, NICs may be configured in a way to increase the fault tolerance of the server's LAN link. If one NIC fails, the fail-over software deactivates the faulty NIC and switches LAN traffic to an alternate card. The rerouting takes place almost instantaneously, without human intervention. The software also gives Simple Network Management Protocol (SNMP) alerts on the failed NICs. When the failed NIC begins working again, the software brings it back into the array automatically and starts balancing traffic across it again.

Many server NICs provide asymmetric port aggregation—also referred to as "asymmetric load balancing" or "asymmetric trunking." This technology distributes outbound server traffic between two or more cards, providing a wider data pipe. The NICs operate together and appear as a single device with one network address. Asymmetric port aggregation is especially useful for Web servers, e-mail servers, and other applications where most of the traffic flows in one direction from the server to the client PCs. This method also provides fault tolerance in that if one of the NICs fails, the others take its load.

A companion technology is symmetric port aggregation (or symmetric load balancing or symmetric trunking). This method combines two or more connections into a wider pipe that can transmit data in both directions. Since combining several 100-Mbps connections does not require replacing hubs or switches, it can put off having to invest in upgrades to gigabit LAN technologies. A dual homing capability allows the server NIC to connect to different switches for additional redundancy, ensuring the server remains available even if one of the attached switches fails.

There are also server NICs that allow Token Ring traffic to run over a 100-Mbps Fast Ethernet backbone connection, eliminating congestion across the Token Ring network backbone without the costs associated with ATM and FDDI. The NIC tunnels Token Ring traffic in Fast Ethernet frames, delivering high-speed performance to Token Ring clients. When installed in a Fast Ethernet server, the NIC allows Token Ring clients to communicate with the server at Fast Ethernet speeds via a special module installed in a Token Ring switch.

Summary

NICs are used to connect computers and other devices to the LAN. Although many users can get by with inexpensive "dumb" NICs costing under $20 (U.S.), enterprise networks require "intelligent" NICs that ensure high availability to support mission-critical applications and management software to facilitate monitoring and control from a central location.

See also

Media Converters

Transceivers

NETWORK MANAGEMENT SYSTEMS

The task of keeping multivendor networks operating smoothly with a minimum of downtime is an ongoing challenge for most organizations. While many companies prefer to retain total control of their network resources, others rely on computer vendors and carriers to find and correct problems on their networks or depend on third-party service firms. Wherever these responsibilities ultimately reside, the tool set used for monitoring the status of the network and initiating corrective action is the network management system (NMS).

With an NMS, technicians can remotely diagnose and correct problems associated with each type of device on the network. Although today's network manager is concerned primarily with diagnosing failures, the likelihood of problems to occur also can be predicted so that traffic may be diverted from failing lines or equipment with little or no inconvenience to users.

Network management begins with such basic hardware components as modems, data sets (CSUs/DSUs), multiplexers, and dial backup units (Figure N-5). Each component typically has the ability to monitor, self-test, and diagnose problems regarding its own operation and report problems to a central management station. The management station operator can initiate test procedures on systems at the other end of a point-to-point line. On more complex multipoint and multidrop configurations, the capability to test and diagnose problems from a central location greatly facilitates problem resolution. This capability also minimizes the need to dispatch technicians to remote locations and reduces maintenance costs.

A minimal NMS consists of a central processing unit, system controller, operating system software, storage device, and operator's console. The central processor may consist of a minicomputer or microcomputer. The system controller, the heart of the NMS, continuously monitors the network and generates status reports from data received from various network components. The system controller also isolates network faults and restores segments of the network that have failed or which are in the process of degrading. The controller usually runs on a powerful platform such as UNIX or Windows NT/2000.

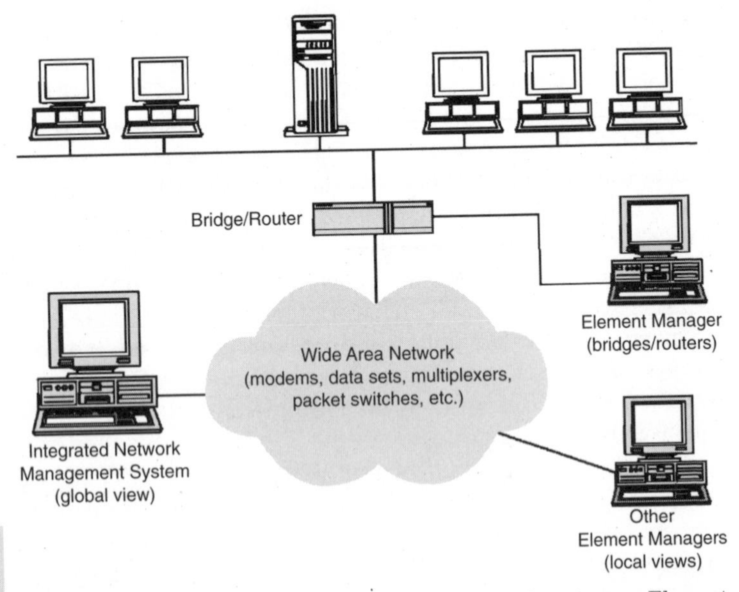

Figure N-5 Each type of device on the network may have its own Element Management System (EMS), which reports to an Integrated Network Management System (INMS).

NMS Functions

Although differing by vendor, the basic functions that most NMSs have in common include topology mapping, administration, performance measurement, control and diagnostics, configuration management, applications management, and security. Some NMSs include other functions such as network modeling, for example, that would enable the operator to simulate aggregate node or circuit failures to test various disaster recovery scenarios.

Topology Mapping

Many NMSs have an automatic discovery capability that finds and identifies all devices or nodes connected to the network. Based on the discovered information, the NMS automatically draws the required topology maps. Nodes that cannot be discovered automatically can be represented by manually adding custom or standard icons to the appropriate map views or by using the NMS's SNMP-based application programming interfaces (APIs) for building map applications without having to manually modify the configuration to accommodate non-SNMP devices.

A network map is useful for ascertaining the relationships of various equipment and connections, for keeping accurate inventory of network components, and for isolating problems on the network. The network map is updated automatically when any device is added or removed from the network. Device status is displayed via color changes to the map. Any changes to the network map are carried through to the relevant submaps.

Administration

The administration element allows the user to take stock of the network in terms of what hardware is deployed and where it is located. It also tells the user what facilities are serving various locations and what lines and equipment are available with which to implement alternate routing. The vehicle for storing and using this information is the relational database management system.

For administrative tasks, multiple specialized databases are used that relate to each other. One of these databases accumulates trouble ticket information. A trouble ticket contains such information as the date and time the problem occurred, the specific devices and facilities involved, the vendor from which it has been purchased or leased, and the service contact. It also contains the name of the operator who initially responded to the alarm, any short-term actions taken to resolve the problem, and space for recording follow-up information. This information may include a record of visits from the vendor's service personnel, dates on which parts were returned for repair, serial numbers of spares installed, and the date of the problem's final resolution.

A trouble ticket database can be used for long-term planning. The network manager can call up reports on all outstanding trouble tickets, trouble tickets involving particular segments of the network, trouble tickets recorded or resolved within a given period, trouble tickets involving a specific type of device or vendor, and even trouble tickets over a given period not resolved within a specific timeframe. The user may customize report formats to meet unique needs.

Such reports provide network managers with insight on the reliability of a given network management station operator, the performance record of various network components, the timeliness of on-site vendor maintenance and repair services, and the propensity of certain segments of the network to fail. And with information on both active and spare parts, network managers can readily support their decisions on purchasing and expansion. In some cases, cost and depreciation information on the network's components is also provided.

Performance Measurement

Performance measurement refers to network response time and network availability. Many NMSs measure response time at the local end, from the time the monitoring unit receives a start-of-transmission (STX) or end-of-transmission (EOT) signal from a given unit. Other systems measure end-to-end response time at the remote unit. In either case, the NMS displays and records response time information and generates operator-specified statistics for a particular terminal, line, or network segment or for the network as a whole. This information may be reported in real time or stored for a specified time frame for future reference.

Network personnel can use this information to track down the cause of the delay. When an application exceeds its allotted response time, for example, personnel can decide whether to reallocate terminals, place more restrictions on access, or install faster communications equipment to improve response time.

Availability is a measure of actual network uptime, either as a whole or by segments. This information may be reported as total hours available over time, average hours available within a specified time, and mean time between failure (MTBF).

With response time and availability statistics, calculated and formatted by the NMS, managers can establish current trends in network usage, predict future trends, and plan the assignment of resources for specific present and future locations and applications.

Control and Diagnostics

With control and diagnostic capabilities, the NMS operator can determine from various alarms (i.e., an audio or visual indication at the operator's terminal) what problems have occurred on the network and pinpoint the sources of those problems so that corrective action can be taken. Alarms can be correlated to certain events and triggered when a particular event occurs. For example, an alarm can be set to go off when a line's bit error rate (BER) approaches a predefined threshold. When that event occurs and the alarm is issued, automated procedures can be launched without operator involvement. In this case, traffic can be diverted from the failing line and routed to an alternative line or service. If the problem is equipment-oriented, another device on "hot" standby can be placed into service until the faulty system can be repaired or replaced.

Configuration Management

Configuration management gives the NMS operator the ability to add, remove, or rearrange nodes, lines, paths, and access devices as business circumstances change. If a T1 link degrades to the point that it can no longer handle data reliably, for example, the NMS may automatically reroute traffic to another private line or through the public network. When the quality of the failed line improves, the system reinstates the original configuration. Some integrated NMSs—those which unify the host (LAN) and carrier (WAN) environments under a single management umbrella—are even capable of rerouting data but leaving voice traffic where it is.

Voice and data traffic can even be prioritized. This NMS capability is very important because failure characteristics for voice and data are very different—voice is more delay-sensitive, and data are more line error–sensitive. On networks that serve multiple business entities and on statewide networks that serve multiple government agencies, the ability to differentiate and prioritize traffic is very important.

On a statewide network, for example, state police have critical requirements 24 hours a day, 7 days a week, whereas motor vehicle branch offices use the network to conduct relatively routine administrative business only 8 hours a day, 5 days a week. Consequently, the response-time objectives of each agency are different, as would be their requirements for restoring the network in case of an outage. On the high-capacity network, there can be two levels of service for data and another for voice. Critical data will have the highest priority in terms of response time and error thresholds and will take precedence over other classes of traffic during restoral. Since routine data will be able to tolerate a longer response time, the point at which restoral is implemented can be prolonged. Voice is more tolerant than data with regard to errors, so restoral may not be necessary at all. The capability to prioritize traffic and reroute only when necessary ensures maximum channel fills, which affects the efficiency of the entire network and, consequently, the cost of operation.

Configuration management applies not only to the links of a network but also to equipment. In the WAN environment, the features and transmission speeds of software-controlled modems may be changed. If a nodal multiplexer fails, the management system can call its redundant components into action or invoke an alternate configuration. And when nodes are added to the network, the management system can devise the best routing plan for the traffic it will handle.

Applications Management

Applications management is the capability to alter circuit routing and bandwidth availability to accommodate applications that change by time of day. Voice traffic, for example, tends to diminish after normal business hours, while data traffic may change from transaction-based to wideband applications that include inventory updates and remote printing tasks.

Applications management includes having the ability to change the interface definition of a circuit so that the same circuit can alternatively support both asynchronous and synchronous data applications. It also includes having the ability to determine appropriate data rates in accordance with response-time objectives or to conserve bandwidth during periods of high demand.

Security

NMSs have evolved to address the security concerns of users. Although voice and data can be encrypted to protect information against unauthorized access, the management system represents a single point of vulnerability to security violations. Terminals employed for network management may be password-protected to minimize disruption to the network through database tampering. Various levels of access may be used to prevent accidental damage. Senior technicians, for example, may have passwords that allow them to make changes to the various databases, whereas less experienced technicians' passwords allow them to only review the databases without making any changes. Other possible points of entry such as gateways, bridges, and routers may be protected with hardware- or software-defined partitions that restrict internal access.

Individual users, too, may be given passwords that permit them to make use of certain network resources and deny them access to others. A number of methods are even available to protect networks from intruders who may try to access network resources with dial-up modems. For instance, the NMS can request a password and hang up if it does not obtain one within 15 seconds. Or it can hang up and call back over an approved number before establishing the connection. To frustrate persistent hackers, the system can limit unsuccessful call attempts before denying further attempts. All successful and unsuccessful attempts at entry are automatically logged to monitor access and to aid in the investigation of possible security violations.

Summary

Today's NMSs have demonstrated their value in permitting technicians to control individual segments or the entire network remotely. In automating various capabilities, NMSs can speed up the process of diagnosing and resolving problems with equipment and lines. The capabilities of NMSs permit maximum network availability and reliability, thus enhancing the management of geographically dispersed operations while minimizing revenue losses from missed business opportunities that may occur as a result of network downtime. Some NMSs offer customizable Web-based reporting, which provides managers with convenient, on-demand insight into network performance from virtually any location. This capability also can be used by service providers to offer their customers personalized insight into their outsourced managed environments.

See also

Network Integration
Network Restoral
Simple Network Management Protocol

NETWORK RESTORAL

Network restoral refers to the processes that bring carrier-provided equipment and lines back to normal operation in response to a failure or condition that disrupts service. Businesses are increasingly relying on their communication networks to interconnect LANs at far-flung locations, improve customer service, exploit new market opportunities, and secure strategic competitive advantages. Thus, when these networks become severely congested or fail, effective solutions must be implemented to restore affected systems as soon as possible.

Although local and long-distance carriers build reliability into their networks at the design stage and monitor performance of the network on a continuing basis through one or more network operations centers (NOCs), there is always the chance that unforeseen problems will occur. When they do, automated processes perform such functions as raise alarms, reroute traffic, activate redundant systems, perform diagnostics, isolate the cause of the problem, generate trouble tickets and work orders, dispatch repair technicians, and return primary facilities and systems to their original service configuration.

Network Redundancy

Most networks are designed with a certain amount of redundancy built in. There usually is a duplicate or backup system that can be called into service immediately on failure of the primary system.

For example, central office switches are equipped with dual processors so that if one processor fails, the second one can take over automatically. The switches are designed to run self-diagnostic tests periodically to help ensure proper operation. If a problem occurs, systems often can automatically fix themselves by rebooting software, for example, or switching automatically to a backup system so that the primary system can reinitialize itself. These systems also have the ability to alert technicians and network managers if the problem cannot be corrected automatically.

Redundancy also applies to signal transfer points (STPs). These are the computers used to route messages over the carrier's packet-based signaling network, which is used to set up calls and implement call-handling features. Each STP has two computers that operate at just under 50 percent capacity. The STP pairs are not collocated but are usually many miles away from each other. If something happens to one STP, its mate can pick up the full load and operate until the repair or replacement of the damaged STP can be accomplished. Should both halves of a mated pair of STPs fail, the switch that normally relies on them can access the signaling network through helper switches that use a different STP pair.

Network control points (NCPs), the customer databases for advanced services such as 800 or virtual private network (VPN) services, not only have dual processors but also, if the second processor should fail, provide a backup NCP for the protection of all customer configuration information.

Digital interface frames (DIFs) provide access to and from interoffice switches for processing long-distance calls. The digital interface units that actually handle this work have spares that take over immediately when a problem occurs. Guiding the overall work of the DIF are two controllers running simultaneously so that if one experiences a problem, the backup controller can take over without customers even noticing that a problem has occurred. As an option, however, a customer's traffic can be sent to another DIF at another interexchange office if the primary switch encounters a problem. This ensures that the customer's calls continue to flow should the primary DIF experience a prolonged outage.

The power systems used to operate the carrier's network also have backup protection. In normal operation, the carrier's power system provides direct current from redundant rectifiers fed by commercial power. If commercial power fails, batteries, which are charged by the rectifiers, provide backup power. An additional level of redundancy is provided by diesel oil–powered generators, which can replace commercial power during prolonged outages.

Network Diversity

Diversity is the concept of providing as many alternative paths as possible to ensure survival of the network when some kind of natural or human-made disaster strikes. Like redundancy, diversity is built into the network during the design stage.

One way in which carriers ensure diversity is to arrange transmission lines as a series of circles or loops to form an interconnecting grid. Should any particular loop be cut—when a backhoe operator hits a buried cable, for example—traffic can be sent over another facility on one or more adjacent loops. In the case of fiberoptic rings around major metropolitan areas, the use of dual counterrotating fiber ring configurations enables carriers to offer disaster-recovery services. In the event of a node failure on the fiberoptic network, traffic is automatically routed to the other ring in a matter of milliseconds.

In larger metropolitan areas, carriers often offer their business customers path diversity (Figure N-6). In being able to reach the carrier's network from two distinct points, businesses can enhance the reliability of their mission-critical applications. Diverse paths protect against cable cuts that normally would bring down the entire trunk group.

To further ensure uninterrupted service, carriers also offer route diversity for their signaling systems. Each pair of STPs is connected to every other pair of STPs by multiple links. To ensure that connectivity will always be available,

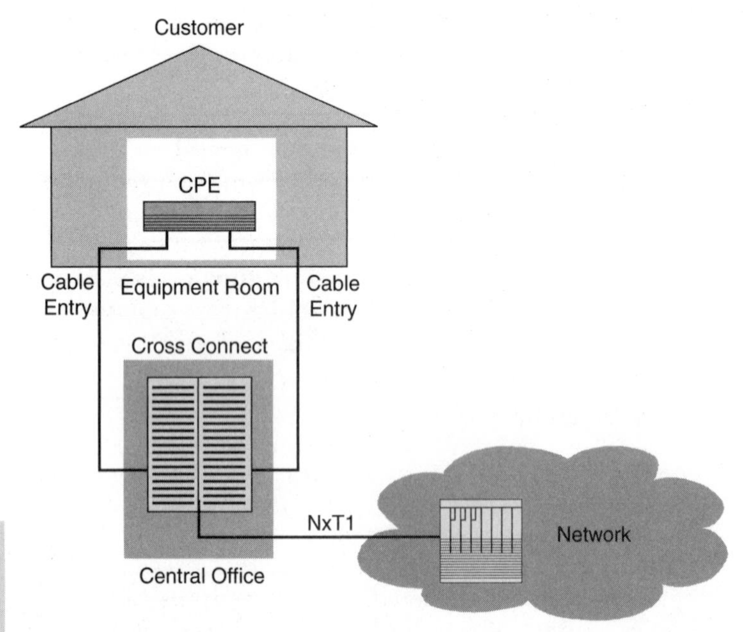

Figure N-6 In this scenario, the business customer has two entry points into the building for cable runs. Traffic is split between the two paths so that if one experiences an outage, such as from a cable cut in a construction site, traffic continues to go out over the other path.

these links are laid out over multiple geographically separated routes. Should something happen to disrupt service along one route, the other routes remain available to keep the carrier's signaling system operational.

Optional Restoral Services

For most businesses, a temporary interruption of service lasting only a few minutes does not present a problem. For businesses that need a much shorter recovery period, carriers offer optional services that can be tailored to meet specific requirements. These services can range from having the carrier plan and build a complete private network to meet certain reliability and performance specifications to selecting one or more of the following lower-cost alternatives:

- For businesses with toll-free 800 services and virtual private networks, the ability to receive calls is of primary importance. The carrier provides routes from two separate switches to the corporate location. In the event of a network disruption, calls are automatically directed to the working switch.
- For businesses with toll-free 800 services, if traffic is blocked at one location for any reason, calls are automatically sent to another corporate location.
- For businesses that use digital services, the carrier can provide a geographically separate backup facility enabling traffic to be switched to the standby links within milliseconds of a service interruption on the primary link.
- For the access portion of a circuit, the carrier can mitigate the effects of certain network failures by automatically transferring service to a dedicated, separately routed access circuit.

Customer-controlled reconfiguration (CCR) is a carrier-provided service that gives businesses a way to organize and manage their own circuits from an on-premises terminal that issues instructions to the carrier's digital cross-connect system (DCS). If a circuit drops to an unacceptable level of performance or fails entirely, the network manager can issue rerouting instructions to the DCS. If several circuits have failed, the network manager can upload a pretested rerouting program to the DCS to restore the affected portion of the network.

Some carriers offer businesses a reservation service in which one or more dedicated digital facilities are brought online after the customer verbally requests it with a phone call. This restoral solution requires that the customer presubscribe to the service and that access facilities already be in place with the local carriers at each end.

Site Recovery Options

Some carriers offer optional site recovery options. This type of service is meant to deal with the loss of a primary data center that runs mission-critical applications. If a customer's data center suffers from a catastrophic fire or natural disaster, for example, traffic will be quickly rerouted to another comparably equipped site. This service is far more economical than having to set up and maintain another data center and links.

To offer site recovery services, the carrier typically partners with an established firm such as Comdisco Disaster Recovery Systems. When disaster strikes, the customer calls the carrier and requests activation of links to the alternate site, a process that may take about 2 hours to complete and which may entail the uploading of new routing tables to each router to reflect the changes.

Summary

Traditionally, most organizations relied on their long-distance carrier for maintaining acceptable network performance. More often than not, the carriers were not up to the task. This led to the emergence of private networks in the 1980s that allowed companies to exercise close control of leased lines with an in-house staff of network managers and technicians. In their eagerness to recapture lost market share, the carriers have made great strides in improving their response to network congestion and outages. This has gone a long way toward restoring lost confidence that once prompted companies to set up and maintain their own networks. Today, companies are once again comfortable in relying on the carriers for maintaining acceptable network performance. Many carriers now back their performance claims with service-level agreements (SLAs) and credit the customer's next invoice if performance falls below certain thresholds.

See also

LAN Restoration
Network Management Systems

NETWORK SECURITY

Protecting vital information has always been a high-priority concern among companies. While access to distributed data networks improves productivity by making applications, processing power, and mass storage readily available to a large and growing user population, it also makes those resources more vulnerable to abuse and misuse. Among the risks are unauthorized access to mission-critical data, information theft, and malicious file tampering that can result in immediate financial loss and, in the long term, loss of customer confidence and damage to competitive position. However, various protective measures can be taken to safeguard information in transit as well as information stored at various points on the network, including servers and desktop computers.

Physical Security

Protecting data in distributed environments starts with securing the premises. Such precautions as locking office doors and wiring closets, restricting access to the data center, and having employees register when they enter sensitive areas can greatly reduce risk. Issuing badges to visitors, installing electronic locks on doors, providing visitor escorts, and having a security guard station in the lobby can reduce risk even further.

Other measures such as keyboard and disk drive locks are also effective in deterring unauthorized access to unattended workstations. These are important security features, especially since some workstations may provide management access to wiring hubs, LAN servers, bridge/routers, and other network access points. In addition, locking down workstations to desks can help protect against equipment theft. Some companies even collect unattended laptops in the office after business hours when building maintenance crews start their shift.

Access Controls

Access controls can prevent unauthorized local access to the network and control remote access through dial-up ports. Network administrators can assign multiple levels of access to different users based on need: public, private, and shared access. Public access allows all users to have read-only access to file information. Private access gives specific users read and write file access, while shared access allows all users to read and write to files.

When a company offers network access to one or more databases, it should restrict and control all user query operations. Each database should have a protective "key," or series of steps, known only to those individuals entitled to access the data. To ensure that intruders cannot duplicate the data from the system files, users should first have to sign on with passwords and then prove that they are entitled to access the data by responding to a challenge with a predefined response known only to that person. This is the basis of a security procedure known as "personal authentication."

Log-on Security

Network operating systems or add-on software can offer effective log-on security that requires that the user enter a log-on ID and password to access local or remote systems. Passwords not only can identify the user, they also can associate the user with a specific workstation, as well as a designated shift, workgroup, or department. The effectiveness of this measure hinges on users' ability to maintain password confidentiality.

A user ID should be suspended after a certain number of passwords have been entered to thwart trial-and-error attempts at access. Changing passwords frequently—especially when key personnel leave the company—and using a multilevel password-protection scheme can enhance security. With multilevel passwords, users can gain access to a designated security level, as well as all lower levels. With specific passwords, on the other hand, users can access only the intended level and not the others above or below. Finally, users should not be allowed to make up their own passwords; they should be assigned using a random password generator or compared to a dictionary to weed out guessable passwords. Although such schemes entail an increased administrative burden, the effort is usually worthwhile.

The effectiveness of passwords can be enhanced by using them in combination with other control measures, such as keyboard lock or card reader. Biometric devices also may be used that identify an authorized user based on such characteristics as a handprint, voice pattern, or the layout of capillary blood vessels in the retina of the eye. Of course, the choice of control measure will depend on the level of security desired and budgetary considerations.

Data Encryption

To protect data (and voice) as they traverse the network requires that they be scrambled with an encryption algorithm. One of the most effective encryption algorithms is that offered by PGP (Pretty Good Privacy), a method that uses a public key to protect computer and e-mail data. The program generates two keys that belong uniquely to the user. One PGP key is secret and stays in the user's computer. The other key is public and is given out to people the user wants to communicate with. The public key can be distributed as part of the message.

PGP does more than encrypt. It has the ability to produce digital signatures, allowing the user to "sign" and authenticate messages. A "digital signature" is a unique mathematical function derived from the message being sent. A message is signed by applying the secret key to it before it is sent. By checking the digital signature of a message, the recipient can make sure that the message has not been altered during transmission. The digital signature also can prove that a particular person originated the message. The signature is so reliable that not even the originator can deny creating it.

Firewalls

A firewall is a method of protecting one network from another untrusted network. The actual mechanism whereby this is accomplished varies widely, but in principle, the firewall can be thought of as a pair of mechanisms: one that blocks traffic and another that permits traffic. Some firewalls place a greater emphasis on blocking traffic, while others emphasize permitting traffic.

One way firewalls protect networks is through packet filtering, which can be used to restrict access from or to certain machines or sites. It also can be used to limit access based on time or day or day of week and by the number of simultaneous sessions allowed, service host(s), destination host(s), or service type. In addition to dedicated firewall systems, this kind of functionality can be set up on various network routers, communications servers, or front-end processors.

Transparent proxies are also used to provide secure outbound communication to the Internet from the corporation's internal network. The firewall software achieves this by appearing to be the default router that provides access to the internal network. However, when packets hit the firewall, the software does not route the packets but immediately starts a dynamic, transparent proxy. The proxy connects to a special intermediate host, which actually connects to the desired service.

Proxies are often used instead of router-based traffic controls to prevent traffic from passing directly between trusted and untrusted networks. Many proxies contain extra logging or support for user authentication. Since proxies must "understand" the application protocol being used, they also can implement protocol-specific security (e.g., an FTP proxy might be configurable to permit incoming traffic and block outgoing traffic).

Remote Access Security

With an increasingly decentralized and mobile workforce, organizations are coming to rely on remote access arrangements that enable telecommuters, traveling executives, salespeople, and branch offices to dial into the corporate network with an 800 number or a set of regional or local access numbers. Appropriate security measures can prevent unauthorized access to corporate resources from the remote access server. One or more of the following security methods can be employed:

- *Authentication* This involves verifying the remote caller by user ID and password, thus controlling access to the server. Security is enhanced if the ID and password is encrypted before going out over the communications link.
- *Access restrictions* This involves assigning each remote user a specific location (i.e., directory or drive) that can be accessed in the server. Access to specific servers also can be controlled.
- *Time restrictions* This involves assigning each remote user a specific amount of connection time, after which the connection is dropped.
- *Connection attempts* This involves limiting the number of consecutive connection attempts and/or the number of times connections can be established on an hourly or daily basis.

Among the most popular remote access security schemes are Remote Access Dial-n User Service (RADIUS) and Terminal Access Controller Access Control System+ (TACACS+). Of the two, RADIUS is the more popular. Users are authenticated through a series of communications between the client and the server. When the client initiates a connection, the communications server puts the name and password into a data packet called the "authentication request," which also includes information identifying the specific server sending the authentication request and the port that is being used for the connection. For added protection, the communications server, acting as a RADIUS client, encrypts the password before passing it on to the authentication server.

When an authentication request is received, the authentication server validates the request and decrypts the data packet to access the user name and password information. If the user name and password are correct, the authentication server sends back an authentication acknowledgment that includes information on the user's network system and service requirements. The acknowledgment can even contain filtering information to limit the user's access to specific network resources.

The older security system is TACACS, which has been updated by Cisco into a version called TACACS+. Although the protocols are different, the proprietary TACACS+ offers many of the same features as RADIUS but is used mainly on networks consisting of Cisco remote access servers and related products. Companies with mixed-vendor environments tend to prefer the more open RADIUS.

Callback Systems

Callback security systems are useful in remote access environments. When a user dials into the corporate network, the answering modem requests the caller's identification, disconnects the call, verifies the caller's identification against a directory, and then calls back the authorized modem at the number matching the caller's identification. This scheme is an effective way to ensure that data communication occurs only between authorized devices, more so when used in combination with data encryption.

Security procedures can even be implemented before the modem handshaking sequence rather than after it, as is usually the case. This effectively eliminates the access opportunity from potential intruders. This method uses a precision high-speed analog security sequence that is not detectable even by advanced line-monitoring equipment.

While these callback techniques work well for branch offices, most callback products are not appropriate for mobile users whose locations vary on a daily basis. Newer products accept roving callback numbers. This feature allows mobile users to call into a remote access server or host computer, type in their user ID and password, and then specify a number where the server or host should call them back. The callback number is then logged and may be used to help track down security breaches.

To safeguard very sensitive information, there are third-party authentication systems that can be added to the server. These systems require a user password and also a special credit card–sized device that generates a new ID every 60 seconds, which must be matched by a similar ID number-generation process on the remote user's computer.

Link-Level Security

When peers at each end of a serial link support the Point-to-Point Protocol (PPP) suite, link-level security features can be implemented. This is so because PPP can integrally support the Password Authentication Protocol (PAP) and Challenge Handshake Authentication Protocol (CHAP) to enforce link security. PPP is a versatile WAN connection standard that can be used for tying dispersed branch offices to the central backbone via dial-up serial links. It is actually an enhanced version of the older Serial Line Internet Protocol (SLIP). SLIP is limited to the IP-only environment, while PPP is used in multiprotocol environments. Since PPP is protocol-insensitive, it can be used to access AppleTalk, IPX and TCP/IP networks, for example.

PAP uses a two-way handshake for the peer to establish its identity. This handshake occurs only on initial link establishment. An ID-password pair is repeatedly sent by the peer to the authenticator until verification is acknowledged or the connection is terminated. However, passwords are sent over the circuit in text format, which offers no protection from interception and playback by network intruders.

CHAP periodically verifies the identity of the peer using a three-way handshake. This technique is employed throughout the life of the connection. With CHAP, the server sends a random token to the remote workstation. The token is encrypted with the user's password and sent back to the server. Then the server does a lookup to see if it recognizes the password. If the values match, the authentication is acknowledged; otherwise, the connection is terminated. Every time remote users dial in, they are given a different token. This provides protection against playback because the challenge value changes in every token.

Some vendors of remote-node products support both PAP and CHAP, while low-end products tend to support only PAP, which is the less robust of the two authentication protocols.

Policy-Based Security

With today's LAN administration tools, security goes far beyond mere password protection to include implementation of a policy-based approach characteristic of most mainframe systems. Under the policy-based approach to security, files are protected by their description in a relational database. This means that newly created files are automatically protected, not at the discretion of each creator, but consistent with the defined security needs of the organization.

Some products use a graphical calendar through which various assets can be made available to select users only during specific hours of specific days. For each asset or group of assets, a different permission type may be applied: permit, deny, or log. Permit allows a user or user group to have access to a specified asset. Deny allows an exception to be made to a permit, not allowing writes to certain files, for example. Log allows an asset to be accessed but stipulates that such access will be logged.

Although the LAN administrator usually has access to a full suite of password controls and tracking features, today's advanced administration tools also provide the ability to determine whether or not a single login ID can have multiple terminal sessions on the same system. Through the console, the LAN manager can review real-time and historical violation activity on line, along with other system activity.

Summary

To protect valuable information, companies must establish a sound security policy before an intruder has an opportunity to violate the network and do serious damage. This means identifying security risks, implementing effective security measures, and educating users on the importance of following established security procedures. Despite advancements in security hardware and software, there are some threats no system can protect against, such as insider attacks or people taking sensitive information out of the building with floppy disks. According to some industry reports, 80 percent of attacks on corporate networks originate with employees.

See also

Firewalls

Proxy Servers

NETWORK STATISTICS

Network Statistics (Netstat) is a utility that displays useful performance information about network connections and activity. It can be run from the command line of any machine with a network operating system that supports TCP/IP (Figure N-7). Netstat includes options that allow the user to

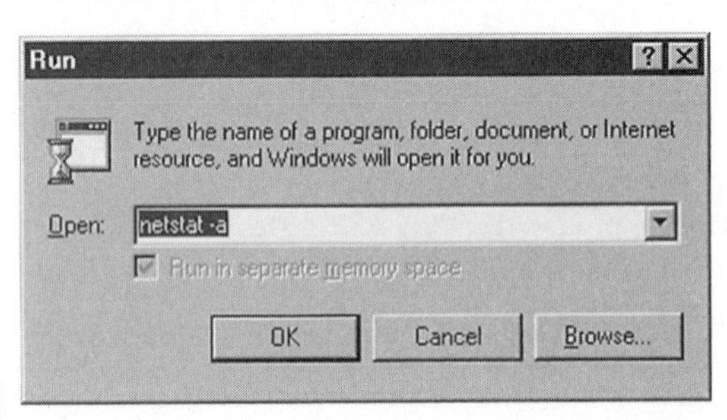

Figure N-7 The Netstat utility with the -a option set, as run from the command line in Windows NT.

Table N-1 Selected Options for the Netstat Utility, Which Can Be Run from the Command Line of Windows NT

Option	Description
-a	Listens for all active ports and displays connection information.
-e	Displays Ethernet statistics. This may be combined with the -s option.
-n	Displays addresses and port numbers in numerical form (rather than attempting name lookups).
-p	Shows connections for the specified protocol specified by Proto; Proto may be TCP or UDP. If used with the -s option to display per-protocol statistics, Proto may be TCP, UDP, ICMP, or IP.
-r	Displays the contents of the routing table.
-s	Displays per-protocol statistics. By default, statistics are shown for TCP, UDP, ICMP, and IP; the -p option may be used to specify a subset of the default.
Interval	Redisplays selected statistics, pausing interval seconds between each display. Press CTRL+C to stop redisplaying statistics. If this parameter is omitted, netstat prints the current configuration information once.

specify the type of information to have displayed. Selected options are described in Table N-1.

To launch Netstat from the command line, the user simply types in the word *netstat* or *netstat* followed by one or more options as in the following examples:

netstat–a   or   netstat–ps

When netstat–a is run, for example, the following type of information is returned:

Proto	Local Address	Foreign Address	State
TCP	ws1-williamsg1:135	0.0.0.0:0	Listening
TCP	ws1-williamsg1:1029	0.0.0.0:0	Listening
TCP	ws1-williamsg1:1034	0.0.0.0:0	Listening
TCP	ws1-williamsg1:1035	0.0.0.0:0	Listening
TCP	ws1-williamsg1:3684	PNTEXCH01.hq.abc.net:1396	Established
TCP	ws1-williamsg1:3689	PNTEXCH01.hq.abc.net:1450	Established
TCP	ws1-williamsg1:3700	PNTEXCH01.hq.abc.net:1396	Established

The column labeled "Proto" indicates the type of protocol used on the connection. The column labeled "Local Address" refers to the local host and its sockets. The column labeled "Foreign Address" indicates the server connections, if any. The column labeled "State" refers to the status of the port as either listening for a connection or having established a connection.

Summary

Netstat is a useful tool for checking the connections and activity on a network. By setting the appropriate options (more are available than listed above), administrators and technicians can monitor the status of network connections, inspect interface configuration information, examine the routing table, and retrieve operational statistics for various network protocols.

See also

Ping

NETWORK SUPPORT

Today's networks have increased in functionality and complexity, pushing support issues into the forefront of management concerns. Whether problems are revealed through network management tools—alarms, diagnostics, and predictive methods—or through user notification, the need for timely and qualified network support services is of critical importance. Recognizing these concerns, local and long-distance carriers now offer network support options in conjunction with their services and facilities.

Types of Services

The support concept encompasses dozens of individual activities from which the business customer may select. Generally, these activities include but are not limited to

- Site engineering, utilities installation, cabling, and rewiring.
- Performance monitoring of the system or network, alarm interpretation, and initiation of diagnostic activities.
- Identification and isolation of system faults and degraded facilities on the network.
- Notification of the appropriate hardware vendor or carrier for restoral action following an outage.
- Testing of the restoral action to verify proper operation of the system or network before it is put back into service.

- The repair or replacement of the faulty system or component.
- Monitoring of the repair/replacement process and the escalation of problems.
- Trouble ticket and work order administration, inventory tracking, maintenance histories, and cost control.
- Administration of equipment moves, adds, and changes.
- Network design, tuning, and optimization.
- Systems documentation and training.
- Preventive maintenance.
- Management reports.

A number of other types of support are also available, such as 24-hour telephone ("hotline") assistance, short-term equipment rental, fast equipment exchange, and guaranteed response time to trouble calls. In addition, the carrier or vendor may offer customized cooperative maintenance plans that qualify the organization for premium reductions if an internal help desk is established to weed out routine problems, many of which are applications-related and beyond the support purview of the carrier or vendor. An increasingly popular support offering is remote diagnostics and network management from the vendor's or carrier's network operations center. Some carriers also offer remote management of firewalls from their network operations center as well.

Levels of Support

Carriers also offer multiple levels of technical support. The most basic form of technical support is toll-free telephone access to technical specialists during normal business hours. This type of service assists customers in resolving hardware or software problems. Typically, there is no charge for this service, and calls are handled on a first-come, first-served basis. There is usually no expiration date for this service—it is available to customers for as long as they use the carrier's services or facilities.

Extended or priority technical assistance is provided via phone 24 hours a day, 7 days a week to assist customers in resolving hardware or software problems. As an extra-cost service, it ensures that customers are called back within 30 minutes during normal business hours and within 1 hour after normal business hours.

Some carriers offer subscription services that provide the most up-to-date technical product information on maintaining network efficiency and reliability. Written by the carrier's own engineers and field service personnel, this kind of service usually emphasizes how to more effectively operate and manage various data communications and network access products. This information can come in a variety of forms, including technical bulletins, product application notes, software release notes, user guides, and field bulletins—in print or on CD-ROM. Increasingly, the Web is being used to distribute such information. Since access is limited to customers, a valid user ID and password usually are required.

Remote dial-in software support addresses the needs of customers operating mission-critical networks. Technical specialists remotely dial-in to the customer's network to resolve software problems via diagnostic testing or by modifying a copy of the system configuration and then downloading the revised configuration file directly to the affected equipment.

Carriers also can assume single-point responsibility for remote network management, providing customers with a proactive approach to service delivery. Technical staff at a central control facility continuously monitor network performance and immediately respond to and resolve any fault resulting from hardware, software, or circuits. From the control facility, network faults are identified and alarm conditions resolved through continuous end-to-end diagnostics. Once a problem is recognized, the latest diagnostic equipment and isolation techniques are used to identify the source of the problem and provide effective resolution. Often problems are identified and corrected before they become apparent to network users.

If the problem originates with the carrier, it assumes ownership until it is resolved. If the problem originates from a local telephone company or competitive access provider, the long-distance carrier reports the problem, makes appropriate status inquiries, and if necessary, escalates the problem within the other company's organization.

Summary

The local and long-distance carriers are competing with equipment vendors and third-party service firms in the provision of network support services, providing customers with a broad range of plans to choose from that encompass just about every aspect of problem identification, diagnostics, and resolution. New support services include contingency planning to assess the risks faced by a business and determine ways that mission-critical processes can be better protected and vulnerability assessment to determine how exposed a company's network is to break-in attempts by hackers.

See also

Help Desks

Network Design Tools

Network Integration

O

OPEN SYSTEMS INTERCONNECTION

The seven-layer Open Systems Interconnection (OSI) reference model was first defined in 1978 and described in International Standards Organization (ISO) Standard 7498. The lower layers (1 to 3) represent local communications, while the upper layers (4 to 7) represent end-to-end communications (Figure O-1). Each layer contributes protocol functions that are necessary to establish and maintain the error-free exchange of information between network users.

The model provides a useful framework for visualizing the communications process and comparing products in terms of standards conformance and interoperability potential. This layered structure not only aids users in visualizing the communications process but also provides vendors with the means for segmenting and allocating various communications requirements within a workable format. This can reduce much of the confusion normally associated with the complex task of supporting successful communications.

Layers

Each layer of the OSI model exchanges information with a comparable layer at the other side of the connection, a

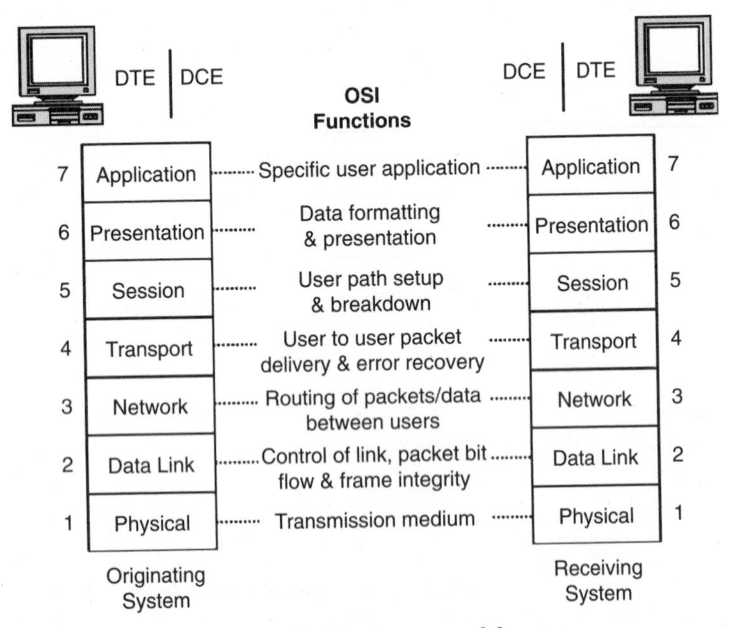

Figure O-1 The seven-layer OSI reference model.

process known as "peer-protocol communications." The functionality associated with each layer is as follows:

Application Layer The highest layer in the OSI reference model is the Application Layer, which serves as the window for users and application processes to access network services. This level applies to the actual meaning rather than the format or syntax (as in Layer 6) of applications and permits communication between users. According to the model, each type of application must employ its own Layer 7 protocol, and with the wide variety of available application types, Layer 7 offers definitions for each, including

- Resource sharing and device redirection
- Remote file access
- Remote printer access

- Interprocess communication
- Network management
- Directory services
- Electronic messaging (such as mail)
- Network virtual terminals

Presentation Layer Layer 6 deals with the format and representation of data that applications use; specifically, it controls the formats of screens and files. Layer 6 defines such things as syntax, control codes, special graphics, and character sets. Additionally, this level determines how variable alphabetic strings will be transmitted, how binary numbers will be presented, and how data will be formatted.

The Presentation Layer formats the data to be presented to the Application Layer. It can be viewed as the translator for the network. This layer may translate data from a format used by the Application Layer into a common format at the sending station and then translate the common format to a format known to the Application Layer at the receiving station. Specifically, the Presentation Layer provides

- *Character code translation* For example, ASCII to EBCDIC.
- *Data conversion* Bit order, carriage return (CR) or carriage return/line feed (CR/LF), integer-floating point, and other functions.
- *Data compression* Reduces the number of bits that need to be transmitted on the network.
- *Data encryption* Encrypts data for security purposes.

Session Layer The Session Layer manages communications; for example, it sets up, maintains, and terminates virtual circuits between sending and receiving devices. It sets boundaries for the start and end of messages and establishes how messages will be sent: half duplex, with each computer taking turns sending and receiving, or full duplex, with each computer sending and receiving at the same time. These details are negotiated during session initiation.

The Session Layer allows session establishment between processes running on different stations. Specifically, the Session Layer provides

- *Session establishment, maintenance, and termination* Allows two application processes on different machines to establish, use, and terminate a connection, called a "session."
- *Session support* Performs the functions that allow these processes to communicate over the network, performing security, name recognition, logging, and other functions.

Transport Layer Layer 4 handles end-to-end transport. If there is a need for reliable, end-to-end sequenced delivery, then the Transport Layer performs this function. For example, each packet of a message might have followed a different route through the network toward its destination. The Transport Layer reestablishes packet order through a process called "sequencing" so that the entire message is received exactly the way it was sent. At this layer, lost data are recovered, and flow control is implemented. With flow control, the rate of data transfer is adjusted to prevent excessive amounts of data from overloading network buffers.

Layer 4 also may support datagram transfers—that is, transactions that need not be sequenced. This is required for voice and video, which may tolerate loss of information but needs to have low delay and low variance in transmittal time. This flexibility is the result of the protocols implemented in this layer, ranging from the five OSI protocols—TP0 to TP4—to the Transmission Control Protocol (TCP) and the User Datagram Protocol (UDP) in the TCP/IP suite and many others in proprietary suites. Some of these protocols do not perform retransmission, sequencing, checksums, and flow control. Specifically, the Transport Layer provides

- *Message segmentation* Accepts a message from the (session) layer above it, splits the message into smaller units (if not already small enough), and passes the smaller units down to the network layer. The Transport Layer at the destination station reassembles the message.
- *Message acknowledgment* Provides reliable end-to-end message delivery with acknowledgments.
- *Message traffic control* Tells the transmitting station to "back off" when no message buffers are available.
- *Session multiplexing* Interleaves several message streams, or sessions, onto one logical link and keeps track of which messages belong to which sessions (see session layer).

Network Layer Layer 3 formats the data into packets, adds a header containing the packet sequence and the address of the receiving device, and specifies the services required from the network. The network does the routing to match the service requirement. Sometimes a copy of each packet is saved at the sending node until it receives confirmation that it has arrived at the next node undamaged, as is done in X.25 packet-switched networks. When a node receives the packet, it searches a routing table to determine the best path for that packet's destination without regard for its order in the message. In a network where not all nodes can communicate directly, this layer takes care of routing packets through the intervening nodes. Intervening nodes may reroute the message to avoid congestion or node failures. Specifically, the Network Layer provides

- *Routing* Routes frames among networks.
- *Subnet traffic control* Routers (Network Layer intermediate systems) can instruct a sending station to scale back its frame transmission when the router's buffer fills up.
- *Frame fragmentation* If it determines that a downstream router's maximum transmission unit (MTU) size is less than the frame size, a router can fragment a frame for transmission and reassembly at the destination station.

- *Logical-physical address mapping* Translates logical addresses, or names, into physical addresses.
- *Subnet usage accounting* Has accounting functions to keep track of frames forwarded by subnet intermediate systems to produce billing information.

Data Link Layer All modern communications protocols use the services defined in Layer 2. The Data Link Layer provides the lowest level of error control. It detects errors and requests the sending node to retransmit the data. This layer has assumed a greater role as communications lines have become less noisy through the replacement of analog lines with digital lines, while end stations have become more intelligent through the use of more powerful processors and high-capacity memory. Combined, these factors have lessened the need for high-level information protection mechanisms in the network, moving them to the end systems. Layer 2 does not know what the information or packets it encapsulates mean or where they are headed. Networks that can tolerate this lack of information are rewarded by low transmission delays. Specifically, the functions provided by the Data Link Layer include

- *Link establishment and termination* Establishes and terminates the logical link between two nodes.
- *Frame traffic control* Tells the transmitting node to back off when no frame buffers are available.
- *Frame sequencing* Transmits/receives frames sequentially.
- *Frame acknowledgment* Provides/expects frame acknowledgments. Detects and recovers from errors that occur in the physical layer by retransmitting unacknowledged frames and handling duplicate frame receipt.
- *Frame delimiting* Creates and recognizes frame boundaries.
- *Frame error checking* Checks received frames for integrity.

- *Media access management* Determines when the node has the right to use the physical medium.

Physical Layer The lowest OSI layer is the Physical Layer. This layer represents the actual interface, electrical and mechanical, that connects a device to a transmission medium. Because the physical interface has become so standardized, it is usually taken for granted in discussions of OSI connections. Yet physical connections—cables and connectors—with their pin-outs and transmission characteristics can still be a problem in designing a reliable network if they do not conform to a common model. Specifically, the functions provided at the Physical Layer include

- *Data encoding* Modifies the simple digital signal pattern (1s and 0s) used by the PC to better accommodate the characteristics of the physical medium and to aid in bit and frame synchronization.
- *Physical medium attachment* Accommodates various possibilities in the medium, such as the number of pins a connector has and what is each pin used for.
- *Transmission technique* Determines whether the encoded bits are to be transmitted by digital or analog signaling.
- *Physical medium transmission* Transmits bits as electrical or optical signals appropriate for the physical medium and determines how many volts/decibels should be used to represent a given signal state, using a given physical medium.

Conformance versus Interoperability

About a dozen laboratories are accredited by the National Institute of Standards and Technology (NIST) to run a suite of tests that certifies vendors' products for conformance to the OSI reference model. However, while the products of different vendors may conform to the OSI model, this does not necessarily mean that they are interoperable.

Conformance testing is the process of comparing a vendor's protocol implementation against a model of the protocol. Conformance test results are sent to NIST for approval. Approved results for each product are then entered into NIST's registry of OSI-conformant products.

However, by itself, conformance does not guarantee that the product of one vendor will work with the product of another vendor, even though both products have passed the same conformance test. OSI product conformance testing only increases the probability of successful interoperability in a customer's multivendor OSI network. To ensure that the products of both vendors do indeed work together on the network, they must be specifically tested for interoperability at the highest level of OSI—the Application Layer. This involves running both vendors' protocol implementations of FTAM or X.400, for example, to see if they work properly across their respective products.

Summary

Throughout the 1980s, the prediction often was made that OSI would replace TCP/IP as the preferred technique for interconnecting multivendor networks. It is now clear that this will not happen in the United States. There are several reasons for this, including the slow pace of OSI standards progress in the 1980s, as well as the expense of implementing complex OSI software and having products certified for OSI interoperability. Furthermore, TCP/IP was already widely available, and plug-in protocols continue to be developed to add functionality. The situation is different in Europe, where OSI compliance was mandated early on by the regulatory authorities in many countries.

See also

Simple Network Management Protocol

Transmission Control Protocol/Internet Protocol (TCP/IP)

P

PEER-TO-PEER NETWORKS

In a peer-to-peer network, computers are linked together for resource sharing. If there are only two computers to link together, networking can be done with a Category 5 crossover cable that plugs into the RJ45 jack of the network interface card (NIC) on each computer. If the computers do not have NICs, they can be connected with either a serial or parallel cable. Once connected, the two computers function as if they were on a local area network (LAN), and each computer can access the resources of the other. If three or more computers must be connected, a wiring hub is required.

Regardless of exactly how the computers are interconnected, each is an equal or "peer" and can share the files and peripherals of the others. For a small business doing routine word processing, spreadsheets, and accounting, this type of network is the low-cost solution to sharing resources such as files, applications, and peripherals. Multiple computers can even share an external cable or Digital Subscriber Line (DSL) modem, allowing them to access the Internet at the same time.

Networking with Windows

Windows 95/98 and Windows NT/2000 are often used for peer-to-peer networking. In addition to peer-to-peer network access, both provide network administration features and memory management facilities, support the same networking protocols—including the Transmission Control Protocol/Internet Protocol (TCP/IP) for accessing intranets, virtual private networks (VPNs), and the public Internet—and provide such options as dial-up networking and fax routing.

One difference between the two operating systems is that in Windows 95/98 the networking configuration must be established manually, whereas in Windows NT/2000 the networking configuration is part of the initial program installation, on the assumption that NT/2000 will be used in a network. Although Windows 95/98 is good for peer-to-peer networking, Windows NT is more suited for larger client/server networks.

Windows supports Ethernet, Token Ring, Asynchronous Transfer Mode (ATM), and Fiber Distributed Data Interface (FDDI) data-frame types. Ethernet is typically the least expensive network to implement. The NICs can cost as little as $20 each, and a five-port hub can cost as little as $40. Category 5 cabling usually costs less than 50 cents per foot in 100-foot lengths with the RJ45 connectors already attached at each end. Snap-together wall-plate kits cost about $6 each.

Configuration Details

When setting up a peer-to-peer network with Windows, each computer must be configured individually. After installing an NIC and booting the computer, Windows will recognize the new hardware and automatically install the appropriate network-card drivers. If the drivers are not already available on the system, Windows will prompt the user to insert the manufacturer's disk containing the drivers, and they will be installed automatically (Figure P-1).

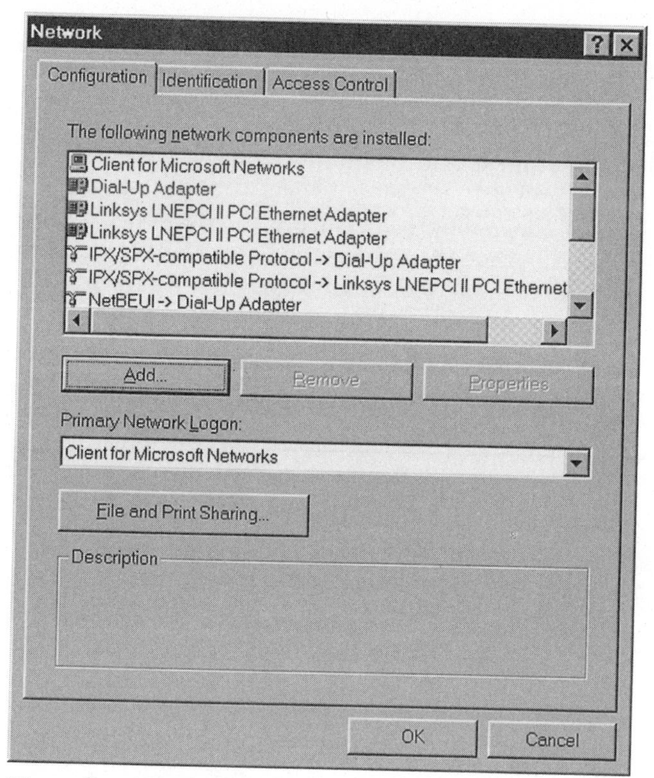

Figure P-1 To verify that the right drivers have been installed, the user opens the Network Control Panel to check the list of installed components. In this case, a Linksys LNEPCI II Ethernet adapter has been installed.

Next, the user must select the client type. If a Microsoft peer-to-peer network is being created, the user must add "Client for Microsoft Networks" as the primary network logon (Figure P-1). Since the main advantage of networking computers is resource sharing, it is important to enable the sharing of both printers and files. The user does this by clicking on the "File and Print Sharing" button and choosing one or both of these capabilities (Figure P-1). Through file and printer sharing, each workstation becomes a potential server.

Identification and security are the next steps in the configuration process. From the "Identification" tab of the dialog box, the user must select a unique name for the computer and the workgroup to which it belongs, as well as a brief description of the computer (Figure P-2). When others use Network Neighborhood to browse the network, they will see the menu trees of all active computers on the network.

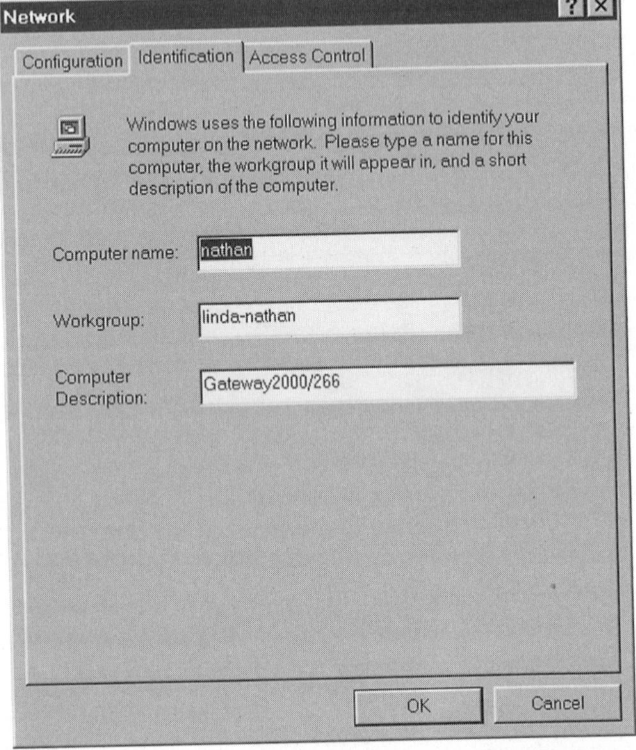

Figure P-2 A unique name for the computer and the workgroup to which it belongs and a brief description of the computer identify it to other users when they access Network Neighborhood to browse the network.

From the “Access Control” tab of the dialog box, the user selects the security type. For a small peer-to-peer network, share-level access is adequate (Figure P-3). This allows printers, hard drives, directories, and other resources to be shared and enables the user to establish password access for each of these resources. In addition, read-only access allows users only to view (not modify) a file or directory.

To allow a printer to be shared, for example, the user right-clicks on the printer icon in the Control Panel and selects “Sharing” from the drop-down list (Figure P-4). Next,

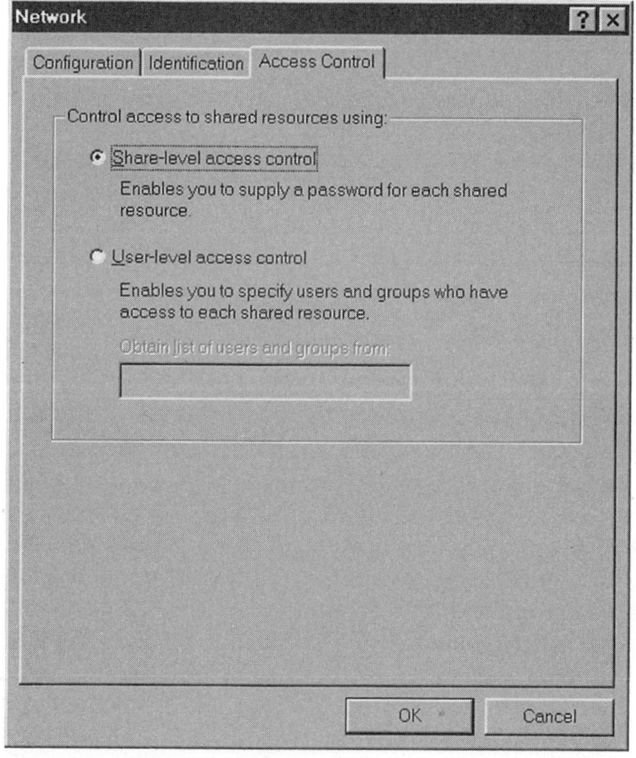

Figure P-3 Choosing share-level access allows the user to password-protect each shared resource.

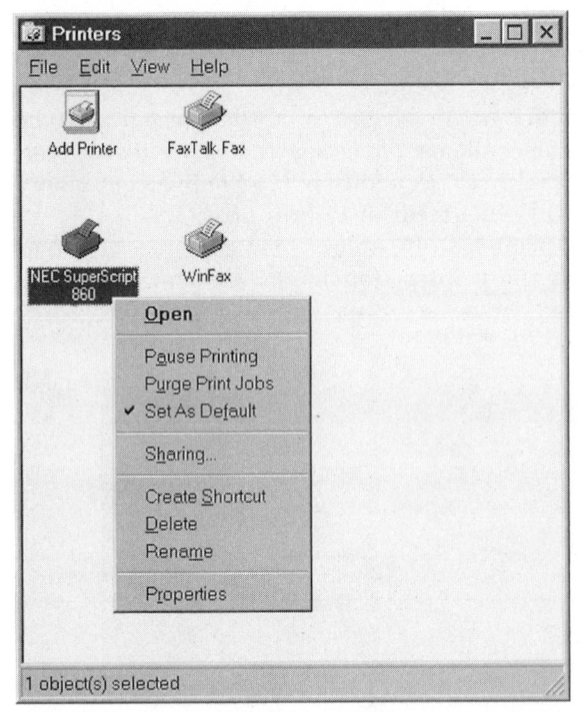

Figure P-4 A printer can be configured as the default printer and for sharing by other computers on the network.

the user clicks on the "Shared As" radio button and enters a unique name for the printer (Figure P-5). If desired, this resource can be given a password as well. When another computer tries to access the printer, the user will be prompted to enter a password. If a password is not necessary, the password field is left blank.

Another security option in the "Access Control" tab is user-level access, which is used to limit resource access by user name. This function eliminates the need to remember passwords for each shared resource. Each user simply logs onto the network with a unique name and password; the network administrator governs who can do what on the net-

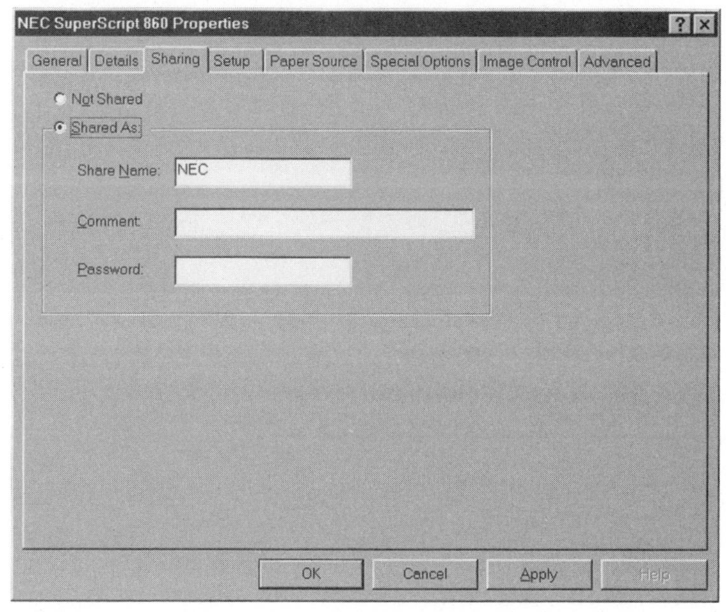

Figure P-5 A printer can be password-protected if necessary.

work. However, this requires the computers to be part of a larger network with a central server—perhaps running Windows NT/2000 server—which maintains the access-control list for the whole network. Since Windows 95/98 and Windows NT/2000 workstations support the same protocols, Windows 95/98 computers can participate in a Windows NT/2000 server domain.

Peer services can be combined with standard client/server networking. For example, if a Windows 95/98 computer is a member of a Windows NT/2000 network and has a color printer to share, the resource "owner" can share that printer with other computers on the network. The server's access-control list determines who is eligible to share resources.

Once the networking infrastructure is in place, the NIC of each computer is individually connected to a hub with Category 5 cable. This cable has connectors on each end that insert into the RJ45 jacks of the hub and NICs. For small networks, the hub usually will not be manageable with the Simple Network Management Protocol (SNMP), so no additional software is installed. Once the computers are configured properly and connected to the hub, the network is operational.

Summary

Peer-to-peer networking is an inexpensive way for small companies and households to share resources among a small group of computers. This type of network provides most of the same functions as the traditional client/server network, including the ability to run network versions of popular software packages. Peer-to-peer networks also are easy to install. Under ideal conditions, installation of the cards, software, hub, and cabling for five users would take only a few hours.

See also

Client/Server Networks

Hubs

PERFORMANCE BASELINING

Performance baselining is a procedure for understanding the behavior of a properly functioning network so that deviations can help identify the cause of problems that may occur in the future. The only way to know a network's normal behavior is to analyze it while it is operating properly. Later, technicians and network managers can compare data from the properly functioning network with data gathered after conditions have begun to deteriorate. This comparison often points to the right steps that lead to a corrective solution.

Information Requirements

The first step in baselining performance is to gather appropriate information from a properly functioning network. Much of this information may already exist, and it is just a matter of finding it.

Topologic Map For example, many enterprise management systems have the capability of automatically discovering devices on the network and creating a topologic map. This kind of information is necessary for knowing what components exist on the network and how they interact physically and logically. For wide area networks (WANs), this means locations, descriptions, and cable plant maps for equipment such as routers, bridges, and network access devices (LAN to WAN).

In addition, information on transmission media, physical interfaces (T1/E1, V series, etc.), line speeds, encoding methods and framing, and access points to service provider equipment should be assembled. Although it is not always practical to map individual workstations in the LAN portions of the WAN or to know exactly what routing occurs in the WAN cloud, knowing the general topology of the WAN can be useful in tracking down problems later.

WAN and LAN Protocols To fully understand how a network behaves, it is necessary to know what protocols are in use. Later, during the troubleshooting process, the presence of unexpected protocols may provide clues as to why network devices appear to be malfunctioning or why data transfer errors or failures are occurring.

Logs Some network problems begin to occur after new devices or applications are installed. The addition of new devices, for example, can cause network problems that have a ripple effect throughout the network. A new end-user device with a duplicate IP address, for instance, could make it impossible for other network elements to communicate. Or a badly configured router added to the network could produce congestion and connection problems. Other problems occur when new data communications are enabled or existing topologies and configurations are changed. A log of these activities can help pinpoint causes of network difficulty. In addition, previous network trouble—and its resolution—is sometimes recorded, and this too can lead to faster problem identification and resolution.

Statistics Often, previously gathered data can provide valuable context for newly created baselines. Previously assembled baselines also may contain event and error statistics and examples of decoded traffic based on network location or time of day. These logs may have been gathered over long periods, yielding valuable information about the history of network performance.

Usage Patterns A profile of users and their typical usage patterns also can speed fault isolation. This entails having several types of information, including what kind of LAN traffic is carried over the WAN.

LAN Traffic on the WAN With knowledge of what kind of LAN traffic to expect on the WAN, technicians and network managers will have a better idea of the analysis that might have to be performed later. In addition, knowing how LAN frames might be handled at end stations can help troubleshooters make a distinction between WAN problems and end-station processing problems.

Traffic Content Knowing the WAN traffic type (voice, data, video, etc.) can help troubleshooters estimate when network traffic is most likely to be heavy, what level of transmit errors can be tolerated, and whether it makes sense to even use a protocol analyzer. For example, an analyzer may incorrectly report errored frames and corrupt data when attempting to process voice or video traffic based on data communication protocols.

Peak Usage Knowing when large data transfers will occur—such as scheduled file backups between LANs connected across the WAN—can help network managers predict and plan for network slowdowns and failures. It also can help technicians schedule repairs so that WAN performance is minimally affected. Some of this information can be obtained from interviews with network administrators or key users. Other times it must be gathered with network analysis tools.

Hard Stats and Decodes

After gathering information on topology, devices, protocols, and typical users of the WAN, hard statistics and examples of decoded network traffic should be gathered. Getting comprehensive baseline data may entail gathering it at regular intervals at numerous points throughout the network.

Statistics Logs To understand usage trends and normal error levels over time, a statistics log is created. Many protocol analyzers let technicians specify the period over which these kinds of data are logged, the interval between log entries, and the type of statistics to log. The log file can be exported to a spreadsheet or other application program for offline analysis.

Frame or Packet Data To see details about typical WAN traffic, frame or packet data can be collected and saved to a file for later examination. Data collection can be done at specific periods during the day or week to find differences between peak and off-peak usage. Saved network traffic also provides insight into device configurations for use later during routine upgrades or repairs.

Targeted Statistics Using configurable traffic filters and counters, selected blocks of data or statistics based on specific network events can be captured, which might include error count thresholds, specific frame types, and in-channel alarms. A comprehensive collection of such data provides a benchmark for comparison if the network begins to malfunction. New protocols on the network, unexpected line and channel utilization levels, and increases in normal errors and in-channel alarms can be isolated according to physical link location, helping narrow the search for the problem.

Applying the Baseline

If network performance and reliability problems occur, the information gathered during baselining can be used to help identify the nature and source of the problem through comparison analysis and historical trends.

Comparison Analysis Baseline information is compared with current information to see network changes. For example, to isolate failing devices or connections, the number of errors recorded during baselining is compared to the current number of errors that occur over a similar time interval.

Historical Trends Current network problems can result from subtly changing conditions that are detected only after examining a series of baselines gathered over time. For example, congestion problems may become apparent only as new users are added to a particular part of the network. Examining historical trends can help isolate these situations.

Summary

Performance baselining provides a profile of normal network behavior, making it easier for technicians and network managers to identify deviations so that appropriate corrective action can be taken. This "snapshot" of the current network also can be used as the input data for subsequent performance modeling. For example, network administrators and operations managers can use the baseline data to conduct "what if" scenarios to assess the impact of proposed changes. A wide variety of changes can be evaluated, such as adding routers, increasing WAN bandwidth or application workloads, and relocating user sites. During analysis, performance thresholds can be customized to highlight network conditions of interest. These capabilities enable users to plan and quantify the benefits of feature migrations, such as different routing protocols, and to make more accurate and cost-effective decisions regarding the location and timing of upgrades.

See also

Network Design Tools
Protocol Analyzers

PING

Ping is a simple test function that allows the user to check if a local or remote system on an IP network is currently up and running. Ping can be run on the command line on UNIX machines or within a client application in the Windows or Macintosh environment. The Ping command can be run using plain language domain names or IP addresses. The general command line syntax for implementing Ping is

ping abc.com or ping 192.168.100.1

This will indicate whether the host at ABC Company is currently online or, if Ping is launched from an internal workstation, whether its internal proxy server is in service. The Ping command sends one datagram per second and prints (or displays) one line of output for every echo response returned. No output is produced if there is no response. A count option can be used in the command-line syntax to specify the number of requests to be sent. Many implementations of Ping also include an option that measures the roundtrip time of the sent packet in milliseconds (ms) as well as the packet loss between two hosts on the network. When Ctrl-C is pressed on the keyboard, Ping provides a brief statistical summary, as in the following example:

PING abc.com: 56 data bytes

64 bytes from 132.58.68.1: icmp_seq=0 ttl=251 time=66 ms

64 bytes from 132.58.68.1: icmp_seq=1 ttl=251 time=45 ms

64 bytes from 132.58.68.1: icmp_seq=2 ttl=251 time=46 ms

64 bytes from 132.58.68.1: icmp_seq=3 ttl=251 time=55 ms

64 bytes from 132.58.68.1: icmp_seq=4 ttl=251 time=48 ms

—- abc.com ping statistics —-

5 packets transmitted, 5 packets received, 0% packet loss

round — trip min/avg/max = 45/52/66 ms

Pinging once or twice is generally enough to provide a reliable indication of a remote system's current state. Ping also can be used to continuously monitor the state of the connection. For Windows and Macintosh machines, there are feature-rich graphical utilities offered as shareware that implement Ping (Figure P-6). Among other things, they let the user specify

- The ping data packet size
- The number of hosts to ping simultaneously
- The ping interval
- The amount of milliseconds to wait for echo reply
- Time to wait until next ping if the last ping succeeded
- Time to wait until next ping if the last ping failed
- The number of failed pings before the utility considers the host is down

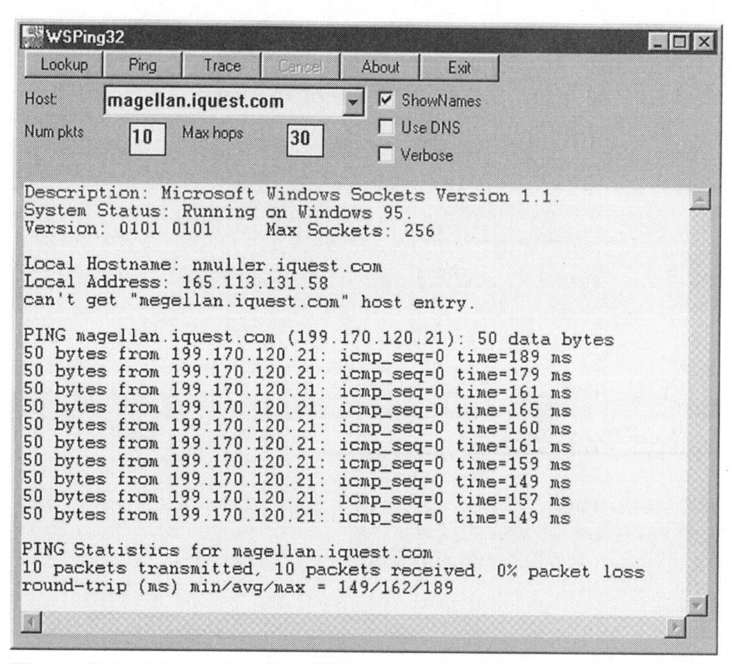

Figure P-6 John A. Junod's ICMP Ping for Windows, a shareware utility.

Monitoring network activities is made even easier by setting the Ping utility to take a specific action whenever the remote host changes its state from up to down and vice versa. Among the actions the Ping utility can implement in response to changes in network activity are the playing of audio messages and the running of custom programs.

For example, the user can specify that an audio file be played to indicate when a remote host crashes or recovers from a crash. The Ping utility also can be configured to run a program as soon as the remote host recovers from a crash.

Summary

Ping is a simple but useful troubleshooting tool for tracking down the source of a failure on an internal network or on the Internet. Basically, when Ping is run, packets are issued to one or more designated hosts and echoed back to provide performance information. Various statistics are displayed that tell the user about the state of the local or remote host and the connection. If the target host is out of service, there is no echo. Instead, the Ping function times out, indicating a problem.

See also

Jitter
Latency
Network Statistics
Quality of Service

PROTOCOL ANALYZERS

A category of test equipment known as the protocol analyzer is used to monitor and diagnose performance problems on LANs and WANs by decoding upper-layer protocols. There are protocol analyzers for all types of communications circuits, including Frame Relay, X.25, T1 and ISDN, and ATM. There are also protocol analyzers that offer full seven-layer decodes of NetBIOS, SNA, SMB, TCP/IP, DEC LAT, XNS/MS-NET, NetWare, and VINES, as well as the various LAN cabling, signaling, and protocol architectures, including those for AppleTalk, ARCnet, Ethernet, StarLAN, and Token Ring. There are even analyzers for wireless services like Bluetooth, which monitor the frames that are transmitted through the air, as well as capture and analyze Bluetooth serial data as they travel between a host and a host controller.

In the case of a LAN, the protocol analyzer connects directly to the cable as if it was just another node or to the test port of data terminal equipment (DTE) or data communication equipment (DCE) where trouble is suspected (Figure P-7).

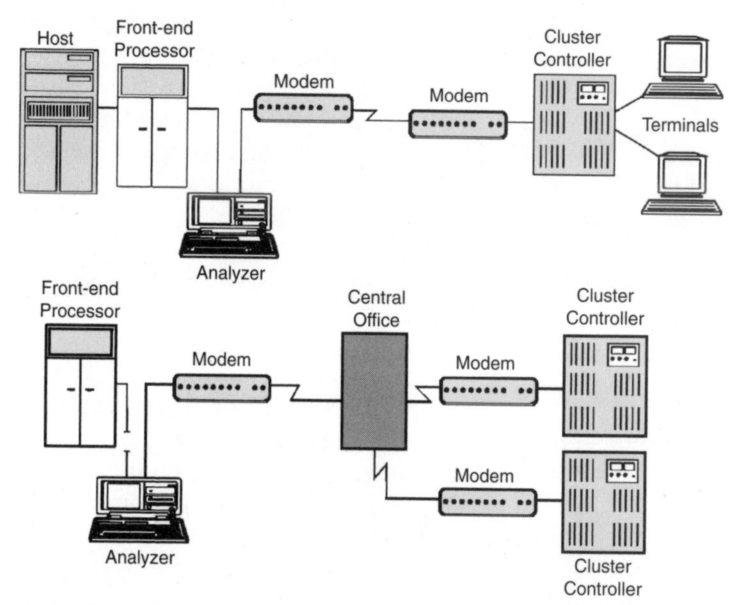

Figure P-7 A protocol analyzer in the monitor mode (*top*) allows the user to check events taking place between front-end processor and local modem over a synchronous link. A protocol analyzer in the simulation mode (*bottom*) allows the user to run a program that exhibits proper operation of the suspect front-end processor over a synchronous multidrop link, which includes two cluster controllers.

Troubleshooting Features

Many protocol analyzers have sophisticated features such as data capture to RAM or disk, automatic configuration, counters, timers, traps, masks, and statistics. These features can dramatically shorten the time it takes to isolate a problem. Some sets are programmable and offer simulation features.

Monitoring and Simulation Protocol analyzers are generally used in either a passive monitoring application or a simulation application. In the monitoring application, the analyzer sits passively on the network and monitors both the integrity of the cabling and the level of data traffic, logging such things as excessive packet collisions and damaged packets that can tie up an Ethernet LAN, for example. The information on troublesome nodes and cabling are compiled for the network manager. In a monitoring application, the protocol analyzer merely displays the protocol activity and user data (packets) that are passed over the cable, providing a window into the message exchange between network nodes.

In the simulation application, the protocol analyzer is programmed to exhibit the behavior of a network node, such as a gateway, communications controller, or front-end processor (FEP). This makes it possible to replace a suspect device on the network with a simulator that is running a program to simulate proper operation. This also enhances the ability to do fault isolation. For example, a dual-port protocol analyzer can monitor a gateway while running a simulation. More sophisticated protocol analyzers can run simulations designed to stress-test individual nodes to verify their conformity to standards. Protocol simulation is most often used to verify the integrity of a new installation.

Trapping The trapping function allows the troubleshooter to command the protocol analyzer to start recording data into its buffer or onto disk when a specific event occurs. For example, the protocol analyzer could be set to trap the first errored frame it receives. This feature permits the capture of only essential information. Some protocol analyzers allow the user to set performance thresholds according to the type of traffic on the network. When these performance thresholds are exceeded, an alarm message is triggered, indicating that there is a problem.

Filtering With the protocol analyzer's filtering capability, the user can exclude certain types of information from capture or analysis. For example, the technician might suspect that errors are being generated at the data link layer, so network-layer packets can be excluded until the problem is located. At the data link layer, the analyzer tracks information such as where the data were generated and whether they contain errors. If no problems are found, the user can set the filter to include only network-layer packets. At this layer, the protocol analyzer tracks information such as where the data are destined and the type of application under which they were generated. If the troubleshooter has no idea where to start looking for problems, then all the packets may be captured and written to disk. Various filters can be applied later for selective viewing.

Bit Error Rate Testing Bit error rate testing is used to determine whether data are being passed reliably over a carrier-provided communications link. This is accomplished by sending and receiving various bit patterns and data characters to compare what is transmitted with what is received. The bit error rate is calculated as a ratio of the total number of bit errors divided by the total number of bits received. Any difference between the two is indicated and displayed as an error. Additional information that may be presented includes sync losses, sync loss seconds, errored seconds, error-free seconds, time unavailable, elapsed time, frame errors, and parity errors.

Packet Generation In being able to generate packets, the protocol analyzer allows the user to test the impact of additional traffic on the network. Using a set of configuration screens, the technician can set the following parameters of the packets:

- The source address the packets will be sent from
- The destination address the packets will be sent to
- The maximum and minimum frame size of the packets
- The spacing between the packets, expressed in microseconds
- The number of packets sent out with each burst

The technician also can customize the contents of the data field section of the packets to simulate real or potential applications. When the packets are generated, the real-time impact of the additional traffic on the network can be observed on the monitor of the protocol analyzer. Packets also may be generated to force a suspected problem to reoccur, thereby expediting identification of the problem.

Load Generation A related capability is load generation, whereby varying traffic rates on the network may be created. By loading the network from 1 to 98 percent, network components such as repeaters, bridges, and transceivers can be stressed for the purpose of identifying any weak links on the network before they become serious problems later.

Mapping Some protocol analyzers have a mapping capability. In automatically documenting the physical location of LAN nodes, many hours of work can be eliminated in rearranging the network map when devices are added, deleted, or moved. The mapping software allows the network manager to name nodes. Appropriate icons for servers and workstations are included. The icon for each station also provides information about the type of adapter used, as well as the node's location along the cable. When problems arise on the network, the network manager can quickly locate the problem by referring to the visual map. Some protocol analyzers can depict network configurations according to the usage of network nodes, arranging them in order of highest to lowest traffic volume.

Programmability The various tasks of a protocol analyzer may be programmed, allowing performance information to be collected automatically. While some analyzers require the use of programming languages, others employ a setup screen, allowing the operator to define a sequence of tests to be performed. Once preset thresholds are met, an appropriate test or sequence of tests is performed automatically. This capability is especially useful for tracking down intermittent problems. An alternative to programming or defining analyzer operation is to use off-the-shelf software that can be plugged into the data analyzer in support of various test scenarios.

Automatic Configuration Some protocol analyzers have an autoconfigure capability that enables the device to automatically configure itself to the protocol characteristics of the line under test. This eliminates the need to go through several manually established screens for setup, which can save a lot of time and frustration.

Decode Capability Some protocol analyzers are unique in their ability to decode packets and display their contents in character notation, in addition to hexadecimal or binary code. Further details about a specific protocol may be revealed through an analyzer's "drill down" capability, which allows the troubleshooter to display each bit field, along with a brief explanation of its status (Figure P-8). This feature may be applied to virtually any protocol, including SNA, X.25, and TCP/IP.

Editing Some analyzer software includes a text editor that can be used in conjunction with captured data. This allows the user to delete unimportant data, enter comments, print reports, and even create files in common database formats.

Switched Environments Protocol analyzers were designed originally for shared networks. They pick up and examine all traffic as it is broadcast across a shared wire. With LAN switches growing in popularity, diagnosing problems has become a complex undertaking. Switches break shared networks into segments, and traffic is only broadcast over a particular segment. Although this improves performance by cutting down on contention and devoting more bandwidth to each station, it also makes diagnosing problems more difficult. A protocol analyzer usually can listen to traffic only on the segment to which it is connected. This makes it difficult to obtain an overall picture of what is happening through the switch.

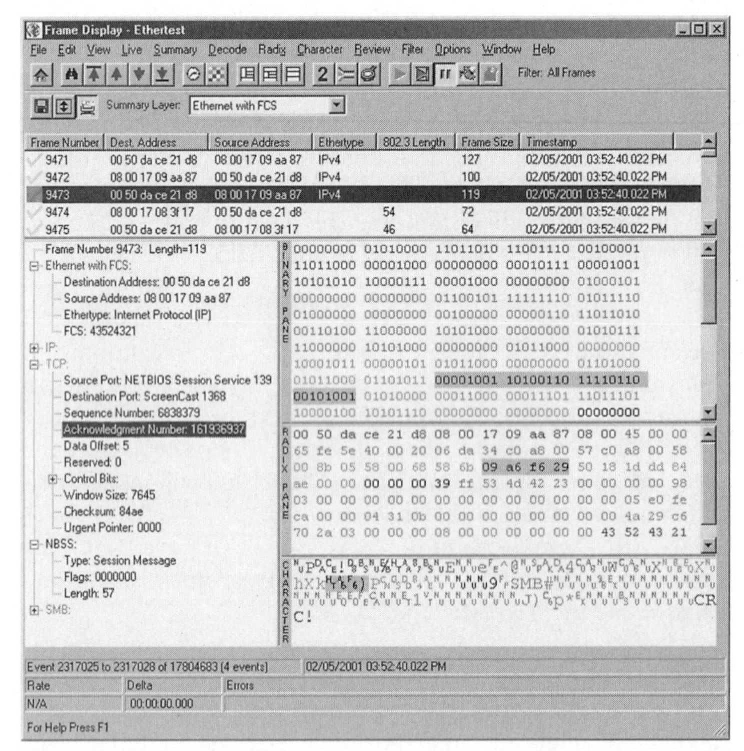

Figure P-8 Frontline Test Equipment offers a protocol analyzer called Ethertest that runs on Windows platforms. The "Frame Decode" pane of Ethertest's Frame Display window provides a comprehensive layered decode of each frame/packet, allowing a problem to be quickly located. Protocol errors are highlighted in red, making them easy to recognize.

One way to deal with this situation is for the technician to set the switch to operate in "promiscuous" mode, which sends all Ethernet packets to all ports on the switch, enabling the analyzer to see all traffic. But this results in a measurement that does not reflect real-world switch conditions. Another technique is port mirroring, which copies traffic going through one port to a port where a protocol analyzer is connected. The problem with this approach is that it limits the analyzer's view to one segment at a time.

The functionality of protocol analyzers has been expanded to monitor switches so that network administrators can get traffic statistics across a switch's ports and the switch itself, detect configuration problems in virtual LANs, and track problems between switches and desktop computers. Today's protocol analyzers can discover virtual LAN (VLAN) configurations in a switch, for example, and detect problems in the configuration. Network managers can set thresholds for traffic levels through a switch port. When the threshold is reached, the analyzer takes the traffic going through that port, mirrors it to the port with the analyzer on it, and alerts the help desk.

Another technique for gathering statistics through an entire switch is called "port looping." The analyzer uses port mirroring to look at each port on the switch for only a short time. By sampling traffic through each port, one at a time, the analyzer can build statistics about traffic through the switch.

Summary

Protocol analyzers have long been among the key diagnostic tools of technicians and network managers, helping them quickly isolate trouble spots. However, the use of protocol analyzers was almost always reactive—they captured trace information and network statistics that were later interpreted by network technicians. This required a skilled analyst to interpret the data. Today's protocol analyzers can detect deteriorating conditions that lead to errors and suggest remedial actions to head off problems before they affect performance. There are even tools that track the performance of data offerings covered by service-level agreements (SLA) so that the organization can be assured that it is getting from the carrier a level of performance and availability it is paying for.

See also

Help Desks

PROXY SERVERS

A proxy server implements a variety of complementary tasks for companies and Internet service providers (ISPs), including caching, filtering Web content, and network address translation. The primary function of a proxy server is caching frequently accessed documents to conserve network bandwidth and reduce response times for clients. It also enables network administrators to maintain better control over the use of network resources by blocking access to specific sites by user, document, and other criteria that can be set by a network administrator. The Network Address Translation (NAT) capability allows subnets to be created and protects internal IP addresses from public view on the Internet, conserving IP addresses and enhancing security.

A corporation could deploy a proxy server in a variety of ways. It can deploy a proxy server just behind the firewall to facilitate access to the Internet and reduce response times. It can be used to protect information on the secure Web server behind the firewall and offer load balancing via caching. For companies that have several subnets, a proxy server deployed at each subnet can reduce traffic on the corporate backbone, eliminating the need for more bandwidth. In situations where remote offices are disconnected from the internal network, a proxy server can provide an inexpensive means for quickly replicating content. Outside the United States, where communications bandwidth is typically much more expensive, proxy servers are even more cost-effective for replicating content.

An ISP would deploy one proxy server at each point of presence (POP) and cluster them at the Internet gateway to provide faster, more reliable service and reduce network congestion between the POP and the central Internet gateway. Some ISPs deploy a proxy server only at their gateway to the Internet, which reduces traffic on their link to the Internet but not on their own network from the POP to the Internet gateway.

Caching

A proxy server typically supports HyperText Transfer Protocol (HTTP), File Transfer Protocol (FTP), and Gopher for caching. It also may support the Secure Sockets Protocol (SSL) for the transmission of encrypted traffic and SOCKS, which is a generic way of tunneling protocols (such as Telnet) that are not "proxied." A proxy server uses sophisticated statistical analysis to store the documents most likely to be needed.

Among the many features of proxy servers is dynamic caching, which enables an administrator to schedule batch updates to the cache. This includes the ability to preload documents or sites into the cache in anticipation of user demand and the ability to automatically refresh documents that already reside in the cache. Administrators can schedule batch updates to take place at regular intervals and off-peak hours so that network bandwidth is not tied up caching documents during periods of heavy network use. Administrators can check the proxy access logs to determine whether frequently accessed sites are actually desirable for caching.

A proxy server may support the Cache Array Routing Protocol (CARP) and Internet Cache Protocol (ICP), which are proposed standards for distributed caching. CARP provides a mechanism for routing content requests among an array of proxy servers in a deterministic fashion. CARP enables load balancing, fault tolerance, more efficient caching, and easier management for multiple proxy servers. ICP enables a proxy server to send queries to neighbor caches to determine whether they already have a document.

CARP is appropriate for a group of proxy servers that are serving the same audience of downstream clients or proxies and that are all under common administrative control. ICP is appropriate for proxies that are not under common administrative control and that may be serving different clients.

Filtering

Network administrators can grant or limit access to network resources, including specific sites and documents, through the use of user name and password, IP address, host name, or domain name filtering.

A proxy server allows administrators to ban access to particular sites using a list of Uniform Resource Locators (URLs) or wildcard patterns. For example, an administrator could use http://*.playboy.com/* to prevent access to all pages belonging to the Playboy site.

A proxy server also can filter on the basis of content type, such as specific Multipurpose Internet Mail Extensions (MIME) types, and on the basis of content, such as HyperText Markup Language (HTML) tags. In addition, system administrators can implement their own security policies by stopping transmission of Java and JavaScript and ActiveX components. Many proxy servers now include virus-scanning software to prevent damage to client data and applications.

Network Address Translation

A proxy server can enhance firewall security in a variety of ways, including network address translation, which prevents external users on the Internet from being able to view the corporate network's structure and IP addresses. Blocking this information severely limits the chances of attack from hackers via address spoofing.

In Figure P-9, the proxy server gets a packet from station 135.112.56.52 for a destination on the public Internet. The address is rewritten so that it appears to come from 194.70.71.5, and the packet is sent out with this address.

When a reply packet comes back, it will be addressed to the public address 194.70.71.5. The proxy server maintains a database of outstanding requests and will look up the address of the station that made the original request. It then rewrites the address of the return packet to 135.112.56.52.

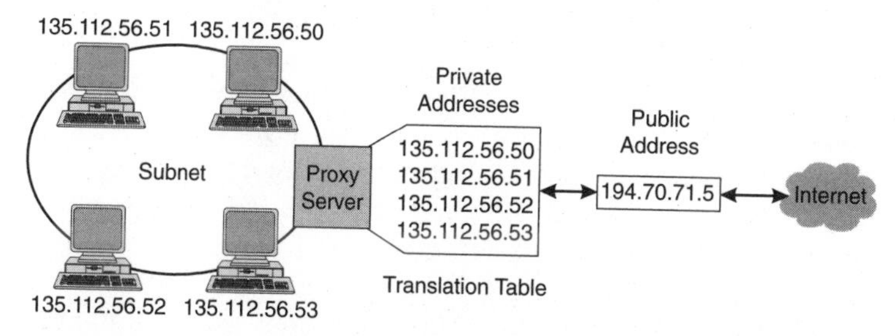

Figure P-9 The network address translation capability of a proxy server allows the creation of subnets with private IP addresses that are locally administered and never exposed to the public Internet. In addition to conserving scarce IP addresses, this capability enhances security by hiding the private IP addresses from public view over the Internet through the use of one or more public IP addresses.

Both static and dynamic address translations are supported. Static address translations explicitly map an external address to an internal address. For incoming packets that have not been specifically requested, such as e-mail, static mapping is used. With dynamic translations, a pool is allocated, and each new IP address to be translated is dynamically mapped to another IP address from the pool in a round-robin fashion. This real-time assignment of IP addresses is implemented with the Dynamic Host Control Protocol (DHCP).

System Log

A proxy server automatically logs all requests using either the common log-file format or an extended log-file format. The extended log-file format includes the referrer field and user agent. Administrators also can create their own log-file format by selecting which HTTP fields they would like to log. A built-in log analysis program includes reports such as total number of requests, total bytes transferred, most common URLs requested, most common IP addresses making requests, performance during peak periods, cache hit rates, and estimated response time reduction.

Summary

For many companies and ISPs, a proxy server is a key element of their overall Internet gateway strategy because it improves the performance and security of communications across the TCP/IP-based Internet and private intranets while permitting more flexibility in the use of IP addresses. The proxy's disk-based caching feature minimizes use of the external network by eliminating recurrent retrievals of commonly accessed documents. This significantly improves interactive response time for locally attached clients. The resulting performance improvements provide a cost-effective alternative to purchasing additional network bandwidth. And since the cache is disk-based, it can be tuned to provide optimal performance based on network usage patterns.

See also

Firewalls

Q

QUALITY OF SERVICE

Quality of Service (QoS) refers to attempts to ensure the delivery of traffic across packet data networks based on the differing performance requirements of the various applications that share the network. There are two approaches to implementing QoS. Packet-by-packet solutions seek to improve the delivery of each individual packet through differentiated treatment at a router, while application-centric QoS solutions focus on the delivery of applications as experienced by users.

With the availability of increasingly cheap bandwidth, there is the temptation to simply add more of it to solve application performance problems. If an application exhibits poor response time, for example, the easiest way for some companies to solve the problem is to purchase more bandwidth in an attempt to ease congestion somewhere in the network. This also saves the up-front cost of purchasing expensive tools or upgrading the routers with software and memory to classify and mark different traffic types to regulate flow in an effort to meet the specific performance requirements of all applications. However, this stopgap approach may not yield the desired results because the faster more bandwidth is made available, the faster it gets used. As a result, most companies never really get ahead of the performance curve.

Another problem with this "fat pipe" approach is that it ignores the need for prioritization schemes to implement service guarantees. Under this approach, certain applications would imply have more bandwidth to hog, still leaving other applications gasping for more. Without careful bandwidth management, routine HyperText Transfer Protocol (HTTP) traffic can make it impossible to implement voice over Internet Protocol (IP), for example. To get ahead of the performance curve requires a combination of intelligence and bandwidth. Intelligence comes in the form of tools that facilitate bandwidth management through such means as partitioning and policy setting.

With partitioning, a certain amount of bandwidth is allocated to a class of traffic. When a traffic class is partitioned, in essence a separate, exclusive channel is created for it within the access link. While partitions control the traffic aggregate for a class, they do not influence individual flows within that aggregate. For example, a partition for all File Transfer Protocol (FTP) traffic can be created, thereby limiting how much of the link all FTP traffic will be allowed to consume. If the bandwidth within that partition is not needed for an FTP session, it can be used by other applications that have traffic to send.

The advantage that bandwidth-management tools provide is the ability to configure the network to support many more users and applications than it would otherwise support. All applications are given a set amount of bandwidth in accordance with their priority, which is determined by each application's performance requirements. Not only does this save on the cost of bandwidth, however cheap, it also eliminates the need to buy more equipment and manage more boxes on the edges of the enterprise network.

Regardless of the particular bandwidth-management tools used, the ability to assign a QoS to each type of traffic ensures the optimal performance of all applications on the network. This also helps to contain operating costs, an important consideration in today's slow-growth economy because it affects competitiveness and investor confidence. Bandwidth-management tools have the added advantage of making the enterprise network easier to manage and administer. They also make the network easier to scale without necessarily having to add network equipment and bandwidth.

Large enterprises usually want to take responsibility for the QoS function and obtain basic transport service from a carrier, while smaller firms that are more resource-constrained might be better off subscribing to the managed services of an Integrated Communications Provider (ICP) that can support IP, Frame Relay, and Asynchronous Transfer Mode (ATM) to handle the growing number of diverse business applications that must be extended to distributed locations. The ICP installs and manages not only the access equipment for the customer but also the access links, ensuring the optimal performance of all applications. This can free smaller firms to focus on core business issues. The ICP's management and life-cycle services ensure that the firm never has to worry about the details of network infrastructure.

Embedded QoS

Ensuring the predictable delivery of increasingly diverse applications and services running over packet-based networks has become the key challenge faced by companies as they adopt the electronic enterprise model of doing business. The ability to set QoS and connection priorities on partitioned bandwidth is inherent in ATM and offers the most effective means of transporting multiple traffic types through the network without affecting the performance of the applications. Table Q-1 summarizes the primary QoS mechanisms built into ATM.

TABLE Q-1 Primary Quality of Service (QoS) Mechanisms Natively Supported by ATM

QoS Mechanism	Description	Applications
Constant Bit Rate (CBR)	Provides always-available bandwidth at a constant bit rate.	Suited for real-time applications such as circuit emulation for PBX-to-PBX trunks, scheduled multicasts, and videoconferencing. Also used for very large bit rate applications such as streaming audio/video and the transfer of medical images and CAD/CAM files.
Variable Bit Rate (VBR)	Provides bandwidth that can fluctuate according to the applications being used.	Suited for compressed voice (VBR real time) or "bursty" LAN connectivity (VBR non–real time).
Available Bit Rate (ABR)	Provides bandwidth to the applications as it becomes available.	Suited for routine applications that are not time-sensitive such as file transfers and e-mail.
Unspecified Bit Rate (UBR)	Provides bandwidth to the applications on a best-effort basis.	Suited for nonessential traffic, such as e-mail "message-delivered" or "message-read" acknowledgments.

Local area network (LAN) and video traffic, for example, can share the same physical link into the ATM network (Figure Q-1). The aggregate bandwidth of a T3 link, for example, may be partitioned between the two applications, with each type of traffic assigned its own QoS. The video stream requires a constant bit rate (CBR) and is assigned to a high-priority (QoS 1) partition in the switching fabric, whereas "bursty" LAN traffic is admitted to the network at a variable bit rate (VBR) and is assigned to a low-priority

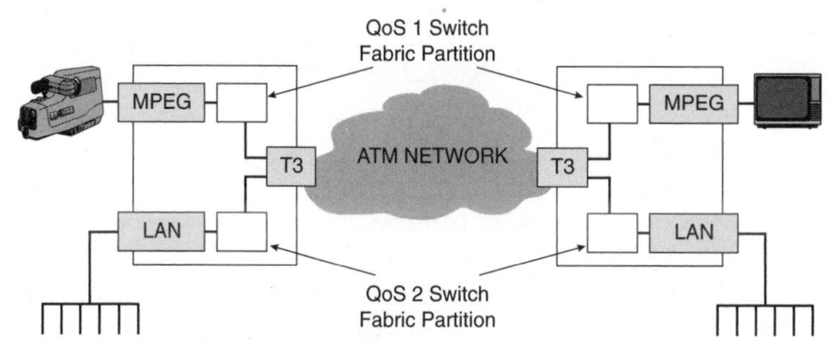

Figure Q-1 In this depiction of QoS reservation, "bursty" LAN traffic tagged as VBR does not cause cell delay for video traffic tagged as CBR because it is assigned a low-priority (QoS2) partition of the switching fabric. In this scheme, time-sensitive video traffic always has the right of way over LAN traffic.

(QoS 2) partition in the switching fabric. This scheme is referred to as "QoS reservation."

The QoS reservation can be set in the following ways:

- When set to 0 percent, there is no distinction made between connections. The system sends ATM cells through the switching fabric on an equal opportunity, round-robin basis.
- When set to 25 percent, the available switch bandwidth is divided so that 25 percent of it is reserved for QoS 1 traffic and 75 percent reserved for QoS 2 traffic.
- When set to 50 percent, the available switch bandwidth is divided so that 50 percent of it is reserved for QoS 1 traffic and 50 percent reserved for QoS 2 traffic.

The ICP configures the QoS parameters and connection priorities for each type of traffic the customer has. The ICP may even supply and manage an integrated access device (IAD) for each customer location to consolidate the different traffic types over virtual circuits that are provisioned over multiple T1 access links. These links can be bonded together to support applications at higher speeds or used separately for such applications as circuit emulation for PBX-to-PBX trunks through the carrier's ATM network.

Policy-Based QoS

Frame Relay and IP do not inherently support QoS with the same granularity as ATM, so protocols must be added in order to prioritize different types of traffic for appropriate handling through the network. Most routers today can add prioritization schemes to expedite the delivery of real-time traffic over Frame Relay. Some of these prioritization schemes, implemented in the router's operating system, are summarized in Table Q-2.

Likewise, there are prioritization schemes that can be added to routers to expedite the delivery of real-time traffic over IP networks. Some of these prioritization schemes are summarized in Table Q-3.

With so many devices on these types of networks, however, information technology (IT) staff easily can get bogged down performing manual QoS configurations to fully optimize the enterprise network. This task can be less tedious and error-prone by using policy-based network management solutions.

To address the QoS challenges of IP networks, policy-based bandwidth management solutions are available from a growing number of vendors. These tools allow network administrators to create assured service levels and deploy security features across enterprise networks, including intranets and virtual private networks (VPNs) based on IP. These tools allow network administrators to set traffic policies designed to guarantee that both mission-critical and routine data traffic are delivered to the network in a timely and consistent manner.

Some of these policy-based tools for IP are available as software solutions that are installed on existing routers located on the edges of the network, while others are implemented in hardware, requiring the purchase of dedicated devices that are also deployed at the edges of the network.

TABLE Q-2 Prioritization Schemes Used for Frame Relay Service, as Implemented in the Router's Operating System

QoS Mechanism	Description	Applications
Rate enforcement on a per-virtual-circuit (VC) basis	The peak rate value for outbound traffic can be set to match the committed information rate (CIR) to provide a constant bit rate.	Suited for real-time applications such as voice, data streaming, and large file transfers.
Dynamic traffic throttling on a per-VC basis	When backward explicit congestion notification (BECN) packets indicate congestion on the network, the outbound traffic rate is automatically stepped down by 25 percent; when congestion eases, the outbound traffic rate is allowed to increase.	This network function ensures that all traffic gets the minimum acceptable incoming or outgoing committed information rate during times of congestion. Routers that do not respond to BECN risk having their traffic discarded.
Enhanced queue support on a per-VC basis	Either custom queue or priority output queue can be configured for individual VCs.	Custom queuing is used in environments that need to guarantee a minimal level of service to all applications. Priority output queuing is used to give mission-critical data the highest priority and hold back less critical traffic during periods of congestion.

The tool vendors typically specialize in IP because intranets based on the TCP/IP protocol suite are very economical and globally available. For these reasons, IP has and will continue to experience higher growth than either ATM or Frame Relay for the foreseeable future.

TABLE Q-3 Select Prioritization Schemes Used for IP Service, as Implemented in the Router's Operating System

QoS Mechanism	Description	Applications
Resource ReSerVation Protocol (RSVP)	Sets up resources through the network to deliver the data stream to each router on the network that has attached subscribers who have preregistered to receive it.	Suited for real-time applications such as scheduled audio/video multicasts, computer-based training (CBT), and distance learning.
Protocol Independent Multicast (PIM)	Sends the data stream only once from the server, which is replicated at a rendezvous point (RP) only as many times as necessary to reach the nearest subscribers who have registered to receive it.	Handles same applications as above but is more bandwidth-efficient than RSVP. PIM also conserves processing resources at the server, since the stream goes out to the network only once.
Real-Time Protocol (RTP)	Sequentially tags IP packets to enable proper reassembly of the packet stream at the receiving end point before conversion to the real-time application.	Handles real-time, multicast, and simulation applications but does not set up network resources as do RSVP and PIM. RTP is augmented by a control protocol (RTCP) to allow monitoring of data delivery and provide minimal control and identification functionality.

IP Precedence	Expedites the handling of IP packets based on the partitioning of the traffic into as many as six classes that can be indicated in the Type of Service (ToS) field of the IPv4 header.	Handles a range of real-time and non-time-sensitive applications on the basis of the class of service they are assigned.
Differentiated Services (Diffserv)	Supersedes the original IP Precedence specification for defining packet priority. Diffserv first prioritizes traffic by class and then differentiates and prioritizes same-class traffic, offering finer priority granularity.	Satisfies differing performance needs of services and applications on the basis of the QoS specified by each packet.

Most policy-based network management systems do not include all the functionality an enterprise really needs. To overcome this limitation, vendors support links with other network management products through integration modules. Among the third-party modules that can be added to core policy tools are those for fault monitoring, billing, performance monitoring, and security.

Summary

Policy-based QoS management tools make it easier to configure and control network resources to accommodate network changes and new applications while staying ahead of the performance curve—things that cannot always be achieved merely by adding cheap bandwidth. Smaller companies that appreciate the value of bandwidth management but that do not have the resources to do it themselves can opt for the services of an ICP. The ICP configures, installs, and manages the customer premises equipment as well as the access and transport links end to end for the optimal performance of all applications in keeping with a service-level agreement that is verified with performance reports and backed up with life-cycle services.

See also

Asynchronous Transfer Mode
Integrated Access Devices
Inverse Multiplexers

R

REDUNDANT ARRAY OF INEXPENSIVE DISKS

A redundant array of inexpensive (or redundant) disks (RAID) is a storage solution that uses multiple disks that function together to provide reliable data recovery. Instead of risking all data on one high-capacity disk, this solution distributes the data across multiple smaller disks. If one of the disks fails, the other disks continue to operate. Companies use RAID technology to safeguard mission-critical data.

RAID solutions are also used to provide real-time services that rely on the uninterrupted flow of content, such as video-on-demand (VOD) programming to cable TV subscribers, which allows multiple viewers to access one copy of the same program stored on a video server. Instead of storing an entire movie on a single hard drive, segments of a movie or program are spread or "striped" across multiple disks in the RAID configuration. In addition to eliminating blocking of a single title to all users, this fault-tolerant storage method prevents video stream interruptions. More than 20 percent of storage disks would have to fail before viewing would be affected. The result is a more efficient, cost-effective, and reliable video-on-demand server that is capable of delivering hundreds of movies to thousands of households simultaneously.

RAID Categories

RAID storage products usually are grouped into categories or levels. RAID Levels 0 to 5 are industry-accepted definitions, while RAID Level 6 and beyond are proprietary storage solutions offered by specific vendors.

- *RAID Level 0* These products are technically not RAID products at all. Although data are striped block by block across all the drives in the array, such products do not offer parity or error correction of the data.
- *RAID Level 1* These products duplicate data, which are stored on separate disk drives. Also called "mirroring," this approach ensures that critical files will be available in case of individual disk drive failures. Each disk in the array has a corresponding mirror disk, and the pairs run in parallel. Blocks of data are sent to both disks at the same time.
- *RAID Level 2* These products distribute the code used for error detection and correction across additional disk drives. The controller includes an error-correction algorithm, which enables the array to reconstruct lost data if a single disk fails. As a result, no expensive mirroring is required. But the code requires that multiple disks be set aside to do the error-correction function. Data are sent to the array one disk at a time.
- *RAID Level 3* These products store user data in parallel across multiple disks. The entire array functions as one large logical drive. Its parallel operation is ideally suited to supporting applications that require high data transfer rates when reading and writing large files. RAID Level 3 is configured with one parity (error-correction) drive. The controller determines which disk has failed by using additional check information recorded at the end of each sector. However, because the drives do not operate independently, every time a file must be retrieved, all the drives in the array are used to fulfill that request. Other user requests are put into a queue.

- *RAID Level 4* These products store and retrieve data using independent writes and reads to several drives. Error-correction data are stored on a dedicated parity drive. In RAID Level 4, data striping is accomplished in sectors, not bytes (or blocks). Sector striping offers parallel operation in that reads can be performed simultaneously on independent drives, which allows multiple users to retrieve files at the same time. While multiple reads are possible, multiple writes are not because the parity drive must be read and written to for each write operation.
- *RAID Level 5* These products interleave user data and parity data, which are then distributed across several disks. Because data and parity codes are striped across all the drives, there is no need for a dedicated parity drive. This configuration is suited for applications that require a high number of input-output (I/O) operations per second, such as transaction-processing tasks that involve writing and reading large numbers of small data blocks at random disk locations. Multiple writes to each disk group are possible because write operations do not have to access a single common parity drive.
- *RAID Level 6* These products improve reliability by implementing drive mirroring at the block level so that data are mirrored on two drives instead of just one. This means that up to two drives in the five-drive disk array can fail without loss of data. If a drive in the array fails with RAID 5, for instance, data must be rebuilt from the parity information that spans the drives. With RAID 6, however, the data simply are read from the mirrored copy of the blocks found on the various striped drives—no rebuilding is required. Although this results in a slight performance advantage, it requires at least 50 percent more disk capacity to implement.
- *RAID Level 7* This proprietary solution of Storage Computer Corp. draws on the concepts of RAID Levels 3 and 4 but with enhancements to overcome the limitations

of those levels. These include a cache arranged into multiple levels and a processor for managing the array asynchronously. This design improves random read and write performance over RAID 3 or 4 even as fault tolerance is maintained. This is made possible by the multilevel cache and processor, which reduce dependence on the dedicated parity disk.

- *RAID Level 10* Some vendors offer hybrid products that combine the performance advantages of RAID 0 with the data availability and consistent high performance of RAID 1 (also referred to as "striping over a set of mirrors"). This method offers high performance and high availability for mission-critical data.

There are other hybrid RAID solutions. RAID 30, for example, is achieved by striping across a number of RAID Level 3 subarrays. RAID 30 generally provides better performance than RAID 3 because of the addition of RAID 0 striping but is not as efficient as RAID 0. RAID 50 is achieved by striping across a number of RAID 5 subarrays. RAID 50 generally provides performance better than RAID 5 because of the addition of RAID 0 striping. Although not as efficient as RAID level 0, it provides better fault tolerance than the single RAID 5.

Summary

Businesses today have multiple data storage requirements. Depending on the application, performance may be valued more than availability; other times, the reverse may be true. Today, it is even common to have different data structures in different parts of the same application. Until recently, the choice among specific RAID solutions involved tradeoffs between cost, performance, and availability—once installed, they cannot be changed to take into account the different storage needs of applications that may arise in the future. Vendors have responded with storage solutions that support a mix of RAID levels (hybrids) simultaneously. Individual disk drives or groups of drives can now be configured via a

PC-based resource manager for high performance, high availability, or as an optimized combination of both. This solves a classic data storage dilemma: meeting the exacting requirements of multiple current applications while staying flexible enough to adapt to changing needs.

See also

Hierarchical Storage Management

REMOTE CONTROL

Remote control is a software-based solution for remotely accessing another computer. In a typical scenario, there are two types of computers—the remote system and the host system. The remote system can be a branch office PC, a PC located at home, or a portable computer whose location varies on a daily basis. The host system can be any computer that a remote user wishes to access, including local area network (LAN)–attached workstations and stand-alone PCs equipped with a modem. Both the remote and host computers must be equipped with the same remote-control software, and the user must be authorized to access a particular host.

With remote control, the remote computer user takes full control of a host system sitting on the corporate network. All the user's keystrokes and mouse movements are sent to the host system, and the image on that screen is forwarded back to the remote PC for display as if the user were sitting in front of the host system. If there is a user at the host system, he or she can watch the video display for the mouse movements initiated by the remote user and all the tasks the remote user performs.

Applications

There are many applications for remote-control software. It allows a mobile user to check e-mail at the office, query a corporate contact database from a hotel room, or access an important file stored on a server or workstation. Help desk operators and technicians can use remote-control software to troubleshoot problems with specific systems. Trainers can periodically monitor the performance of users to determine their need for additional training.

With remote control, however, security can become an important concern. This is so because the host system's monitor displays all the information that is being manipulated remotely, allowing any casual observer to view all screen activity, including electronic mail, financial data, and confidential documents. A possible solution would be to turn off the monitor and leave the computer running, but this is not always possible.

Features

The leading remote-control programs are feature-rich, even permitting the remote user to access the host system through the Internet to save on long-distance charges on direct modem-to-modem connections. Such programs provide such useful features as

- *Callback* This feature enhances security by having a remote host call back to a guest. With fixed callback, the guest is automatically called back at a previously specified number. With a roving callback, the guest has the opportunity to enter the callback number.
- *Chat* Allows two connected users to type messages to each other. Some products allow messaging during a remote-control or file-transfer session.
- *Clipboard* Allows the user to highlight and copy material between the remote computer and host.
- *Compression* Compresses transferred data by a factor of at least 2 to 1 to cut dial-up line costs.

- *Directory synchronization* Synchronizes directories on the host and remote machines, ensuring that they contain the same directory trees.
- *Drag-and-drop* Allows the user to copy or move files from one machine to the other with the drag-and-drop technique.
- *Drive mapping* Makes the disk drives on both the remote and local machines seamlessly accessible to users on both ends. This feature would allow a user on a remote host to open a document stored on a local drive from within a word processor. Without remote drive mapping, the user would have to transfer the document to the remote host before being able to open it.
- *Emulation* Some vendors offer limited modem support in their remote-control software, supporting only TTY terminal emulation and only ASCII and Xmodem file transfer protocols. Others provide extensive emulation support, even allowing users to log on to commercial services such as AOL.
- *Encryption* Encrypts the data stream during transmission so that remote sessions cannot be monitored.
- *File management* Files can be displayed and sorted by name, size, modification date and time, and attribute flags. The directory can be filtered to display only specified types of files.
- *File synchronization* Synchronizes files on the remote machine, ensuring that they contain the same files as the host. File transfers can be expedited by copying only the parts of files that have changed. This also saves on long-distance connect charges.
- *Printer redirection* The ability to reroute printer output from the host to the local printer.
- *Screen data caching* Caches screen data from the host locally so that they all do not have to be retransmitted. The cache can be set to retain data between sessions.

- *Scripting facility* Scripts allow unattended file transfer operations. They allow various tasks such as the selection of connections and file transfers to be automated. A script can even automate logon procedures. Multiple tasks can be automated in a single script.
- *Transfer restart* If a connection is lost during a file transfer, the remote-control program can restart the transfer where it left off.
- *Virus scanning* Automatically scans files for viruses as they are being transferred between the remote computer and host.
- *Voice-to-data switching* Establishes a remote-control connection from a voice call without having to hang up and redial. This feature is especially convenient for technical support applications.

One of the most popular remote-control solutions is Symantec's pcTelecommute. The product's DayEnd Sync feature allows telecommuters to keep their office PC up to date with the most current versions of files when they work from home. It even reminds users at the end of the workday which files have changed and prompts them to start automatic file synchronization with their office PC.

Summary

Remote-control software allows users to dial up a specific modem-equipped or LAN-attached PC to access its files or troubleshoot a problem, even through the Internet. As long as the software is installed on both ends, the remote computer assumes the capabilities of the host. Everything on the host's screen is mirrored on the remote computer's screen. Today's increasingly distributed workforce makes this kind of connectivity a virtual necessity, especially for computer users who divide their work time between home and office. Always-on connections such as Digital Subscriber Line (DSL) and cable provide telecommuters with all the bandwidth they need for remote control and file synchronization between computers at the home office and corporate office.

See also

Remote Node

REMOTE MONITORING

The common platform from which to monitor multivendor networks is Simple Network Management Protocol's (SNMP's) Remote Monitoring (RMON) Management Information Base (MIB). Although a number of SNMP MIBs collect performance statistics to provide a snapshot of events, RMON enhances this monitoring capability by keeping a past record of events that can be used for fault diagnosis, performance tuning, and network planning.

Hardware- and/or software-based RMON-compliant devices (i.e., probes) placed on each network segment monitor all data packets sent and received. The probes view every packet and produce summary information on various types of packets, such as undersized packets, and events, such as packet collisions. The probes also can capture packets according to predefined criteria set by the network manager or test technician. At any time, the RMON probe can be queried for this information by a network management application or an SNMP-based management console so that detailed analysis can be performed in an effort to pinpoint where and why an error occurred.

The original Remote Network Monitoring MIB defined a framework for the remote monitoring of Ethernet. Subsequent RMON MIBs have extended this framework to Token Ring and other types of networks. A map of the RMON MIB for Ethernet and Token Ring is shown in Figure R-1.

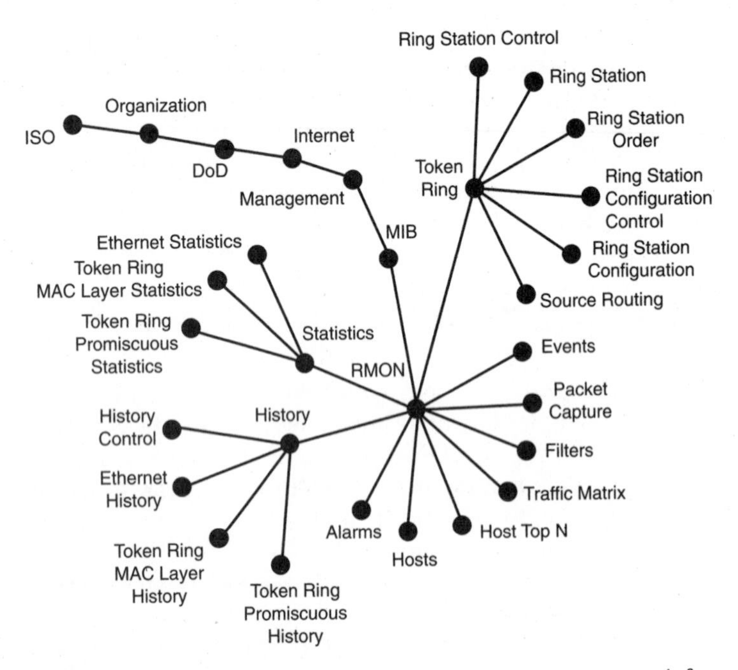

Figure R-1 A map of SNMP's remote monitoring management information base—RMON MIB.

RMON Applications

A management application that views the internetwork, for example, gathers data from RMON agents running on each segment in the network. The data are integrated and correlated to provide various internetwork views that provide end-to-end visibility of network traffic, both LAN and wide area network (WAN). The operator can switch between a variety of views.

For example, the operator can switch between a Media Access Control (MAC) view (which shows traffic going through routers and gateways) and a network view (which shows end-to-end traffic) or can apply filters to see only traffic of a given protocol or suite of protocols. These traffic matrices provide the information necessary to configure or partition the internetwork to optimize LAN and WAN utilization.

In selecting the MAC level view, for example, the network map shows each node of a segment separately, indicating intrasegment node-to-node data traffic. It also shows total intersegment data traffic from routers and gateways. This combination allows the operator to see consolidated internetwork traffic and how each end node contributes to it.

In selecting the network level view, the network map shows end-to-end data traffic between nodes across segments. By connecting source and ultimate destination, without clouding the view with routers and gateways, the operator can immediately identify specific areas contributing to an unbalanced traffic load.

Another type of application allows the network manager to consolidate and present multiple segment information, configure RMON alarms, and provide complete Token Ring RMON information, as well as perform baseline measurements and long-term reporting. Alarms can be set on any RMON variable. Notification via traps can be sent to multiple management stations. Baseline statistics allow long-term trend analysis of network traffic patterns that can be used to plan for network growth.

Ethernet Object Groups

The RMON specification consists of 9 Ethernet/Token Ring groups and 10 specific Token Ring RMON extensions (refer back to Figure R-1).

Ethernet Statistics Group The Statistics Group provides segment-level statistics (Figure R-2). These statistics show packets, octets (or bytes), broadcasts, multicasts, and collisions on the local segment, as well as the number of occurrences of packets dropped by the agent. Each statistic is maintained in its own 32-bit cumulative counter. Real-time packet size distribution is also provided.

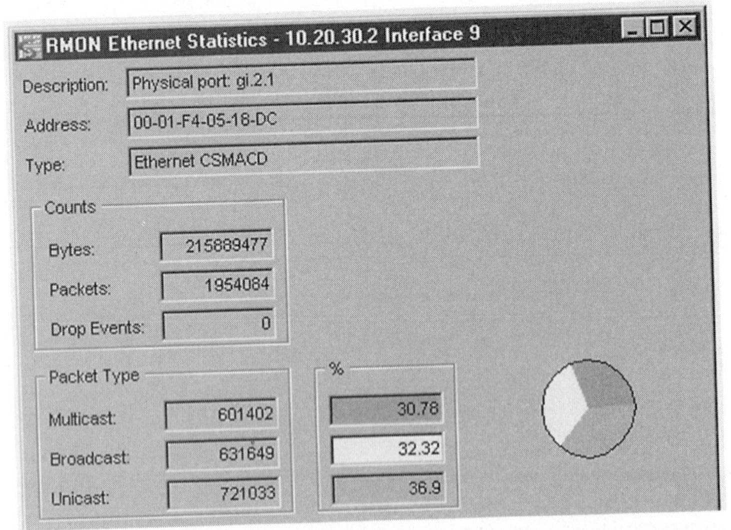

Figure R-2 The Ethernet Statistics window accessed from Enterasys Networks' NetSight Element Manager. This window would be used to view a detailed statistical breakdown of traffic on the monitored Ethernet network segment. The data provided apply only to the interface or network segment.

Ethernet History Group With the exception of packet size distribution, which is provided only on a real-time basis, the History Group provides historical views of the statistics provided in the Statistics Group. The History Group can respond to user-defined sampling intervals and bucket counters, allowing for some customization in trend analysis.

The RMON MIB comes with two defaults for trend analysis. The first provides for 50 buckets (or samples) of 30-second sampling intervals over a period of 25 minutes. The second provides for 50 buckets of 30-minute sampling intervals over a period of 25 hours. Users can modify either of these or add additional intervals to meet specific requirements for historical analysis. The sampling interval can range from 1 second to 1 hour.

Host Table Group The RMON MIB specifies a host table that includes node traffic statistics: packets sent and received, octets sent and received, as well as broadcasts, multicasts, and errored packets sent. In the host table, the classification "errors sent" is the combination of packet undersizes, fragments, CRC/alignment errors, collisions, and oversizes sent by each node.

The RMON MIB also includes a host timetable that shows the relative order in which the agent discovered each host. This feature not only is useful for network management purposes but also assists in uploading those nodes to the management station of which it is not yet aware. This reduces unnecessary SNMP traffic on the network.

Host Top N Group The Host Top N Group extends the host table by providing sorted host statistics, such as the top 10 nodes sending packets or an ordered list of all nodes according to the errors sent over the last 24 hours. The data selected and the duration of the study are both defined at the network management station. The number of studies that can be run depends on the resources of the monitoring device.

When a set of statistics is selected for study, only the selected statistics are maintained in the Host Top *N* counters; other statistics over the same time intervals are not available for later study. This processing—performed remotely in the RMON MIB agent—reduces SNMP traffic on the network and the processing load on the management station, which would otherwise need to use SNMP to retrieve the entire host table for local processing.

Alarms Group The Alarms Group provides a general mechanism for setting thresholds and sampling intervals to generate events on any counter or integer maintained by the agent, such as segment statistics, node traffic statistics defined in the host table, or any user-defined packet match counter defined in the Filters Group. Both rising and falling thresholds can be set, each of which can indicate network faults. Thresholds can be established for both the absolute value of a statistic and its delta value, enabling the manager to be notified of rapid spikes or drops in a monitored value.

Filters Group The Filters Group provides a generic filtering engine that implements all packet capture functions and events. The packet capture buffer is filled with only those packets which match the user-specified filtering criteria. Filtering conditions can be combined using the Boolean parameters "and" or "not." Multiple filters are combined with the Boolean "or" parameter.

Packet Capture Group The types of packets collected depend on the Filter Group. The Packet Capture Group allows the user to create multiple capture buffers and to control whether the trace buffers will wrap (overwrite) when full or stop capturing. The user may expand or contract the size of the buffer to fit immediate needs for packet capturing rather than permanently commit memory that will not always be needed.

Notifications (Events) Group In a distributed management environment, the RMON MIB agent can deliver traps to multiple management stations that share a single community name destination specified for the trap. In addition to the three traps already mentioned—rising threshold and falling threshold (see Alarms Group) and packet match (see Packet Capture Group)—seven additional traps can be specified:

- *coldStart* This trap indicates that the sending protocol entity is reinitializing itself such that the agent's configuration or the protocol entity implementation may be altered.
- *warmStart* This trap indicates that the sending protocol entity is reinitializing itself such that neither the agent configuration nor the protocol entity implementation is altered.
- *linkDown* This trap indicates that the sending protocol entity recognizes a failure in one of the communication links represented in the agent's configuration.

- *linkUp* This trap indicates that the sending protocol entity recognizes that one of the communication links represented in the agent's configuration has come up.
- *authenticationFailure* This trap indicates that the sending protocol entity is the addressee of a protocol message that is not properly authenticated. While implementations of the SNMP must be capable of generating this trap, they also must be capable of suppressing the emission of such traps via an implementation-specific mechanism.
- *egpNeighborLoss* This trap indicates that an External Gateway Protocol (EGP) neighbor for whom the sending protocol entity was an EGP peer has been marked down and the peer relationship is no longer valid.
- *enterpriseSpecific* This trap indicates that the sending protocol entity recognizes that some enterprise-specific event has occurred.

The Notifications (Events) Group allows users to specify the number of events that can be sent to the monitor log. From the log, any specified event can be sent to the management station. The log includes the time of day for each event and a description of the event written by the vendor of the monitor. The log overwrites when full, so events may be lost if not uploaded to the management station periodically.

Traffic Matrix Group The RMON MIB includes a traffic matrix at the MAC layer. A traffic matrix shows the amount of traffic and number of errors between pairs of nodes—one source and one destination address per pair. For each pair, the RMON MIB maintains counters for the number of packets, number of octets, and error packets between the nodes. Users can sort this information by source or destination address.

Applying remote monitoring and statistics-gathering capabilities to the Ethernet environment offers a number of benefits. The availability of critical networks is maximized, since remote capabilities allow for a more timely resolution of the problem. With the capability to resolve problems remotely, operations staff can avoid costly travel to troubleshoot problems on site. With the capability to analyze data collected at specific intervals over a long period of time, intermittent problems can be tracked down that would normally go undetected and unresolved.

Token Ring Extensions

As noted, the first version of RMON defined media-specific objects for Ethernet only. Later, media-specific objects for Token Ring were added.

Token Ring MAC Layer Statistics This extension provides statistics, diagnostics, and event notification associated with MAC traffic on the local ring. Statistics include the number of beacons, purges, and IEEE 803.5 MAC management packets and events; MAC packets; MAC octets; and ring soft error totals.

Token Ring Promiscuous Statistics This extension collects utilization statistics of all user data traffic (non-MAC) on the local ring. Statistics include the number of data packets and octets, broadcast and multicast packets, and data frame size distribution.

Token Ring MAC Layer History This extension offers historical views of MAC layer statistics based on user-defined sample intervals, which can be set from 1 second to 1 hour to allow short-term or long-term historical analysis.

Token Ring Promiscuous History This extension offers historical views of promiscuous (i.e., unfiltered) statistics based on user-defined sample intervals, which can be set from 1 second to 1 hour to allow short-term or long-term historical analysis.

Ring Station Control Table This extension lists status information for each ring being monitored. Statistics include ring state, active monitor, hard error beacon fault domain, and number of active stations.

Ring Station Table This extension provides diagnostics and status information for each station on the ring. The type of information collected includes station MAC address, status, and isolating and nonisolating soft error diagnostics.

Source Routing Statistics The extension for source routing statistics is used for monitoring the efficiency of source-routing processes by keeping track of the number of data packets routed into, out of, and through each ring segment. Traffic distribution by hop count provides an indication of how much bandwidth is being consumed by traffic-routing functions.

Ring Station Configuration Control The extension for station configuration control provides a description of the network's physical configuration. A media fault is reported as a "fault domain," an area that isolates the problem to two adjacent nodes and the wiring between them. The network administrator can discover the exact location of the problem—the fault domain—by referring to the network map. Some faults result from changes to the physical ring—including each time a station inserts or removes itself from the network. This type of fault is discovered through a comparison of the start of symptoms and the timing of the physical changes.

The RMON MIB not only keeps track of the status of each station but also reports the condition of each ring being monitored by a RMON agent. On large Token Ring networks with several rings, the health of each ring segment and the number of active and inactive stations on each ring can be monitored simultaneously. Network administrators can be alerted to the location of the fault domain should any ring go into a beaconing (fault) condition. Network managers also can be alerted to any changes in backbone ring configuration, which could indicate loss of connectivity to an interconnect device such as a bridge or to a shared resource such as a server.

Ring Station Configuration The ring station group collects Token Ring–specific errors. Statistics are kept on all significant MAC level events to assist in fault isolation, including ring purges, beacons, claim tokens, and such error conditions as burst errors, lost frames, congestion errors, frame copied errors, and soft errors.

Ring Station Order Each station can be placed on the network map in a specified order relative to the other stations on the ring. This extension provides a list of stations attached to the ring in logical ring order. It lists only stations that comply with IEEE 802.5 active monitoring ring poll or IBM trace tool present advertisement conventions.

RMON II

The RMON MIB is basically a MAC level standard. Its visibility does not extend beyond the router port, meaning that it cannot see beyond individual LAN segments. As such, it does not provide visibility into conversations across the network or connectivity between the various network segments. Given the trends toward remote access and distributed workgroups, which generate a lot of intersegment traffic, visibility across the enterprise is an important capability to have.

RMON II extends the packet capture and decoding capabilities of the original RMON MIB to Layers 3 through 7 of the Open Systems Interconnection (OSI) reference model. This allows traffic to be monitored via network-layer addresses—which lets RMON "see" beyond the router to the internetwork and distinguish between applications.

Analysis tools that support the network layer can sort traffic by protocol rather than just report on aggregate traffic.

This means that network managers will be able to determine, for example, the percent of Internet Protocol (IP) versus Internet Packet Exchange (IPX) traffic traversing the network. In addition, these higher-level monitoring tools can map end-to-end traffic, giving network managers the ability to trace communications between two hosts—or nodes—even if the two are located on different LAN segments. RMON II functions that will allow this level of visibility include

- *Protocol directory table* Provides a list of all the different protocols a RMON II probe can interpret.
- *Protocol distribution table* Permits tracking of the number of bytes and packets on any given segment that have been sent from each of the protocols supported. This information is useful for displaying traffic types by percentage in graphical form.
- *Address mapping* Permits identification of traffic-generating nodes, or hosts, by Ethernet or Token Ring address in addition to MAC address. It also discovers switch or hub ports to which the hosts are attached. This is helpful in node discovery and network topology applications for pinpointing the specific paths of network traffic.
- *Network-layer host table* Permits tracking of bytes, packets, and errors by host according to individual network-layer protocol.
- *Network-layer matrix table* Permits tracking, by network-layer address, of the number of packets sent between pairs of hosts.
- *Application-layer host table* Permits tracking of bytes, packets, and errors by host and according to application.
- *Application-layer matrix table* Permits tracking of conversations between pairs of hosts by application.
- *History group* Permits filtering and storing of statistics according to user-defined parameters and time intervals.
- *Configuration group* Defines standard configuration parameters for probes that includes such parameters as

network address, serial line information, and SNMP trap destination information.

RMON II is focused more on helping network managers understand traffic flow for the purpose of capacity planning rather than for the purpose of physical troubleshooting. The capability to identify traffic levels and statistics by application has the potential to greatly reduce the time it takes to troubleshoot certain problems. Without tools that can pinpoint which software application is responsible gobbling up a disproportionate share of the available bandwidth, network managers can only guess. Often it is easier just to upgrade a server or a buy more bandwidth, which inflates operating costs and shrinks budgets.

Summary

Applying remote monitoring and statistics-gathering capabilities to the Ethernet and Token Ring environments via the RMON MIB offers a number of benefits. The availability of critical networks is maximized, since remote capabilities allow for timely problem resolution. With the capability to resolve problems remotely, operations staff can avoid costly travel to troubleshoot problems on site. With the capability to analyze data collected at specific intervals over a long period of time, intermittent problems can be tracked down that would normally go undetected and unresolved. And with RMON II, these capabilities are enhanced and extended up to the applications level across the enterprise.

See also

Open Systems Interconnection
Protocol Analyzers
Simple Network Management Protocol

REMOTE NODE

Remote node is a method of remote access that permits users to dial into the corporate network and perform tasks as if they were locally attached to the LAN. With remote node, the remote system performs client functions, while the office-based host systems perform true server functions. This allows remote users to take advantage of a host's processing capabilities. Only the results are transmitted over the connection back to the remote computer.

With remote node, the user's modem-equipped PC dials into the LAN and behaves as if it were a local LAN node. Instead of keystrokes and screen updates, the traffic on the remote node's dial-up line is essentially normal network traffic. The remote PC does not control another PC, as in remote control; rather, it runs regular applications as if it were attached directly to the LAN.

Remote node usually relies on a remote access server (RAS) that is set up and maintained at a corporate location. This allows many remote users, all at different remote locations, to share the same resources. The server usually has Ethernet and/or Token Ring ports and built-in modems for dial-up access at up to 56 kbps. Instead of modems, some vendors offer high-speed asynchronous ports. Depending on vendor, there may be optional support for higher-speed connections over such services as Integrated Services Digital Network (ISDN) or Frame Relay.

Remote node offers many of the same features as remote control and, in some cases, surpasses them in functionality. This is especially true in such key areas as management, event reporting, and security.

Management

Most vendors allow the remote access server's routing software to be configured from a local management console, a remote telnet session, or SNMP management station. SNMP support facilitates ongoing management and integrates the remote access server into the management environment commonly used by network administrators.

Management utilities allow for status monitoring of individual ports, the collection of service statistics, and the viewing of audit trails on port access and usage. Other statistics include port address, traffic type, connect time/break connection, and connect time exceeded. Filters can be applied to customize management reports with only the desired type of information.

The SNMP management station, in conjunction with the vendor-supplied MIB, can display alerts about the operation of the server. The network administrator is notified when an application processor has been automatically reset because of a timeout, for example, or when there is a hardware failure on a processor, which triggers an antilocking mechanism reset. This type of reset ensures that all users are not locked out of the server by the failing processor.

Event Reporting

Remote node products typically include management software that is installed at the server to provide usage statistics, including packets sent and received and transmission errors, as well as who is logged on, how long they have been connected, and what types of modems are attached to the device.

Support for SNMP enables the server to trap a variety of meaningful events. Special drivers pass these traps to any SNMP-based network management platform, such as Hewlett-Packard's OpenView or IBM's NetView.

Security

Remote node products generally offer more levels of security than remote-control products. Depending on the size of the network and the sensitivity of the information that can be remotely accessed, one or more of the following security methods can be employed:

- *User ID and password* These are used routinely to control access to the server. Security is enhanced if ID and password are encrypted before going out over the communications link.
- *Authentication* This involves the server verifying the identity of the remote caller by issuing a challenge to which he or she must respond with the correct answer.
- *Access restrictions* This involves assigning each remote user a specific location (i.e., directory or drive) that can be accessed in the server. Access to specific servers also can be controlled.
- *Time restrictions* This involves assigning each remote user a specific amount of connection time, after which the connection is dropped.
- *Connection restrictions* This involves limiting the number of consecutive connection attempts and/or the number of times connections can be established on an hourly or daily basis.
- *Protocol restrictions* This involves limiting users to a specific protocol for remote access.

There are other options for securing the LAN, such as callback and cryptography. With callback, the remote client's call is accepted, the line is disconnected, and the server calls back after checking that the telephone number is valid. While this works well for branch offices, most callback products are not appropriate for mobile users whose locations vary on a daily basis. There are products on the market, however, that accept roving callback numbers. This feature allows mobile users to call into a remote access server or host computer, type in their user ID and password, and then specify a number where the server or host should call them back. The callback number is then logged and may be used to help track down security breaches.

To safeguard very sensitive information, there are third-party security systems that can be added to the server. Some of these systems require a user password and also a special credit card–sized device that generates a new ID every 60 seconds, which must be matched by a similar ID number-generation process on the remote user's computer.

In addition to callback and encryption, security can be enforced via IP filtering and log-in passwords for the system console and for telnet- and File Transfer Protocol (FTP)–server programs. Many remote node products also enforce security at the link level using the Point-to-Point Protocol (PPP) with the Challenge Handshake Authentication Protocol (CHAP) and Password Authentication Protocol (PAP).

Summary

Companies are being driven to provide effective and reliable remote access solutions—remote control as well as remote node—to meet the productivity needs of the today's increasingly decentralized workforce. While remote control usually relies on dial-up connections established by modems, remote node connections to the remote access server can be established by dial-up routers as well as by modems. Dial-up routers serve branch offices and modems are used by telecommuters and mobile professionals.

See also

Network Security
Remote Control

REPEATERS

A repeater is a device that extends the inherent distance limitations of various transmission media used on LANs and WANs, including wireless links, by boosting signal power so that it stays at the same level regardless of the distance it must travel. As such, the repeater operates at the lowest level of the OSI reference model—the Physical Layer (Figure R-3).

Repeaters are necessary because signal strength weakens with distance: The longer the path a signal must travel, the weaker it gets. This condition is known as "signal attenuation." On a telephone call, a weak signal will cause low volume, interfering with the parties' ability to hear each other. In cellular networks, when a mobile user moves beyond the range of a cell site, the signal fades to the point of disconnecting the call. In the LAN environment, a weak signal can result in corrupt data, which can substantially reduce throughput by forcing retransmissions when errors are detected. When the signal level drops low enough, the chances of interference from external noise increase, rendering the signal unusable.

Repeaters also can be used to link different types of network media—fiber to coaxial cable, for example. Often LANs are interconnected in a campus environment by means of repeaters that form the LANs into connected network seg-

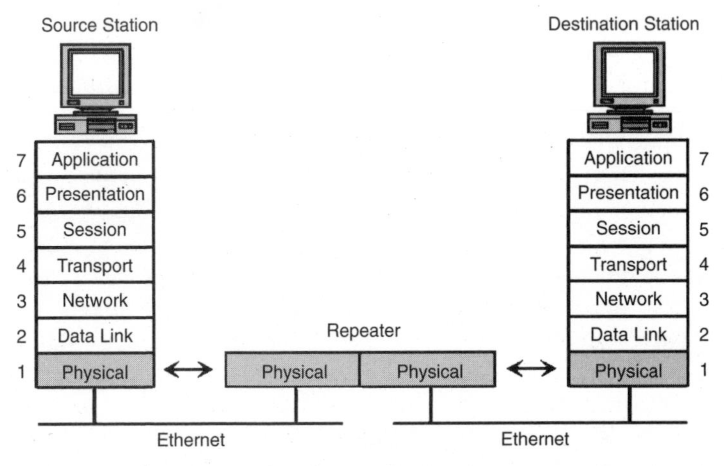

Figure R-3 Repeaters operate at the Physical Layer of the OSI reference model.

ments. The segments may employ different transmission media—thick or thin coaxial cable, twisted-pair wiring, or optical fiber. The cost of media converters is significantly less than that of full repeaters, and they can be used whenever media distance limitations will not be exceeded in the network.

Hubs or switches usually are equipped with appropriate modules that perform the repeater and media conversion functions on sprawling LANs. But the use of hubs or switches also can eliminate the need for repeaters, since most cable segments in office buildings will not run more than 100 feet (about 30 meters), which is well within the distance limitation of most LAN standards, including 1000BaseT Gigabit Ethernet running over Category 5 cable.

Regenerators

Often the terms *repeater* and *regenerator* are used interchangeably, but there is a subtle difference between the two. In an analog system, a repeater boosts the desired signal strength but also boosts the noise level as well. Consequently, the signal-to-noise ratio on the output side of the repeater remains the same as on the input side. This means that once noise is introduced into the desired signal, it is impossible to get the signal back into its original form again on the output side of the repeater.

In a digital system, regenerators are used instead of repeaters. The regenerator determines whether the information-carrying bits are 1s or 0s, on the basis of the received signal on the input side. Once the decision of 1 or 0 is made, a fresh signal representing that bit is transmitted on the output side of the regenerator. Because the quality of the output signal is a perfect replication of the input signal, it is possible to maintain a very high level of performance over a range of transmission impairments. Noise, for instance, is filtered out because it is not represented as a 1 or 0.

Summary

Stand-alone repeaters have transceiver interface modules that provide connections to various media. There are fiberoptic transceivers, coaxial transceivers, and twisted-pair transceivers. Some repeaters contain the intelligence to detect packet collisions and will not repeat collision fragments to other cable segments. Some repeaters also can "deinsert" themselves from a hub or switch when there are excessive errors on the cable segment, and they can submit performance information to a central management station.

See also

Bridges
Gateways
Media Converters
Routers

ROUTERS

A router operates at Layer 3 of the OSI reference model, the Network Layer. The device distinguishes among Network Layer protocols—such as IP, IPX, AppleTalk, and DEC Local Area Transport (LAT)—and makes intelligent packet delivery decisions using an appropriate routing protocol. It can be used to segment a network with the goals of limiting broadcast traffic and providing security, control, and redundant paths. A router also can provide multiple types of interfaces, including those for T1, Frame Relay, ISDN, Asynchronous Transfer Mode (ATM), cable networks, and DSL services, among others. Some routers can perform simple packet filtering to control the kind of traffic that is allowed to pass through them, providing a rudimentary firewall service. Larger routers can perform advanced firewall functions.

A router is similar to a bridge in that both provide filtering and bridging functions across the network. But while bridges operate at the Physical and Data Link Layers of the OSI reference model, routers join LANs at the Network Layer (Figure R-4). Routers convert LAN protocols into WAN protocols and perform the process in reverse at the remote location. They may be deployed in mesh as well as point-to-point networks and, in certain situations, can be used in combination with bridges.

Although routers include the functionality of bridges, they differ from bridges in the following ways: They generally offer more embedded intelligence and, consequently, more sophisticated network management and traffic control capabilities than bridges. Another distinction—perhaps the most significant one—between a router and a bridge is that a bridge delivers packets of data on a "best effort" basis, specifically, by discarding packets it does not recognize onto an adjacent network. Through a continual process of discarding unfamiliar packets, data get to their proper destination—on a network where the bridge recognizes the packets as belonging

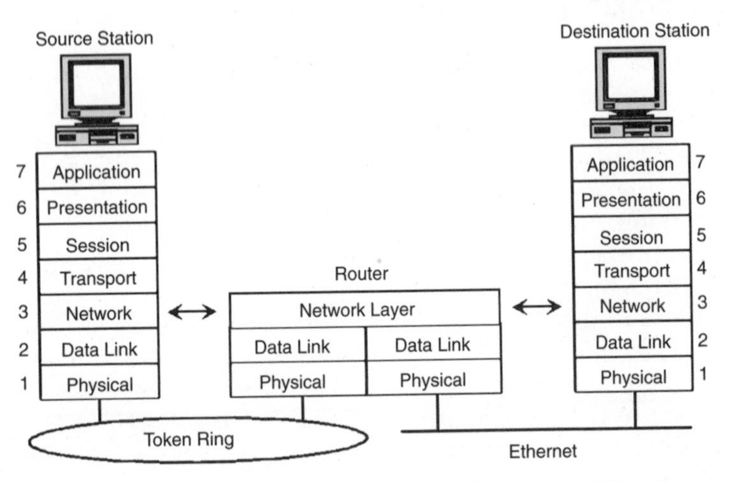

Figure R-4 Routers operate at the Network Layer of the OSI reference model.

to a device attached to its network. By contrast, a router takes a more intelligent approach to getting packets to their destination—by selecting the most economical path (i.e., least number of hops) on the basis of its knowledge of the overall network topology, as defined by its internal routing table. Routers also have flow control and error protection capabilities.

Types of Routing

There are two types of routing: static and dynamic. In static routing, the network manager configures the routing table to set fixed paths between two routers. Unless reconfigured, the paths on the network never change. Although a static router will recognize that a link has gone down and issue an alarm, it will not automatically reroute traffic. A dynamic router, on the other hand, reconfigures the routing table automatically and recalculates the most efficient path in terms of load, line delay, or bandwidth.

Some routers balance the traffic load across multiple access links, providing an $N \times$ T1 inverse multiplexer function. This allows multiple T1 access lines operating at 1.544 Mbps each to be used as a single higher-bandwidth facility. If one of the links fails, the other links remain in place to handle the offered traffic. As soon as the failed link is restored to service, traffic is spread across the entire group of lines as in the original configuration.

Routing Protocols

Each router on the network keeps a routing table and moves data along the network from one router to the next using such protocols as Open Shortest Path First (OSPF) and Routing Information Protocol (RIP).

Although still supported by many vendors, RIP does not perform well in today's increasingly complex networks. As the network expands, routing updates grow larger under

RIP and consume more bandwidth to route the information. When a link fails, the RIP update procedure slows route discovery, increases network traffic and bandwidth usage, and may cause temporary looping of data traffic. Also, RIP cannot base route selection on such factors as delay and bandwidth, and its line-selection facility is capable of choosing only one path to each destination.

The newer routing standard, OSPF, overcomes the limitations of RIP and even provides capabilities not found in RIP. The update procedure of OSPF requires that each router on the network transmit a packet with a description of its local links to all other routers. On receiving each packet, the other routers acknowledge it, and in the process, distributed routing tables are built from the collected descriptions. Since these description packets are relatively small, they produce a minimum of overhead. When a link fails, updated information floods the network, allowing all the routers to simultaneously calculate new tables.

Types of Routers

Multiprotocol nodal, or hub, routers are used for building highly meshed internetworks. In addition to allowing several protocols to share the same logical network, these devices pick the shortest path to the end node, balance the load across multiple physical links, reroute traffic around points of failure or congestion, and implement flow control in conjunction with the end nodes. They also provide the means to tie remote branch offices into the corporate backbone, which might use such WAN services as Transmission Control Protocol/Internet Protocol (TCP/IP), T1, ISDN, and ATM. Some vendors also provide an optional interface for Switched Multimegabit Data Service (SMDS).

Access routers are typically used at branch offices. These are usually fixed-configuration devices available in Ethernet and Token Ring versions that support a limited number of protocols and physical interfaces. They provide connectivity to high-end multiprotocol routers, allowing large and small nodes to be managed as a single logical enterprise network. Although low-cost, plug-and-play bridges can meet the need for branch office connectivity, low-end routers can offer more intelligence and configuration flexibility at comparable cost.

The newest access routers are multiservice devices that are designed to handle a mix of data, voice, and video traffic. They support a variety of WAN connections through built-in interfaces that include dual ISDN Basic Rate Interface (BRI) interfaces, dual analog ports, T1/Frame Relay ports, and an ISDN interface for videoconferencing. Such routers can run software that provides IPSec virtual private network (VPN), firewall, and encryption services.

Midrange routers provide network connectivity between corporate locations in support of workgroups or the corporate intranet, for example. These routers can be stand-alone devices or packaged as modules that occupy slots in an intelligent wiring hub or LAN switch. In fact, this type of router is often used to provide connectivity between multiple wiring hubs or LAN switches over high-speed LAN backbones such as ATM, Fiber Distributed Data Interface (FDDI), and Fast Ethernet.

There is a consumer class of routers capable of providing shared access to the Internet over such broadband technologies as cable and DSL. The Linksys Instant Broadband EtherFast Cable/DSL Router (Figure R-5), for example, is used to connect a small group of PCs to a high-speed Internet connection or to an Ethernet backbone. Configurable through any networked PC's Web browser, the router can be set up as a firewall and Dynamic Host Configuration Protocol (DHCP) server, allowing it to act as an externally recognized Internet device with its own IP address for the home LAN. Unlike a typical router, which can only share 100 Mbps over all its connections, the Linksys device is also equipped with a four-port Ethernet switch that dedicates 100 Mbps to every connected PC.

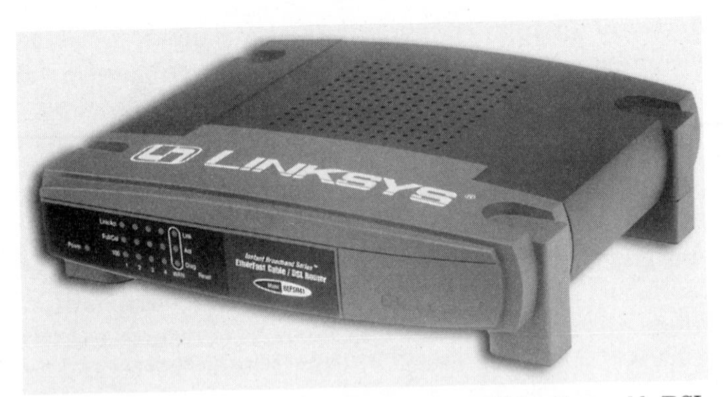

Figure R-5 The Linksys Instant Broadband EtherFast cable/DSL router is representative of the new breed of user-friendly multifunction routers aimed at the emerging home networking market.

The Linksys router also supports Network Address Translation (NAT), a feature that translates one public IP address, given by the cable or DSL Internet provider, and assigns automatically up to 253 private IP addresses to users on the LAN. All the users given an IP address by the router are safe behind the firewall, so incoming and outgoing requests are filtered, keeping unwanted requests off the LAN.

At the same time, the Linksys router supports a feature called "DMZ/Expose Host" that dissembles one of the 253 private IP addresses to become a public IP address so that outside users can access that PC without getting blocked by the firewall. An example would be a gamer playing another gamer via the Internet. They want to access each other's computers so that they can play the game.

Summary

Routers fulfill a vital role in implementing complex mesh networks such as the Internet and private intranets using Layer 3 protocols, usually IP. They also have become an economical means of tying branch offices into the enterprise network and providing PCs tied together on a home network with shared access to broadband Internet services such as cable and DSL. Like other interconnection devices, routers are manageable via SNMP, as well as the proprietary management systems of vendors. Just as bridging and routing functions made their way into a single device, routing and switching functions are being combined in the same way, and even add firewall, DHCP, and NAT capabilities.

See also

Bridges
Firewalls
Gateways
Inverse Multiplexers
Repeaters

S

SIMPLE NETWORK MANAGEMENT PROTOCOL

Since 1988, the Simple Network Management Protocol (SNMP) has been the de facto standard for the management of multivendor Transmission Control Protocol/Internet Protocol (TCP/IP)–based networks. SNMP specifies a structure for formatting messages and for transmitting information between reporting devices and data-collection programs on the network. The SNMP-compliant devices on the network are polled for performance-related information, and results are passed to a network management console. Alarms are also passed to the console. There, the information gathered can be viewed to pinpoint problems on the network or stored for later analysis.

SNMP runs on top of TCP/IP's datagram protocol—the User Datagram Protocol (UDP)—a transport protocol that offers a connectionless-mode service. This means that a session need not be established before network management information can be passed to the central control point. Although SNMP messages can be exchanged across any protocol, UDP is well suited to the brief request/response message exchanges characteristic of network management communications.

SNMP is a very flexible network management protocol that can be used to manage virtually any object. An "object" refers to hardware, software, or a logical association, such as a connection or virtual circuit. An object's definition is written by the equipment vendor and is held in a management information base (MIB). The MIB is simply a list of switch settings, hardware counters, in-memory variables, or files that are used by the network management system to determine the alarm and reporting characteristics of each device on the network, including those connected over local area networks (LANs).

As noted, SNMP is basically a request/response protocol. The management system retrieves information from the agents through SNMP's "get" and "get-next" commands. The "get" request retrieves the values of specific objects from the MIB. The MIB lists the network objects for which an agent can return values. These values may include the number of input packets, the number of input errors, and routing information. The "get-next" request permits navigation of the MIB, enabling the next MIB object to be retrieved, relative to its current position. A "set" request is used to request a logically remote agent to alter the values of variables. In addition to these message types, there are "trap" messages, which are unsolicited messages conveyed from management agent to management stations. Other commands are available that allow the network manager to take specific actions to control the network. Although these commands look like SNMP commands, they are really vendor-specific implementations. For example, some vendors use a "stat" command to determine the status of network connections.

All the major network management platforms support SNMP, including Hewlett-Packard's OpenView and IBM's NetView. In addition, many of the third-party systems and network management applications that plug into these platforms support SNMP. The advantage of using such products is that they take advantage of SNMP's capabilities while providing a graphical user interface (GUI) to make SNMP easier to use (Figure S-1). Even MIBs can be selected for display and navigation through the GUI.

Another advantage of commercial products is that they can use SNMP to provide additional functionality. For example, OpenView and NetView are used to manage network devices that are IP addressable and run SNMP. Their automatic discovery capability finds and identifies all IP nodes on the network, including those of other vendors that support SNMP. On the basis of discovered information, the management system automatically draws a network topology map. Nodes that cannot be discovered automatically can be represented in either of two ways: first, by manually adding custom or standard icons to the appropriate map views, and second, by using SNMP-based Application Programming Interfaces (APIs) for building

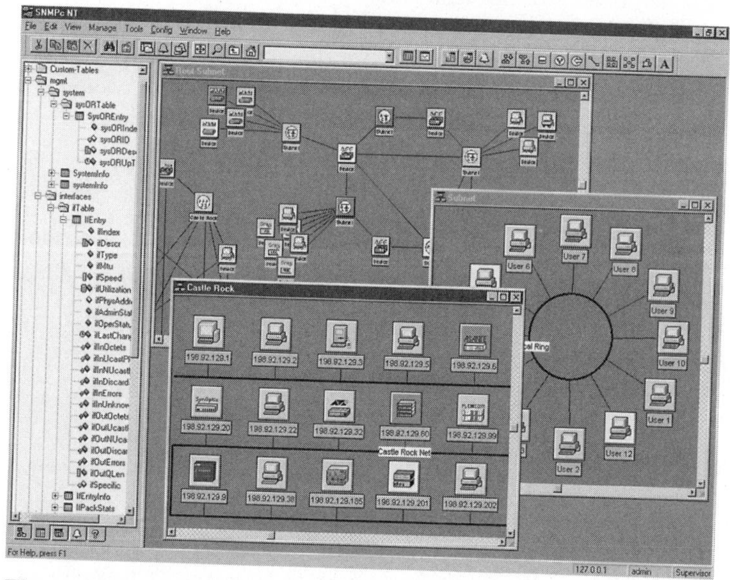

Figure S-1 Castle Rock Computing, Inc., offers SNMPc for Windows NT and Windows 2000, which, among other things, provides a graphical display that supports multilevel hierarchical mapping.

map applications without having to manually modify the configuration to accommodate non-SNMP devices.

Architectural Components

SNMP is one of three components constituting a total network management system (Figure S-2). The other two are the MIB and the network manager (NM). The MIB defines the controls embedded in network components, while the NM contains the tools that enable network administrators to comprehend the state of the network from the gathered information.

Network Manager

The NM is a program that may run on one host or more than one host, with each responsible for a particular subnet. SNMP communicates network management data to a single

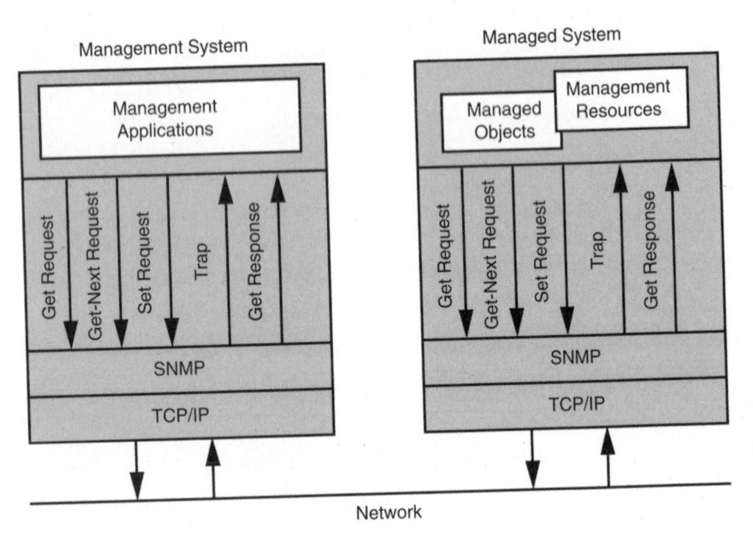

Figure S-2 The SNMP architecture.

site, called a Network Management Station (NMS). Under SNMP, each network segment must have a device, called an "agent," that can monitor devices (called "objects") on that segment and report the information to the NMS. The agent may be a passive monitoring device whose sole purpose is to read the network, or it may be an active device that performs other functions as well, such as bridging, routing, and switching. Devices that are non-SNMP-compliant must be linked to the NMS via a proxy agent.

The NMS provides the information display, communication with agents, information filtering, and control capabilities. The agents and their appropriate information are displayed in a graphical format, often against a network map. Network technicians and administrators can query the agents and read the responses on the NMS display. The NMS also periodically polls the agents, searching for anomalies. Detection of an anomaly results in an alarm at the NMS.

Management Information Base

The MIB is a list of objects necessary to manage an entity on the network. As noted, an "object" refers to hardware, software, or a logical association such as a connection or virtual circuit. The attributes of an object might include such things as the number of packets sent, routing table entries, and protocol-specific variables for IP routing. A basic object of any MIB is sysDescr, which is a textual description of the entity. This value includes the full name and version identification of the system's hardware type, software operating system, and networking software. This object should contain only printable ASCII characters.

The first MIB was concerned primarily with IP routing variables used for interconnecting different networks. There are 110 objects that form the core of the standard SNMP MIB. The latest generation MIB, known as MIB II, defines over 160 objects. It extends SNMP capabilities to a variety of media and network devices, marking a shift from Ethernets and TCP/IP wide area networks (WANs) to all media types used on LANs and WANs. Many vendors want to add value to their products by making them more manageable, so they create private extensions to the standard MIB, which can include 200 or more additional objects.

Many vendors of SNMP-compliant products include MIB tool kits that generally include two types of utilities. One, a MIB compiler, acts as a translator that converts ASCII text files of MIBs for use by an SNMP management station. The second type of MIB tool converts the translator's output into a format that can be used by the management station's applications or graphics. These output handlers, also known as "MIB editors" or "MIB walkers," let users view the MIB and select the variables to be included in the management system. Some vendors of SNMP management stations do not offer MIB tool kits but rather an optional service whereby they will integrate into the management system any MIB a user requires for a given network. This service includes debugging and technical support.

There are also MIB browsers (Figure S-3) that allow network managers, technicians, and engineers to query a remote device for software and hardware configurations via SNMP and make changes to the remote device. The remote device could be a router, switch hub, server, firewall, or any other device that supports SNMP. Another common use for a MIB browser is to find out what MIBs and object IDs (OIDs) are supported on a particular device.

Summary

SNMP's popularity stems from the fact that it works, it is reliable, and it is widely supported. The protocol itself is in the public domain. SNMP capabilities have been integrated into just about every conceivable device that is used on today's LANs and WANs. MIBs contain a list of objects that can be monitored by the SNMP.

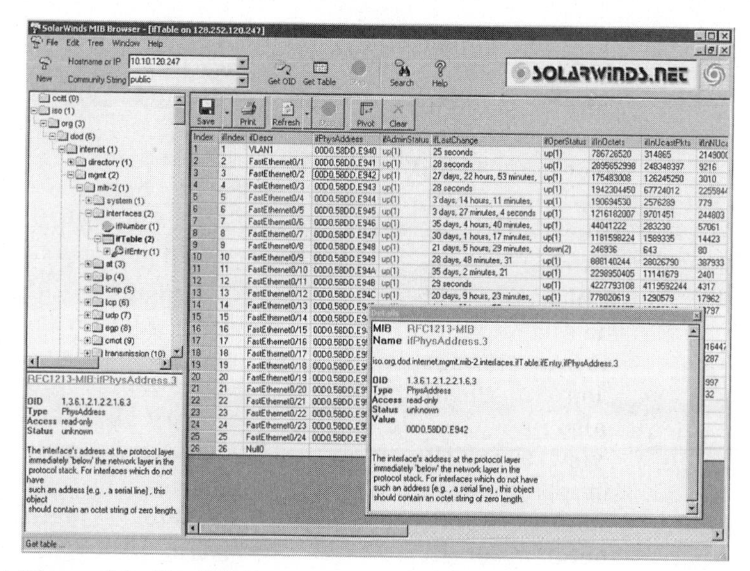

Figure S-3 The MIB browser from SolarWinds.Net, Inc., is capable of reading over a thousand standard and proprietary MIBs and 110,000 unique OIDs.

See also

Network Management Systems

Open Systems Interconnection

Protocol Analyzers

Remote Monitoring

STARLAN

AT&T developed StarLAN in the 1980s to satisfy the need for a low-cost, easy-to-install LAN that would offer more configuration flexibility than Token Ring and higher availability than Ethernet. The hub-based StarLAN was offered in two versions—1 and 10 Mbps. Since it was based on IEEE

802.3 standards, StarLAN offered interoperability with Ethernet and Token Ring through driver software.

The StarLAN architecture was based on the use of one or more hubs. Connectivity between PCs and hubs was achieved through the use of unshielded twisted-pair (UTP) wiring. In multihub networks, up to five levels of hubs could be cascaded, with one hub designated as the header to which one or more intermediate hubs were connected. The maximum distance between adjacent hubs was 250 meters. The maximum span of a five-level network was 2500 meters.

Summary

In 1991, AT&T and NCR merged under the name AT&T Global Information Solutions, where responsibility for StarLAN resided until 1996. That year, AT&T Global Information Solutions changed its name back to NCR Corp. in anticipation of being spun off to AT&T shareholders as an independent, publicly traded company. Around that time, NCR discontinued the StarLAN product line.

See also

ARCnet
Ethernet
Token Ring

SYNCHRONOUS COMMUNICATION

In synchronous or real-time communication, data go out over the link as a continuous bit- or byte-oriented stream. Instead of start and stop bits to bracket each character, as in asynchronous communication, the sending and receiving devices are synchronized with a clock or a signal encoded into the data stream. This mechanism provides the means of extracting individual characters or blocks of information from the stream.

Synchronization of the devices at each end of the link is accomplished by sending control characters, called "synchronization" or "syn characters," before any user data are actually sent. Once the devices become synchronized, transmission can take place with assurance that the data stream will be interpreted accurately by the receiver. To guard against the loss of synchronization, the receiver is periodically brought into synchronization with the transmitter through the use of the control characters embedded in the data stream.

In synchronous communication, characters are separated by time. In dispensing with the need for start and stop bits, synchronous communication is usually much more efficient in the use of bandwidth than asynchronous transmission. Assuming parity is enabled, resulting in an 8-bit byte, the 2 extra bits for the start-stop functions result in a 20 percent loss of useful bandwidth. However, synchronous communication is usually more expensive to implement because of the added complexity.

Summary

While asynchronous communication sends small blocks of data with a lot of overhead bits, synchronous techniques use big blocks of data with control bits only at the start and end of the entire transmission. But because of the minimal use of overhead bits in synchronous communication, the devices must be timed so as to correctly interpret the information in the data stream, and they need a relatively clean line to minimize the chance of errors. A clock signal in the data stream keeps the sending and receiving devices in continuous synchronization during long transmissions. A dial-up modem used over an analog line does not handle synchronous communication very well because line impairments such as noise can easily disrupt the synchronization.

See also

Asynchronous Communication

SYNCHRONOUS DATA LINK CONTROL

Synchronous Data Link Control (SDLC) was introduced by IBM in 1973 and is the preferred link level protocol for its Systems Network Architecture (SNA). It was intended to replace the older Binary Synchronous Communication (BSC) protocol developed in 1965 for wide area connections between IBM equipment. It is equivalent to the High-Level Data Link Control (HDLC) developed by the International Organization for Standardization (ISO) and adapted by many non-IBM vendors.

Like HDLC, SDLC ensures that data passed up to the next layer have been received as transmitted—error-free, without loss, and in the correct order. However, SDLC is not a peer-to-peer protocol like HDLC, and it is used only on leased lines where the connections are permanent. In a point-to-point configuration, there is one primary station that controls all communications and one secondary station. In a point-to-multipoint configuration, there is one primary station and multiple secondary stations arranged as drops along the line.

The primary station can be a mainframe or midrange central computer or a communications controller that acts as a concentrator for a number of local terminals. The primary station is capable of operating in full-duplex mode (simultaneous send and receive), while the secondary stations operate in half-duplex mode (alternately send and receive). The primary station is aware of the transmission status of the secondary stations at all times. The drops can be in different locations. A mainframe in New York, for example, may support a multidrop line with controllers connected to drops in offices in Atlanta, Chicago, Dallas, and Los Angeles.

SDLC uses the same frame format as HDLC. It is a variable-length frame that is bounded by two 8-bit flags, each containing the binary value of 01111110. The 8-bit address field of each SDLC frame always identifies the secondary station on the line. When the primary station invites a secondary station to send data (i.e., polling), it identifies the station being polled. Each secondary station sees all transmissions from the primary but responds only to frames with its own address. In a point-to-multipoint configuration, up to 254 secondary station addresses are possible, with one additional used for testing and another for broadcasting information from the primary station to all secondary stations. The SDLC frame's 8-bit control field is used to indicate whether the frame contains application-specific data, supervisory data, or command data.

The variable-length information field contains application-specific data—in other words, the user's data. The content of this field is always in multiples of 8 bits. This field is optional, however, since control fields containing unnumbered commands do not transmit application data in the frames.

The 16-bit frame check sequence (FCS) field is used to verify the accuracy of the data. As in HDLC, the FCS is the result of a mathematical computation performed on the frame at its source. The same computation is performed at the receive side of the link. If the answer does not agree with the value on the FCS field, this means some bits in the frame have been altered in transmission, in which case the frame is discarded. A window of up to seven frames can be sent from either side before acknowledgment is required. Acknowledgment of received frames is encoded in the control field of data frames so that if data are flowing in both directions, no additional frames are needed for frame acknowledgment.

Summary

SDLC is a full-duplex protocol that enables the primary and secondary stations to send data to each other at the same time. Since SDLC is a bit-oriented protocol, it is insensitive to code, which may be American Standard Code for Information Interchange (ASCII) or IBM's own Extended Binary-Coded Decimal Interchange Code (EBCDIC). SDLC is much more efficient than the older BSC. With the former, the acknowledgment for the data is usually sent with the data themselves, while in the latter, the acknowledgment is a separate transmission.

See also

High-Level Data Link Control

SYNCHRONOUS OPTICAL NETWORK

Synchronous Optical Network (SONET) is an industry standard for high-speed, Time Division Multiplexed (TDM) transmission over optical fiber. Carriers and large companies use SONET facilities for fault-tolerant backbone networks and fiber rings around major metropolitan areas. SONET-based services have a performance objective of 99.9975 percent error-free seconds and an availability rate of at least 99.999 percent. SONET combines bandwidth and multiplexing capabilities, allowing users to fully integrate voice, data, and video over the fiberoptic facility.

The SONET standards were developed by the Alliance for Telecommunications Industry Solutions (ATIS), formerly known as the Exchange Carriers Standards Association (ECSA), with input from Bellcore (now known as Telcordia Technologies), the former research and development arm of the original seven regional Bell operating companies (RBOCs) that were created in 1984 as a result of the breakup of AT&T. The standards for North America are published and distributed by the American National Standards Institute (ANSI). In 1989, the International Telecommunication Union (ITU) published the Synchronous Digital Hierarchy (SDH) standard, which is what the rest of the world uses.

Advantages

SONET provides a highly reliable transport infrastructure, offering numerous benefits to carriers and users.

Bandwidth The enormous amounts of bandwidth available with SONET and its integral management capability permit carriers to create global intelligent networks capable of supporting the next generation of services. SONET-based information superhighways support bit-intensive applications such as three-dimensional computer-aided design (CAD), medical imaging, collaborative computing, interactive virtual reality programs, and multipoint videoconferences, as well as new consumer services such as video on demand and interactive entertainment. Under current SONET standards, bandwidth is scalable from about 52 Mbps to about 13 Gbps, with the potential to go much higher. The SONET standard specifies a hierarchy of electrical and optical rates as summarized in Table S-1.

Even though the SONET standards go up only to OC-256, some equipment vendors support SONET-compliant OC-768 and OC-1536. Their systems offer a switch fabric and backplane that is OC-768/OC-1536–ready. When the optics become available and viable for 40/80 Gbps, carriers will need only to change the line cards in the optical transport switching systems to take advantage of these speeds.

The base signal rate of SONET on both the electrical and optic sides is 51.84 Mbps. On the electrical side, the synchronous transport signal (STS) is what goes to customer premises equipment (CPE), which is electrical. On the optical side, the optical carrier (OC) is what goes to the network. Both the electrical and optical signals can be multiplexed in a hierarchical fashion to form higher rate signals.

The STS-1/OC-1 frame, from which all larger frames are constructed, has a 9 × 90-byte format, which permits efficient packing of data rates in a payload of 783 bytes, plus 27 bytes for transport overhead, for a total of 810 bytes. This results in a usable payload of 48.384 Mbps (Figure S-4). The overhead bytes are used for real-time error monitoring, self-diagnostics,

TABLE S-1 Hierarchy of Electrical and Optical Rates Specified in SONET Standards

Electrical Level	Optical Level	Line Rate
STS-1	OC-1	51.84 Mbps
STS-3	OC-3	155.520 Mbps
STS-9	OC-9	466.560 Mbps
STS-12	OC-12	622.080 Mbps
STS-18	OC-18	933.120 Mbps
STS-24	OC-24	1.244 Gbps
STS-36	OC-36	1.866 Gbps
STS-48	OC-48	2.488 Gbps
STS-192	OC-192	9.95 Gbps
STS-256	OC-256	13.271 Gbps
STS-768*	OC-768 *	39.813 Gbps
STS-1536*	OC-1536 *	79.626 Gbps

*Not yet an official SONET standard.

and fault analysis. The signal is transmitted byte by byte beginning with byte 1, scanning left to right from row 1 to row 9. The entire frame is transmitted in 125 microseconds. Higher-level signals (STS-*n*) are integer multiples of the base electrical rate that are interleaved and converted to optical signals (OC-*n*).

A key feature of SONET framing is its ability to accommodate existing synchronous and asynchronous signal formats. The SONET payload can be subdivided into smaller "envelopes" called "virtual tributaries" (VTs) to transport lower-capacity signals. Because VTs can be placed anywhere on higher-speed SONET payloads, they provide effective transport for existing North American and international formats. Table S-2 highlights some of these VTs.

Standardization

Because the vast installed base of legacy T3 equipment and newer Wave Division Multiplexers (WDM) are proprietary in implementation among equipment vendors, carriers have been limited in terms of product selection and configuration

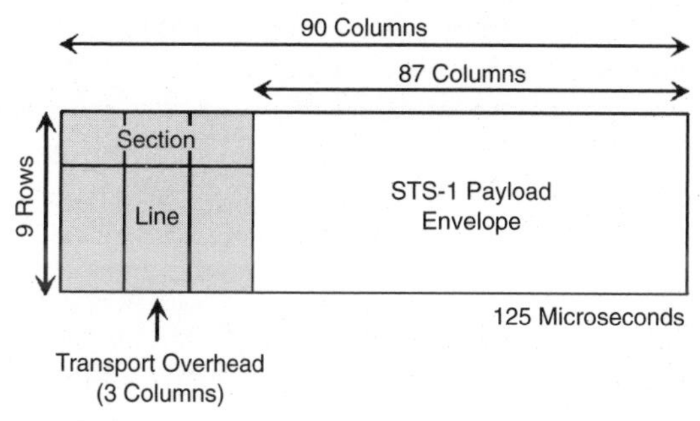

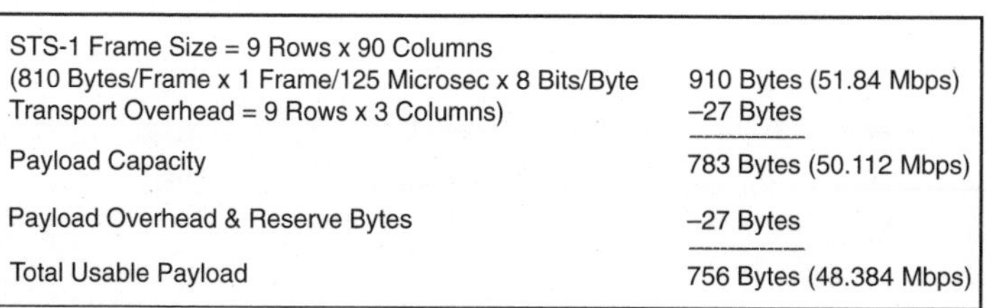

STS-1 Frame Size = 9 Rows x 90 Columns	
(810 Bytes/Frame x 1 Frame/125 Microsec x 8 Bits/Byte	910 Bytes (51.84 Mbps)
Transport Overhead = 9 Rows x 3 Columns)	–27 Bytes
Payload Capacity	783 Bytes (50.112 Mbps)
Payload Overhead & Reserve Bytes	–27 Bytes
Total Usable Payload	756 Bytes (48.384 Mbps)

Figure S-4 A SONET STS-1/OC-1 frame.

TABLE S-2 Select Virtual Tributary (VT) Payload Envelopes Specified in SONET Standards

VT Level	Line Rate (Mbps)	Standard
VT1.5	1.728	DS1
VT2	2.304	CEPT1
VT3	3.456	DS1C
VT6	6.912	DS2
VT6-N	$N \times 6.9$	Future
Async DS3	44.736	DS3

flexibility. SONET standards make possible seamless interconnectivity among compliant equipment, eliminating the need to deploy equipment from the same vendor end to end and making it much easier to interconnect compliant networks, even among international locations.

In eliminating proprietary optics, carriers are freed from the necessity of dealing with a single vendor for all their equipment needs. Instead, they can buy equipment based on price and performance and mix and match hardware from multiple vendors. SONET also gives corporate users flexibility in choosing CPE instead of being locked into the carriers' preferred vendor.

Bandwidth Management

A major advantage of SONET is its ability to easily manage huge amounts of bandwidth. Within the SONET infrastructure, carriers can tailor the width of information highways in a standard way. Carriers can parcel out specific amounts of this bandwidth to meet the needs of a broad and diverse array of user applications. Such parceling can be accomplished without adding equipment to the network or manually reconnecting cables. SONET eliminates central office reliance on metallic DSX technology with its cumbersome manual cabling and jumpers and replaces it with remotely configurable optical cross-connects. SONET provides more efficient switching and transport by eliminating the need for such midlevel network elements as back-to-back M13 multiplexers, for example, that normally cross-connect T1 facilities.

With remotely configurable SONET equipment, carriers can more expeditiously support the connectivity requirements of their customers. For example, a DS3 signal, with a rate of 44.736 Mbps, can be mapped directly into an STS-1, at 51.84 Mbps. The remaining STS-1 bytes are used for overhead and stuffing. Or two 51.84-Mbps channels can be combined to support LAN traffic between Fiber Distributed Data Interface (FDDI) backbones operating at 100 Mbps.

Real-Time Monitoring

SONET permits sophisticated self-diagnostics and fault analysis to be performed in real time, making it possible to identify problems before they disrupt service. Intelligent network elements, specifically the SONET Add-Drop Multiplexer (ADM), can automatically restore service in the event of failure via a variety of restoral mechanisms.

SONET's embedded control channels enable the tracking of end-to-end performance and identification of elements that cause errors. With this capability, carriers can guarantee transmission performance, and users can readily verify it without having to go offline to implement various test procedures. For network managers, these capabilities allow a proactive approach to problem identification, which can prevent service disruptions. Along with the self-healing capabilities of ADMs, these diagnostic capabilities ensure that properly configured SONET-compliant networks experience virtually no downtime.

Survivable Networking

SONET offers multiple ways to recover from network failures, including

- *Automatic protection switching* The capability of a transmission system to detect a failure on a working facility and to switch to a standby facility to recover the traffic. One-to-one protection switching and "one to *n*" protection switching are supported.
- *Bidirectional line switching* Requires two fiber pairs between each recoverable node. A given signal is transmitted across one pair of fibers. In response to a fiber facility failure, the node preceding the break loops the signal back toward the originating node, where the data traverse a different fiber pair to its destination.
- *Unidirectional path switching* Requires one fiber pair between each recoverable node. A given signal is transmitted in two different paths around the ring. At the receiving end, the network determines and uses the best path. In response to a fiber facility failure, the destination node switches traffic to the alternate receive path.

Universal Connectivity

As the foundation for future high-capacity backbone networks, SONET carries a variety of current and emerging traffic types, including Internet Protocol (IP), Asynchronous Transfer Mode (ATM), Frame Relay, Switched Multimegabit Data Services (SMDS), and Ethernet. To ensure universal connectivity down to the component level, equipment vendors support common product specifications, including pin-for-pin compatible transmitters and receivers. Uniformity and compatibility are achieved by using common specifications for module pin-out, footprint, logic interface, optical performance parameters, and power supplies.

Channelization

Channelized interfaces provide network configuration flexibility and contribute to lower telecommunications costs. For example, channelized T1 delivers bandwidth in economical 56/64-kbps DS0 units, each of which can be used for voice or data and routed to different locations within the network or aggregated as needed to support specific applications. Likewise, channelized DS3 delivers economical 1.544-Mbps DS1 units, which can be routed separately or aggregated as needed. Channelized interfaces also apply to the SONET world. OC-48, for example, can be channelized as follows:

- Four OC-12 tributaries, all configured for IP (packet over SONET) framing or all configured for ATM framing.
- Four OC-12 tributaries, with two configured for IP (packet over SONET) framing and the other two for ATM framing.
- Two OC-12, with one configured for IP framing and the other for ATM framing; 8 OC-3, with four configured for IP framing and four for ATM framing.
- Two OC-12, with one configured for IP framing and the other for ATM framing; 6 OC-3, with three configured for IP framing and three for ATM framing; 6 DS3, with three configured for IP framing and three for ATM framing.
- Forty-eight DS3s, with 24 configured for IP framing and 24 for ATM framing.

Multiservice channelized SONET may be implemented on a single OC-48 line card, which provides IP packet and ATM cell encapsulation to support business data and Internet services from the same hardware platform. Some vendors offer products that act as a native IP router as well as a native ATM switch in a single modular platform, enabling any port to be configured for either packet over SONET (PoS) or ATM service. The configurations are implemented through keyboard commands at service time rather than permanently assigned at network build out time.

A new generation of optical systems has become available that support both SONET and WDM. They allow wavelengths to be shared by a number of services. Carriers are no longer forced to choose between SONET and WDM; it is now possible to have both functioning together so that a wavelength provisioned through a WDM system can carry a SONET payload.

Summary

The continued deployment of SONET-compliant equipment on public and private networks will have a significant impact on telephone companies, interexchange carriers, and corporate users. SONET offers scalable bandwidth, integral fault recovery and network management, interoperability between public services and corporate enterprise networks, and multivendor equipment interoperability. For carriers, timely and effective deployment of SONET determines the types of broadband services that can be offered to subscribers, since SONET comprises the physical layer on which broadband services are built.

See also

Fiber Distributed Data Interface
Fibre Channel

T

T1 LINES

T1 lines are a type of T-carrier facility that provides a transmission rate of up to 1.544 Mbps using digital signal level 1 (DS1) signaling. Among other types of applications, T1 lines can be used to interconnect local area networks (LANs) across the wide area network (WAN). Two pairs of wires are used to achieve full-duplex transmission—one pair for the send path and one pair for the receive path. The available bandwidth is divided into 24 channels operating at 64 kbps each plus an 8-kbps channel for basic supervision and control. Voice is sampled and digitized via Pulse Code Modulation (PCM) before being placed on the line. T1 digital lines are used for more economical and efficient transport over the WAN.

Cost savings are the result of consolidating multiple lower-speed voice and data channels via a multiplexer or channel bank and sending the traffic out over the higher-speed T1 line. This is more cost-effective than dedicating a separate lower-speed line to each terminal device connected to the WAN. The economics are such that only five to eight analog lines are needed to cost-justify the move to T1.

Greater bandwidth efficiency can be obtained by compressing voice and data to make room for even more channels over the available bandwidth. This can result in even more cost savings. Individual channels also can be dropped or inserted at various destinations along the line's route. Network management information can be embedded in each channel for enhanced levels of supervision and control.

Usually a T1 multiplexer provides the means for companies to realize the full benefits of T1 lines, while channel banks offer a low-cost alternative. The difference between the two devices is that T1 multiplexers offer higher line capacity, support more types of modules and interfaces, and provide more management features than channel banks.

D4 Framing

T1 multiplexers and channel banks transmit voice and data in frames that are called "D4 frames." The frames are bounded on each side by framing bits that perform two functions: They identify the beginning of each frame and help locate the signaling information. For voice, this bit is carried in the eighth bit position of frames 6 and 12 (Figure T-1).

D4 frames consist of 193 bits, which equates to 24 channels of 8 bits each, plus a single framing bit. Each frame contains a framing bit or signaling bit in the last position (193), which permits the management of the DS1 facility itself. This is done by robbing the least significant bit in the data stream, which otherwise carries information or signaling data. Another bit is used to mark the start of a frame. Twelve D4 frames compose a superframe.

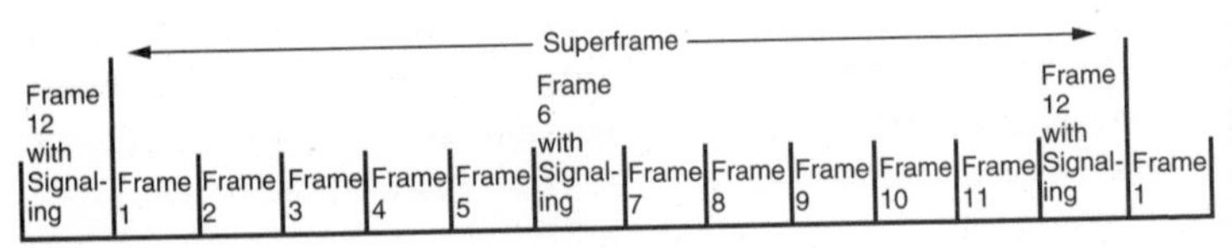

Figure T-1 D4 framing (i.e., superframe).

Extended Superframe Format

Extended Superframe Format (ESF) is an enhancement to T-carrier that specifies methods for error monitoring, reporting, and diagnostics. Use of ESF allows technicians to maintain and test the T1 line while it is in service and often fix minor troubles before they affect service. ESF extends the normal 12-frame superframe structure of the D4 format to 24 frames. By doubling the number of bits available, more diagnostic functionality also becomes available (Figure T-2).

Of the 8 kbps of bandwidth (repetition rate of bit 193, the framing bit) allocated for basic supervision and control, 2 kbps is used for framing, 2 kbps for Cyclic Redundancy Checking (CRC-6), and 4 kbps for the Facilities Data Link (FDL). With CRC, the entire circuit may be segmented so that it can be monitored for errors without disrupting normal data traffic. In this manner, performance statistics can be generated to monitor T1 circuit quality. Via FDL, performance report messages are relayed to the customer's equipment, usually a channel service unit (CSU), at 1-second intervals. Alarms also can use the FDL, but performance report messages always have priority.

ESF diagnostic information is collected by the CSU at each end of the T1 line for access by both the carrier and customer. CSUs gather statistics on such things as clock synchronization errors and framing errors, as well as errored seconds, severely errored seconds, failed seconds, and bipolar violations. A supervisory terminal connected to the CSU displays this information, furnishing a record of circuit performance.

Originally, the CSU compiled performance statistics every 15 minutes. This information would be kept updated for a full 24 hours so that a complete performance history for the day could be available to the carrier. The carrier would have to poll each CSU to retrieve the collected data and clear its storage register. By equipping the CSU with dual registers—one for the carrier and one for the user—both carrier and user alike have full access to the T1 line's performance history.

Frame Number	D4 Format: Value of 193rd Bit	D4 Format: Use	ESF Format: Value of 193rd Bit	ESF Format: Use
1	1	F_T	X	FDL
2	0	F_S	X	CRC
3	0	F_T	X	FDL
4	0	F_S	0	F_S
5	1	F_T	X	FDL
6	1	F_S	X	CRC
7	0	F_T	X	FDL
8	1	F_S	0	F_S
9	1	F_T	X	FDL
10	1	F_S	X	CRC
11	0	F_S	X	FDL
12	0	F_T	1	F_S
13	First 12 Frames Repeated		X	FDL
14			X	CRC
15			X	FDL
16			0	F_S
17			X	FDL
18			X	CRC
19			X	FDL
20			1	F_S
21			X	FDL
22			X	CRC
23			X	FDL
24			1	F_S

Frames 1–12: 1 Superframe (D4). Frames 1–24: 1 Extended Superframe (ESF).

Notes:

- F_T Terminal framing bit } (F bit)
- F_S Multiframe alignment bit } (F bit)
- FDL 4 Kbps data link bit (M bit)
- CRC Cyclic redundancy check bit (C bit)
- X Data dependent

Figure T-2 A comparison of bit 193 bit in D4 and ESF formats.

Today, the CSU is not required to store performance data for 24 hours. Also, the CSU no longer responds to polled requests from carriers but simply transmits ESF performance messages every second.

ESF also allows end-to-end performance data and sectionalized alarms to be collected in real time. This allows the customer to narrow down problems between carrier access points and on interoffice channels and to find out in which direction the error is occurring.

E1 Frame Format

In today's increasingly global economy, more companies are expanding their private networks beyond the United States and Canada to European, Asian, and South American locations. In doing so, the first thing they notice is that the primary bit rate service may not be T1 but E1. Whereas T1 has a maximum bit rate of 1.544 Mbps, E1 has a maximum bit rate of 2.048 Mbps. Between the two, there is only one common characteristic: the 64-kbps channel, or DS0. A T1 line carries 24 DS0s, while an E1 carries 32 DS0s. Despite this commonality, the DS0s of a T1 line and the DS0s of an E1 line are not compatible.

Although each uses PCM to derive a 64-kbps voice channel, the form of PCM encoding differs. T1 uses a PCM encoding technique based on mulaw companding, while E1 uses a PCM coding technique based on A-law companding. This difference is not as great as it may seem; most multiplexers and carrier switches have the integral capability to convert between the two. Conversion includes both the signaling format and the companding method.

In E1, as in T1, there is the need to identify the DS0s to the receiver. The E1 format uses framing for this function, as does T1; i.e., there are 8000 frames per second, with each frame containing one sample from each time slot, numbered 0 to 31.

The frame synchronization in E1 uses half of time slot 0, while signaling occupies time slot 16 in the 0 to 31 sequence, resulting in 30 channels left for user information. However, it is not possible for all 30 channels to signal within the 8 bits available in time slot 16. The channels therefore must take turns using time slot 16. Two channels send their signaling bits in each frame. The 30 user channels then take 15 frames to cycle through all the signaling bits. One additional frame is used to synchronize the receiver to the signaling channel, so the full multiframe ends up having 16 frames. This multiframe corresponds to the T1 superframe.

Summary

In recent years, T1 lines and services have become the basic building blocks of digital networks. They can support voice, data, and video channels. The individual channels can be added to or dropped from the aggregate bandwidth to improve the efficiency and economy of a private network. In addition, these channels can traverse the public switched telephone network (PSTN) to bring off-net locations into the private network. Alternatively, the channels can go through a carrier's Frame Relay network or Asynchronous Transfer Mode (ATM) network, allowing even greater efficiencies and economies for certain applications. The channels of a T1 can even be bonded together to support bandwidth-intensive applications on private networks.

See also

Data Compression
Multiplexers

THIN CLIENT ARCHITECTURE

The thin client architecture originated with Oracle Corp. in 1995 as part of its concept of network computing. In this model of computing, applications are deployed, managed, supported, and executed from servers on the LAN. This allows organizations to deploy low-cost client devices on the desktop and, in the process, overcome the critical application deployment challenges of management, access, performance, and security. Since then, Oracle has abandoned its original network computing model and replaced it with the Internet computing model, in which all applications and databases reside on the Internet. Regardless of environment, the user devices remain the same—thin clients.

Fat versus Thin Clients

The terms *fat* and *thin* refer primarily to the amount of processing being performed at the client. Terminals are the ultimate thin clients because they rely exclusively on the host for applications and processing. Stand-alone PCs are the ultimate fat clients because they have the resources to run all applications locally and handle the processing themselves. Spanning the continuum from all-server processing to all-client processing is the client/server environment where there is a distribution of work between both devices.

Client/server was once thought to be the ideal computing solution. Despite the initial promises held out for client/server solutions, today there is much dissatisfaction with their implementation. Client/server solutions are too complex, desktops are too expensive to administer and upgrade, and the applications still are not secure and reliable enough. Furthermore, client/server applications take too long to develop and deploy, and incompatible desktops prevent universal access. The thin client architecture attempts to overcome these limitations.

Benefits of Thin Clients

Businesses that have embraced thin clients are using them for a variety of applications. Most thin clients are used to access an office suite like Microsoft Office, but some are used to run mission-critical applications, such as accounting, transaction processing, and order-entry applications. Thin clients are also running engineering, enterprise resource planning (ERP), and medical applications.

Users of thin clients are usually task-oriented and prefer to do their work without being distracted by technology issues. These are front-line professionals, such as doctors in

HMOs, accountants, engineers, and salespeople of big-ticket items, such as industrial equipment and real estate. Thin clients are also used for back-office operations supported by clerical and administrative staff, low-level salespeople, and workers on the shop floor. The human resources departments of large companies can use thin clients installed solely with a browser to allow job applicants to fill out forms.

Thin client computing has several compelling benefits that are of key concern to organizations dealing with escalating information technology (IT) costs:

- *Cost of ownership* Thin client computing lowers the total cost of ownership for the IT infrastructure.
- *Platform independence* The use of thin clients allows applications to be written and deployed without regard for the desktop platform they will run on.
- *Flexibility* The use of thin clients eases the deployment of new applications, since they are installed and maintained at the server rather than every desktop.
- *Security* In terms of administration and overall protection of the network, some security features, such as virus filtering, are best implemented at the server rather than every desktop. In addition, centralizing security at the server allows easier control of access to files, applications, and networks.

If there is one disadvantage to thin client computing, it is that many companies may have to upgrade their servers to accommodate the increased load that inevitably will occur when they are forced to support more clients. In addition to adding ports and interfaces to other networks, the servers may have to be upgraded with redundancy features to ensure continuous availability. However, even with these one-time costs, over a span of 3 to 5 years, companies can achieve significant savings in total cost of ownership. The savings accumulate primarily in easier administration of the computing environment.

Operation of Thin Clients

The operation of thin clients is fairly simple: They are dependent on servers for boot-up, applications, processing, and storage. Since most thin clients may not have a hard drive, the server provides booting service to the network computers when they are turned on. The server can be a suitably equipped PC, a RISC-based workstation, a midrange host like the IBM AS/400, or even a mainframe. The server typically connects to the LAN with an Ethernet or Token Ring adapter and supports Transmission Control Protocol/Internet Protocol (TCP/IP) for WAN connections to the public Internet or a private intranet.

Since all applications reside on the server, installation is done only once—not hundreds or thousands of times at individual desktops—via electronic software distribution tools. Periodic updates to applications and bug fixes are conducted on the server. This ensures that every network computer uses the same version of the application every time it is accessed.

Network computers can access both Java and Windows applications on the server, as well as various terminal emulations for access to legacy data. Users accessing Java applications do so through a Java-enabled Web browser, which also gives them access to applications on the Internet or intranet. For Windows applications, the server typically allows access to multiple users in accordance with a network license from the software provider. The server's operating system also may include terminal support for 3270, 5250, and X-Windows servers.

Role of Java

Java plays a key role in the thin client architecture. Developed by Sun Microsystems, Java provides a cleaner, simpler language that can be processed faster and more efficiently than C or C++ on nearly any microprocessor.

Whereas C or C++ source code is optimized for a particular model of processor, Java source code is compiled into a universal format. It writes for a virtual machine in the form of simple binary instructions. Compiled byte code is executed by a Java run-time interpreter, performing all the usual activities of a real processor, but within a safe, virtual environment instead of a particular computer platform.

Much of the Java applications development at major corporations hinges around the Web because the Internet has become an economical way to access corporate information, applications, and business tools from remote locations. A Java-enabled Web browser is used as the interface for this access.

With the advent of thin clients has come the concept of "thin servers," which are dedicated, special-purpose devices that are optimized for supporting thin clients. In addition to supporting a narrow range of network applications, the thin server supports localized services to reduce network traffic congestion and provides fast access to routine applets used for many database and spreadsheet programs.

Unlike traditional server-based systems that require an investment in separate hardware and software, the thin server comes complete with hardware and software at a fraction of the cost. These servers usually come ready to set up, install, and configure out of the box. Typically, all the user needs to do is plug in an Ethernet cable and set the IP address. A thin server usually comes with a browser graphical user interface (GUI), a Java-based management application, and an embedded HyperText Transfer Protocol (HTTP)–compliant operating system.

Businesses that have embraced thin clients are using them for a variety of applications. Most thin clients are used to access an office suite like Microsoft Office, but some are used to run mission-critical applications, such as accounting, transaction processing, and order-entry applications. Thin clients are also running engineering, ERP, and medical applications.

Summary

The promise of the thin client architecture is that it allows organizations to more quickly realize value from the applications and data required to run their businesses, receive the greatest return on computing investment, and accommodate both current and future enterprise computing needs. This does not mean thin clients will replace fat-client PCs. The two are really complementary architectures that can be managed centrally. In some cases, it may even be difficult to distinguish between the two—the line of demarcation seems to be quite fluid. Overall, thin clients should provide the economies and efficiencies organizations are looking for, but these advantages will come primarily from among the user population that is task-oriented.

See also

Client/Server Networks

Network Computing

TOKEN RING

Token Ring is a type of LAN that was introduced by IBM in 1985. It had a top speed of 4 Mbps and was developed as a response to the commercial availability of Ethernet, which was developed jointly by Digital Equipment, Intel, and Xerox. When Ethernet was introduced, IBM did not endorse it, mainly because its equipment would not work in that environment. Later, in 1989, the speed of Token Ring was boosted to 16 Mbps.

The ring is essentially a closed loop, although various wiring configurations that employ a multistation access unit (MAU)[1] and patch panel may cause it to resemble a star

[1]This is a nonintelligent concentrator that can be used as the basis for implementing Token Ring LANs.

topology (Figure T-3). In addition, today's intelligent wiring hubs and Token Ring switches can be used to create dedicated pipes between rings and provide switched connectivity between users on different rings.

The cable distance of a 4-Mbps Token Ring is limited to 1600 feet between stations, while the cable distance of a 16-Mbps Token Ring is 800 feet between stations. Because each node acts as a repeater in that data packets and the token

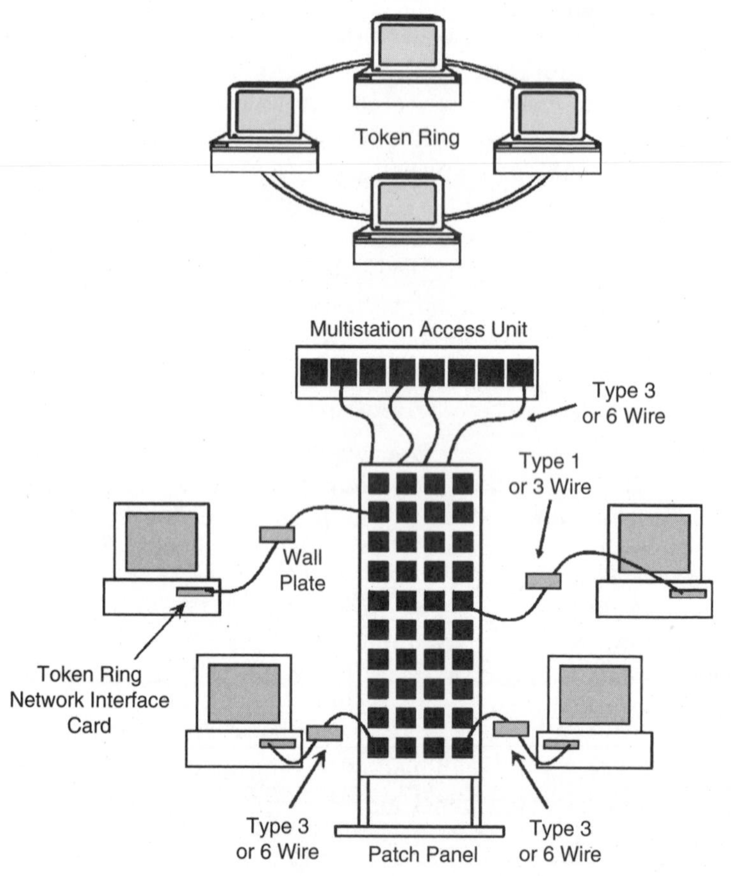

Figure T-3 Token Ring topologies: closed ring and star-wired.

are regenerated at their original signal strength, Token Ring networks are not as limited by distance as are bus-type networks. Like its nearest rival, Ethernet, Token Ring networks normally use twisted-pair wiring, shielded or unshielded.

Advantages of Token Ring

The ring topology offers several advantages:

- Since access to the network is not determined by a contention scheme, as is Ethernet, a higher throughput rate is possible in heavily loaded situations, limited only by the slowest element—sender, receiver, or link speed.
- With all messages following the same path, there are no routing problems to contend with. Logical addressing may be accommodated to permit message broadcasting to selected nodes.
- Adding terminals is accomplished easily—one connector is unplugged, the new node is inserted, and both nodes are plugged into the network. Other nodes are updated with the new address automatically.
- Control is simple, requiring little in the way of additional hardware or software to implement.
- The cost of network expansion is proportional to the number of nodes.

Another advantage of Token Ring is that the network can be configured to give high-priority traffic precedence over lower-priority traffic. Only if a station has traffic equal to or higher in priority than the priority indicator embedded in the token can it transmit data onto the network.

The Token Ring in its pure configuration is not without liabilities, however. Failed nodes and links can break the ring, preventing all the other terminals from using the network. At extra cost, a dual-ring configuration with redundant hardware and bypass circuitry is effective in isolating faulty nodes from the rest of the network, thereby increasing reliability. Through the use of bypass circuitry, physically adding or deleting terminals to the Token Ring network, is accomplished without breaking the ring. Specific procedures must be used to ensure that the new station is recognized by the others and is granted a proportionate share of network time. The process for obtaining this identity is referred to as "neighbor notification." This situation is handled quite efficiently, since each station becomes acquainted with the address of its predecessor and successor on the network on initialization (power-up) or at periodic intervals thereafter.

Frame Format

The frame size used on 4-Mbps Token Rings is 4048 bytes, while the frame size used on 16-Mbps Token Rings is 16,192 bytes. The IEEE 802.5 Standard defines two data formats—tokens and frames (Figure T-4). The token, three octets in length, is the means by which the right to access the medium is passed from one station to another. The frame format of Token Ring differs only slightly from that of Ethernet. The following fields are specified for IEEE 802.5 Token Ring frames:

- *Start Delimiter (SD)* Indicates the start of the frame.
- *Access Control (AC)* Contains information about the priority of the frame and a need to reserve future tokens, which other stations will grant if they have a lower priority.
- *Frame Control (FC)* Defines the type of frame, either Media Access Control (MAC) information or information for an end station. If the frame is a MAC frame, all stations on the ring read the address, but only the destination station can read the user data.
- *Destination Address (DA)* Contains the address of the station that is to receive the frame. The frame can be addressed to all stations on the ring.
- *Source Address (SA)* Contains the address of the station that sent the frame.

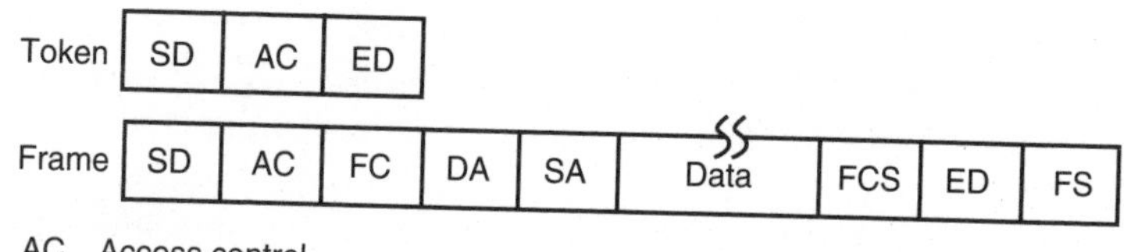

AC Access control
DA Destination address
ED Ending delimiter
FC Frame control
FCS Frame check sequence
FS Frame status
SA Source address
SD Starting delimiter

Figure T-4 Format of IEEE 802.5 token and frame.

- *Data* Contains the data "payload." If the frame is a MAC frame, this field may contain additional control information.
- *Frame Check Sequence (FCS)* Contains error-checking information to ensure the integrity of the frame to the recipient.
- *End Delimiter (ED)* Indicates the end of the frame.
- *Frame Status (FS)* Provides indications of whether one or more stations on the ring recognized the frame, whether the frame was copied, or whether the destination station is not available.

Operation

A token is circulated around the ring, giving each station in sequence a chance to put information on the network. The station seizes the token, replacing it with an information frame. Only the addressee can claim the message. At the completion of the information transfer, the station reinserts the token on the ring. A token-holding timer controls the maximum amount of time a station can occupy the network before passing the token to the next station.

A variation of this token-passing scheme allows devices to send data only during specified time intervals. The ability to determine the time interval between messages is a major advantage over the contention-based access method used by Ethernet. This time-slot approach can support voice transmission and videoconferencing, since latency is controllable.

To protect the Token Ring from potential disaster, one terminal is typically designated as the control station. This terminal supervises network operations and does important housecleaning chores, such as reinserting lost tokens, taking extra tokens off the network, and disposing of "lost" packets. To guard against the failure of the control station, every station is equipped with control circuitry so that the first station detecting the failure of the control station assumes responsibility for network supervision.

Dedicated Token Ring

Dedicated Token Ring (DTR), also known as "full-duplex Token Ring," lets devices directly connected to a Token Ring switch transmit and receive data simultaneously at 16 Mbps, effectively providing each station with 32 Mbps of throughput.

Under the IEEE 802.5r Standard for DTR, which defines the requirements for end stations and concentrators that operate in full-duplex mode, all new devices will coexist with existing Token Ring equipment and will adhere to the token-passing access protocol. The DTR concentrator consists of C-Ports and a data transfer unit (DTU). The C-Ports provide basic connectivity from the device to Token Ring stations, traditional concentrators, or other DTR concentrators. The DTU is the switching fabric that connects the C-Ports within a DTR concentrator. In addition, DTR concentrators can be linked to each other over a LAN or WAN via data transfer services such as Asynchronous Transfer Mode (ATM).

High-Speed Token Ring

With 16-Mbps Token Ring, connections between switches easily become congested at busy times, and high-performance servers become less able to deliver their full bandwidth potential. The need for a high-speed solution for Token Ring has become readily apparent in recent years. Other high-speed technologies are available—FDDI, Fast Ethernet, and ATM—but they are inadequate for the Token Ring environment.

In 1997, several Token Ring vendors teamed up to address this situation by forming the High Speed Token Ring Alliance (HSTRA). A year later, the alliance issued a specification for High Speed Token Ring (HSTR) that offers 100 Mbps and preserves the native Token Ring architecture. However, to keep costs to a minimum and to shorten its development time, HSTR is based on the IEEE 802.5r Standard for Dedicated Token Ring, adapted to run over the same 100-Mbps physical transmission scheme used by dedicated Fast Ethernet. HSTR links can be run in either half-duplex or full duplex mode, just like DTR.

HSTR uses existing switches, hubs, bridges, routers, network interface cards (NIC), and cabling. This introduces greater throughput where the enterprise needs it most—at the server and backbone. Upgrading these connections with HSTR requires only that an HSTR uplink be plugged into a Token Ring switch and that the existing 16-Mbps server network NIC be replaced with a 100-Mbps HSTR NIC. To complete the upgrade, the two devices are connected with appropriate cabling. The 100-Mbps HSTR operates over both Category 5 UTP and IBM Type 1 STP cable, as well as multimode fiberoptic cabling.

It is also possible to connect desktop systems to Token Ring switches on dedicated 100-Mbps HSTR connections. Token Ring vendors offer 4/16/100-Mbps adapter cards that enable companies to standardize on a single network adapter and prepare their infrastructure for the eventual move to HSTR. While the HSTR standard does not define an autonegotiation algorithm, individual vendors have a number of ways to implement the feature while adhering to the standard. With this feature, HSTR products operate at the maximum connection speed, automatically determining whether to transmit at 4, 16, or 100 Mbps. Many corporations install autonegotiating 4/16/100-Mbps NICs in today's desktops, even if there is no immediate need for 100 Mbps throughput to the desktop. When the hub or switch at the other end of the connection is later upgraded to 100-Mbps HSTR, the Token Ring desktop will automatically adjust transmission to 100 Mbps.

Since Ethernet packets can be carried over Token Ring links, HSTR makes a good backbone medium for the mixed-technology LAN. With support for the maximum Token Ring frame size, an HSTR backbone segment is able to handle Ethernet and Token Ring frames on the same virtual LAN (VLAN) connection, which Fast Ethernet would not be able to do without a lot of processing to break down the larger Token Ring frames.

Summary

Token Ring is a stable technology with proven capacity for handling today's applications. At the same time, network managers can protect their current investments in Token Ring by understanding application performance and the capacity of the network and tuning it accordingly. The DTR standard prolongs the useful life of Token Ring networks while meeting the increased bandwidth requirements of emerging applications such as document imaging, desktop videoconferencing, and multimedia. Nevertheless, Token Ring has been overtaken by Ethernet in terms of both technology and market share. Not only is Ethernet cheap to implement, it also offers a migration path to higher speeds that Token Ring standards lack. While Ethernet has reached gigabit-per-second speeds, Token Ring has not been standardized beyond 100 Mbps.[2]

[2]Although some vendors such as Cisco Systems offer a form of Gigabit Token Ring, critics claim that they are proprietary products and not authentic Token Ring.

See also

ARCnet
Ethernet
Fiber Distributed Data Interface
StarLAN

TRANSCEIVERS

A transmitter-receiver (transceiver) connects a computer, printer, or other device to a LAN (Figure T-5). The transceiver may be a component integrated on the network interface card (NIC), or it can be a separate device that connects to the NIC with a drop cable. In the latter case, the transceiver cable and connectors form the attachment unit interface (AUI) and the transceiver is the medium attachment unit (MAU).

In the Ethernet environment, the MAU has four basic functions:

- *Transmit* Transmits serial data onto the medium.

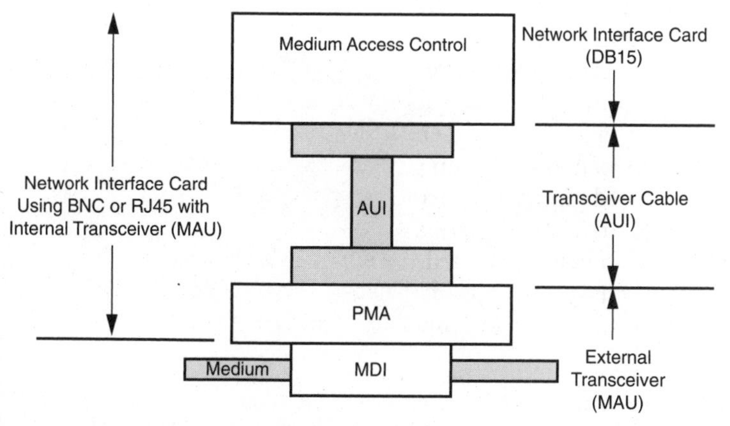

Figure T-5 Transceiver architecture.

- *Receive* Receives serial transmission and passes these signals to the attached station.
- *Collision detection* Detects the presence of simultaneous signals on the network and alerts the station.
- *Jabber function* Automatically interrupts the transmit function to inhibit abnormally long data stream output.

The MAU consists of the physical medium attachment (PMA), which provides the functions and two connectors. On the network side, the MAU attaches to the medium-dependent interface (MDI). The specific interface depends on the type of medium used. For example, 10BaseT (twisted-pair) uses a RJ-45 connector and 10Base2 (thin coax) uses a BNC connector (Figure T-6), while the older 10Base5 (thick coax) implements a special "vampire" tap that pierces the coaxial cable and makes contact with both the center conductor and shield.

Status Indicators

Transceivers offer several indicators to keep the user informed of performance status at any given time:

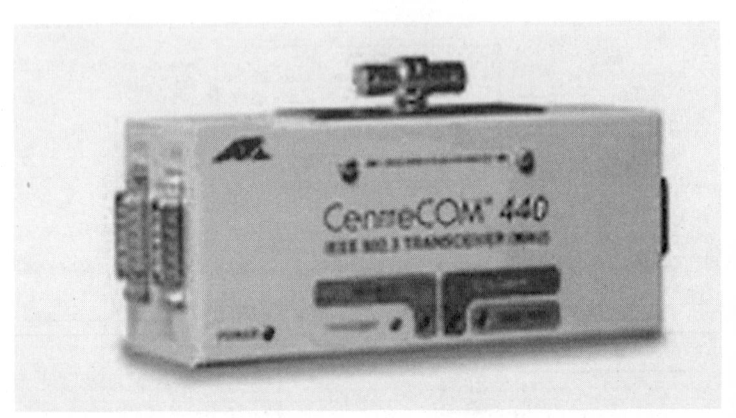

Figure T-6 This four-port transceiver from Allied Telesyn has a BNC connector (*top*) that is used for connecting the unit to a 10Base2 (thin coax) Ethernet LAN.

- *Transmit* Indicates that packets are being transmitted onto the media.
- *Receive* Indicates that packets are being received from the media.
- *SQE* Indicates that a signal quality error (SQE) test signal is present.
- *Collision* Indicates that a collision has occurred.

A user-selectable switch is provided, permitting the network manager to choose between enabling and disabling the SQE test function. This feature permits the transceiver to be used with repeaters that cannot support the heartbeat function.

Summary

Transceivers are available in a variety of configurations to support different LAN types and media. There are transceivers for all versions of Ethernet, as well as FDDI and ATM networks. There are transceivers for coaxial cable (thick and thin), twisted-pair wiring (shielded and unshielded), optical fiber (single- and multimode), and wireless (spread spectrum and infrared).

See also

Media Converters

Network Interface Cards

TRANSMISSION CONTROL PROTOCOL/INTERNET PROTOCOL (TCP/IP)

Transmission Control Protocol/Internet Protocol (TCP/IP) is a suite of networking protocols that is valued for its ability to interconnect diverse computing platforms—from PCs,

Macintoshes, and UNIX systems to mainframes and supercomputers. The protocol suite originated from the work done by four key individuals over 30 years ago: Vinton Cerf, Robert Kahn, Leonard Kleinrock, and Lawrence Roberts (Figure T-7). Each disagrees on who deserves the lion's share of credit in the development of the Internet. Although the early experiments of Kleinrock made his computer the first node on the early Advanced Research Projects Agency

Figure T-7 Vinton Cerf (*top left*), Robert Kahn (*top right*), Leonard Kleinrock (*bottom left*), and Lawrence Roberts (*bottom right*) are the four individuals generally credited with the initial development work that led to today's Internet.

Network (ARPANET), Cerf and Roberts generally get the credit for designing the network architecture that eventually became known as the Internet.

The U.S. government's Advanced Research Projects Agency (ARPA) funded further development of the TCP/IP suite in the 1970s. As noted, the protocol suite was developed to enable different networks to be joined to form a virtual network known as an "internetwork." The original Internet was formed by converting an existing conglomeration of networks belonging to ARPANET over to TCP/IP, which evolved to become the backbone of today's Internet. Today, the Internet Engineering Task Force (IETF) oversees the development of the TCP/IP protocol suite and related protocols.

Several factors have driven the acceptance of TCP/IP for mainstream business and consumer use over the years. These include the technology's ability to support LAN and WAN connections, its open architecture, and a set of specifications that are freely available in the public domain. Although not the most functional or robust transport available, TCP/IP offers a mature, dependable environment for corporate users who need a common denominator for their diverse and sprawling networks.

Key Protocols

The key protocols in the suite include the Transmission Control Protocol (TCP), the Internet Protocol (IP), and the User Datagram Protocol (UDP). There are also application services that include the telnet protocol, providing virtual terminal service, the File Transfer Protocol (FTP), the Simple Mail Transfer Protocol (SMTP), and the Simple Network Management Protocol (SNMP).

Transmission Control Protocol TCP forwards data delivered by IP to the appropriate process at the receiving host. Among other things, TCP defines the procedures for breaking up the data stream into packets and reassembling them in the proper order to reconstruct the original data stream at the receiving end. Since the packets typically take different routes to their destination, they arrive at different times and out of sequence. All packets are stored temporarily until the missing packets arrive so that they can be put in the correct order. If a packet arrives damaged, it is simply discarded and another one resent.

To accomplish these and other tasks, TCP breaks the messages or data stream down into a manageable size and adds a header to form a packet. The packet's header (Figure T-8) consists of

- *Source port (16 bits)/destination port (16 bits) address* The source and destination ports correspond to the calling and called TCP applications. The port number is usually

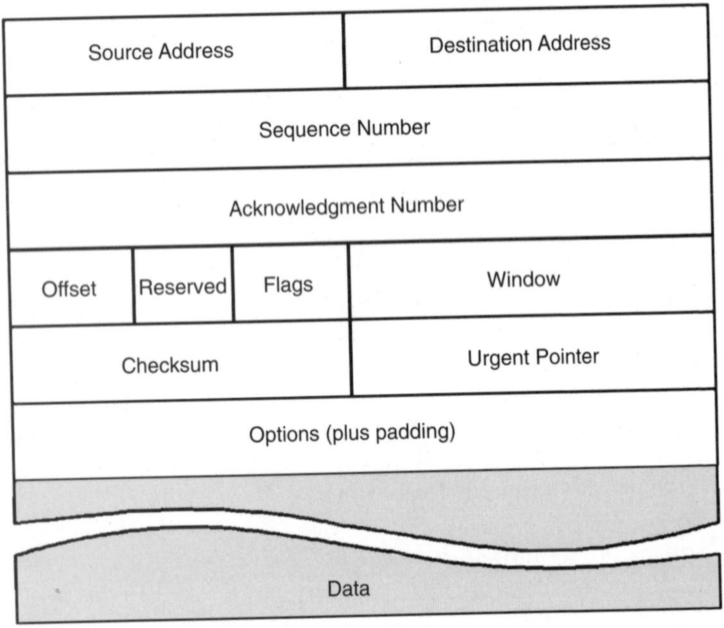

Source Address
Destination Address
Sequence Number
Acknowledgment Number
Offset
Reserved
Flags
Window
Checksum
Urgent Pointer
Options (plus padding)
Data

Figure T-8 TCP packet header.

assigned by TCP whenever an application makes a connection. There are well-known ports associated with standard services such as telnet, FTP, and SMTP.

- *Sequence number (32 bits)* Each packet is assigned a unique sequence number that lets the receiving device reassemble the packets in sequence to form the original data stream.
- *Acknowledgment number (32 bits)* The acknowledgment number indicates the identifier or sequence number of the next expected packet. Its value is used to acknowledge all packets transmitted in the data stream up to that point. If a packet is lost or corrupted, the receiver will not "acknowledge" that particular packet. This negative acknowledgment triggers a retransmission of the missing or corrupted packet.
- *Offset (4 bits)* The offset field indicates the number of 32-bit words in the TCP header. This is required because the TCP header may vary in length, according to the options that are selected.
- *Reserved (6 bits)* This field is not currently used but may accommodate some future enhancement of TCP.
- *Flags (6 bits)* The flags field serves to indicate the initiation or termination of a TCP session, to reset a TCP connection, or to indicate the desired type of service.
- *Window (16 bits)* The window field, also called the "receive window size," indicates the number of 8-bit bytes that the host is prepared to receive on a TCP connection. This provides precise flow control.
- *Checksum (16 bits)* The checksum is used to determine whether the received packet has been corrupted in any way during transmission.
- *Urgent Pointer (16 bits)* The urgent pointer indicates the location in the TCP byte stream where urgent data end.
- *Options (0 or more 32-bit words)* The options field is typically used by TCP software at one host to communicate

with TCP software at the other end of the connection. It passes such information as the maximum TCP segment size that the remote machine is willing to receive.

The bandwidth and delay of the underlying network impose limits on throughput. Poor transmission quality causes packets to be discarded, which in turn results in retransmissions and causes poor throughput.

Internet Protocol The Internet is composed of a series of autonomous systems, or subnetworks, each of which is locally administered and managed. These subnets may consist of Ethernet LANs, Integrated Services Digital Network (ISDN), Frame Relay networks, and ATM networks, over which IP runs. IP delivers data between these different networks through routers that process packets from one autonomous system (AS) to another.

Each node in the AS has a unique IP address. The IP adds its own header and checksum to make sure the data are routed properly (Figure T-9). This process is aided by the presence of routing update messages that keep the address tables in each router current. Several different types of update messages are used, depending on the collection of subnets involved in a management domain. The routing tables list the various nodes on the subnets as well as the paths between the nodes. If the data packet is too large for the destination node to accept, it will be segmented into smaller packets.

The IP header consists of the following fields:

- *IP version (4 bits)* The current version of IP is 4; the next generation of IP is 6.
- *IP header length (4 bits)* Indicates header length; if options are included, the header may have to be padded with extra 0s so that it can end at a 32-bit-word boundary. This is necessary because header length is measured in 32-bit words.

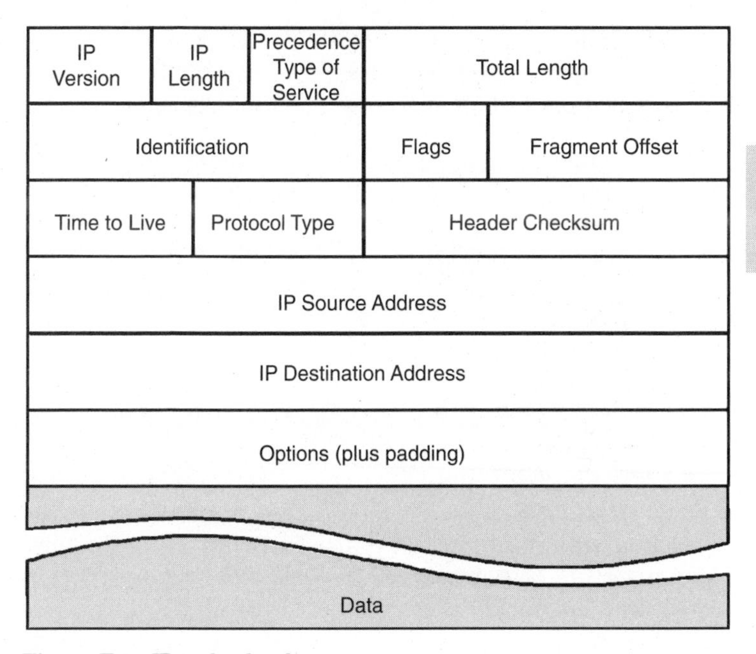

Figure T-9 IP packet header.

- *Precedence and type of service (8 bits)* Precedence indicates the priority of data packet delivery, which may range from 0 (lowest priority) for normal data to 7 (highest priority) for time-critical data (i.e., multimedia applications). Type of service contains quality of service (QoS) information that determines how the packet is handled over the network. Packets can be assigned values that maximize throughput, reliability, or security and minimize monetary cost or delay. This field will play a larger role in the future as the Internet evolves to handle more multimedia applications.[3]

[3]Differentiated Services (Diffserv) supersedes the original IP Precedence/Type of Service specification for defining packet priority. Using this field, Diffserv first prioritizes traffic by class and then differentiates and prioritizes same-class traffic, offering finer priority granularity.

- *Total packet length (16 bits)* Total length of the header plus the total length of the data field of the packet.
- *Identification (16 bits)* A unique ID for a message that is used by the destination host to recognize packet fragments that belong together.
- *Flags (3 bits)* Indicates whether or not the packets can be fragmented for delivery; if a packet cannot be delivered without being fragmented, it will be discarded and an error message will be returned to the sender.
- *Fragmentation offset (13 bits)* If fragmentation is allowed, this field indicates how IP packets are to be fragmented. Each fragment has the same ID. Flags are used to indicate that more fragments are to follow, as well as indicate the last fragment in the series.
- *Time to live (8 bits)* This field indicates how long the packet is allowed to exist on the network in its undelivered state. The hop counter in each host or gateway that receives the packet decrements the value of the time-to-deliver field by one. If a gateway receives a packet with the hop count decremented to zero, it will be discarded. This prevents the network from becoming congested by undeliverable packets.
- *Protocol type (8 bits)* Specifies the appropriate service to which IP delivers the packets, such as TCP or UDP.
- *Header checksum (16 bits)* This field is used to determine whether the received packet has been corrupted in any way during transmission. The checksum is updated as the packet is forwarded because the time-to-live field changes at each router.
- *IP source address (32 bits)* The address of the source host (e.g., 130.132.9.55).
- *IP destination address (32 bits)* The address of the destination host (e.g., 128.34.6.87).
- *Options (up to 40 bytes)* Although seldom used for routine data, this field allows one or more options to be specified.

Option 4, for example, time stamps all stops that the packet made on the way to its destination. This allows measurement of overall network performance in terms of average delay and nodal processing time.

Internet performance depends on the resources available at the various hosts and routers—transmission bandwidth, buffer memory, and processor speed—and how efficiently these resources are used. Although each type of resource is manageable, there are always tradeoffs between cost and performance.

User Datagram Protocol While TCP offers assured delivery, it does so at the price of overhead. UDP, on the other hand, functions with minimum overhead (Figure T-10); it merely passes individual messages to IP for transmission. Since IP is not reliable without TCP, there is no guarantee of delivery.

Nevertheless, UDP is very useful for certain types of communications, such as quick database lookups. For example, the Domain Name System (DNS) consists of a set of distributed databases that provide a service that translates between system names and their IP addresses. For simple messaging between applications and these network resources, UDP does the job. UDP is also well suited to the

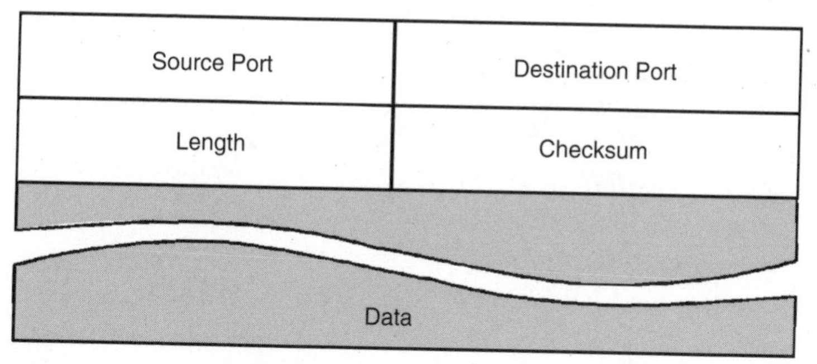

Figure T-10 The UDP header.

brief request/response message exchanges characteristic of SNMP. The UDP header consists of the following fields:

- *Source port (16 bits)* This field identifies the source port number.
- *Destination port (16 bits)* This field identifies the destination port number.
- *Length (16 bits)* Indicates the total length of the UDP header and data portion of the message.
- *Checksum (16 bits)* Validates the contents of a UDP message. Use of this field is optional. If it is not computed for the request, it can still be included in the response.

Applications using UDP communicate through a specified numbered port that can support multiple virtual connections, which are called "sockets." A socket is an IP address and port, and a pair of sockets (source and destination) forms a TCP connection. One socket can be involved in multiple connections (Figure T-11).

Some ports are registered ("well known") and can be found on many TCP/IP implementations. Well-known ports are numbered from 0 to 1023. Telnet, for example, always uses port 23 for communications, while FTP uses port 21. The well-known ports are assigned by the Internet Assigned

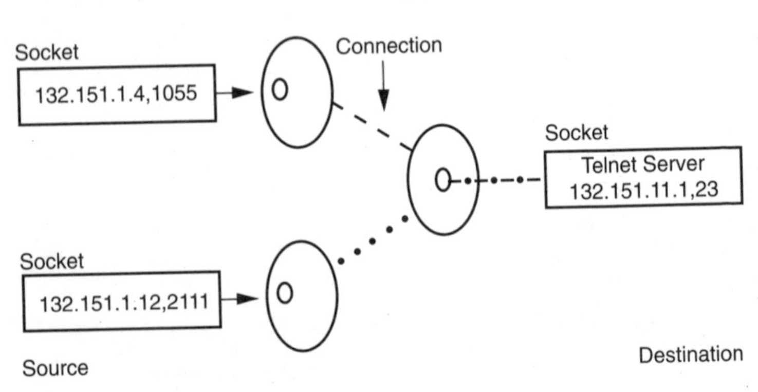

Figure T-11 UDP socket application.

Numbers Authority (IANA) and on most systems can only be used by system (or root) processes or by programs executed by privileged users. Other examples of UDP well-known ports are listed in Table T-1.

In addition to the well-known ports, there are also registered ports numbered from 1024 to 49151 and private ports numbered from 49152 to 65535.

High-Level TCP/IP Services

The TCP/IP model includes three simple types of services for file transfers, electronic mail, and virtual terminal sessions:

- *File Transfer Protocol (FTP)* A protocol used for the bulk transfer of data from one remote device to another. Usually implemented as application-level programs, FTP

TABLE T-1 Examples of UDP Well-Known Ports

Service	Port	Description
Users	11	Shows all users on a remote system.
Quote	17	Returns a "quote of the day."
Mail	25	Used for electronic mail via SMTP.
Domain Name Server	53	Translates system names and their IP addresses.
BOOTpc	68	Client port used to receive configuration information.
TFTP	69	Trivial File Transfer Protocol used for initializing diskless workstations.
World Wide Web	80	Provides access to the Web via the HyperText Transfer Protocol (HTTP).
Snagas	108	Provides access to an SNA Gateway Access Server.
nntp	119	Provides access to a newsgroup via the Network News Transfer Protocol (NNTP).
SNMP	161	Used to receive network management queries via the Simple Network Management Protocol.

uses the telnet and TCP protocols. Most FTP offerings have options to support the unique aspects of each vendor's file structures. Data in the FTP environment consists of a stream of data followed by an end-of-file marker, allowing only entire files to be transferred—not selected records within a file. Sending a file via FTP to a user on another TCP/IP network requires a valid user ID and password for a host on that network.

- *Simple Mail Transfer Protocol (SMTP)* A protocol for exchanging mail messages between systems without regard for the type of user interface or the functionality that is available locally. SMTP sessions consist of a series of commands, starting with both ends exchanging "handshake" messages to identify themselves. This is followed by a series of commands that indicate that a message is to be sent and receipts are needed and by commands that actually transfer the data. Separating the data message from the address field allows a single message to be delivered to multiple users and to verify that there is at least one deliverable addressee before sending the contents. SMTP modifies every message that it receives by adding a time stamp and a reverse-path indicator into each message. This means that a mail message in the SMTP environment usually consists of a fairly long header with information from each node that handled the message. Many user interfaces are able to automatically filter out this kind of information, however.
- *Telnet Virtual Terminal Service* The telnet protocol defines a network-independent virtual terminal through which a user can log in to remote TCP/IP hosts. The user goes through the standard log-in procedure on the remote TCP/IP host and must know the characteristics of the remote operating system to execute host-resident commands. Telnet enables remote terminals to access different hosts by fooling an operating system into thinking that a remote terminal is locally connected. Most telnets operate in the full-duplex mode, meaning that they are

capable of sending and receiving at the same time. There is a half-duplex mode to accommodate IBM hosts, however. In this case, a turnaround signal switches the sending of data to the other side of the connection.

Summary

In its early years of development and implementation, TCP/IP was considered of interest only to research institutions, academia, and defense contractors. Today, corporations have embraced TCP/IP as a platform that can meet their needs for multivendor, multinetwork connectivity. Because it was developed in large part with government funding, TCP/IP code is in the public domain; this availability has encouraged its use by thousands of vendors worldwide, who apply it to support nearly all types of computers and network devices. Because of its flexibility, comprehensiveness, and nonproprietary nature, TCP/IP has captured a considerable and growing share of the commercial internetworking market.

See also

Open Systems Interconnection

Simple Network Management Protocol

U

UNIFIED MESSAGING

While the ways in which we can communicate have diversified, so too has the number of devices people must use to receive all the messages—desk phones, cellular phones, fax machines, alphanumeric pagers, and e-mail systems, to name a few. Unified messaging brings order to this chaos by consolidating the reception, notification, presentation, and management of communications to the local area network (LAN)–connected desktop. The goal of unified messaging is to make individuals, workgroups, and organizations more efficient and responsive.

The unified messaging capability can be provided through message servers connected to a corporate Private Branch Exchange (PBX; Figure U-1) or carrier switch. These servers can have a distributed architecture, allowing unified messaging services to be added incrementally throughout the carrier or corporate network as demand warrants. More recently, unified messaging services have become available over the Internet, allowing users to view communication activity through their Web browser.

A unified messaging service deposits each subscriber's e-mail, fax, and voice messages into a universal messaging inbox so that the subscriber can find all messages in a single

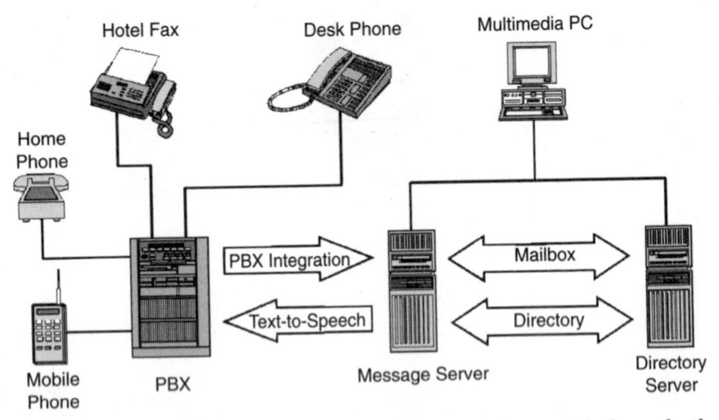

Figure U-1 Unified messaging application implemented through the integration of a corporate LAN and PBX.

place and through a single interface, such as a telephone handset, Web browser, or desktop application like PhoneSoft (Figure U-2).

Notification Options

The unified messaging systems notify subscribers whenever a new message arrives. Depending on the type of system or service that is used, notification methods include an e-mail message, message-waiting indicator light, stutter dial tone, pager, and out-dial.

When a service is used, for example, it is common for voice-messaging subscribers to hear a stutter dial tone on arrival of a new voice message but not when an e-mail or fax message has arrived. With the advent of unified messaging, however, this changes, and the system notifies the subscriber of an incoming message whenever a message arrives, regardless of whether it is a voice, fax, or e-mail message. A stutter dial tone is only one of several potential notification options. The service provider may choose which notification methods it offers to subscribers.

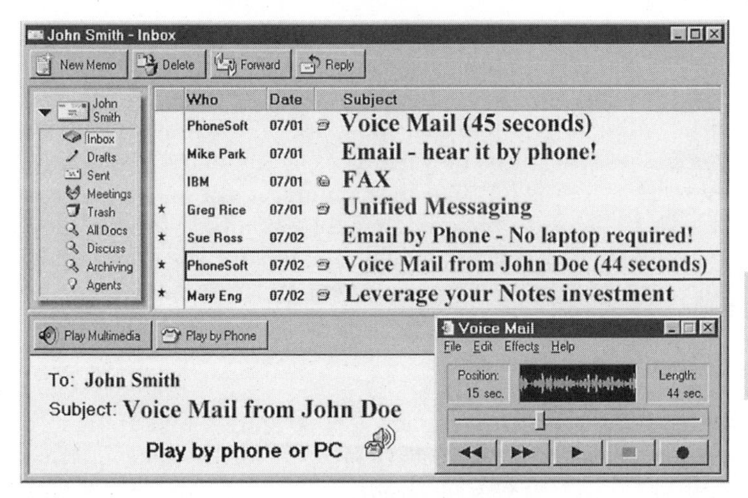

Figure U-2 Active Voice's PhoneSoft Unified Messaging delivers all voice, fax, and e-mail messages to the user's Notes inbox. Voice messages are simply Notes mail messages containing compressed WAV files.

If the specified notification method is e-mail, for example, the unified messaging system sends an e-mail notification message to a subscriber-specified e-mail address. In the body of the notification message, the system embeds a hypertext link that points to the voice or fax message in the system's message store. When the subscriber clicks on the hypertext link, the mail application passes the Uniform Resource Locator (URL) to the subscriber's Web browser, which opens up a window to the subscriber's universal inbox, where the voice or fax message appears.

If a voice message comes in and the subscriber's workstation supports multimedia, the subscriber can listen to the voice message over the workstation's audio system. If the message is a fax, the system presents the fax as a graphic image on the workstation's screen. If the message is itself an e-mail message, the system simply passes it along to the subscriber.

Let's say the subscriber wants to be notified of incoming messages via a telephone's message-waiting indicator. When a new voice, fax, or e-mail message arrives and the subscriber checks for messages, the system indicates how many of each type of message have arrived. The system's synthesized voice might tell the subscriber, "You have two new voice messages, three new fax messages, and five new e-mail messages." Using buttons on the handset, the subscriber can choose to listen to the voice messages, output the fax messages to the nearest fax machine, and save the e-mail messages for viewing at a more convenient time and place.

The Role of Browsers

Because the Web can support text, graphics, and audio efficiently, it provides an effective medium for the consolidation and presentation of e-mail, fax, and voice messages. Some unified messaging systems can take advantage of Web technology to securely handle these as binary objects for large numbers of subscribers.

With a Web browser such as Microsoft Internet Explorer, unified messaging system subscribers can view fax and e-mail messages on their computer screens and listen to voice mail messages over headphones or speakers attached to their workstation. They also can use the message compose and reply features of the browser interface to respond to incoming messages.

For example, a subscriber can send a voice-mail message to another recognized unified messaging subscriber in response to a voice, fax, or e-mail message. By clicking the "Compose Voice" button in the browser window and picking the recipient of the message from a directory window, the subscriber records a response and, when finished, clicks the "Send" button to deliver the voice mail.

A subscriber also can send an e-mail message in response to an incoming voice, fax, or e-mail message. In response to an incoming voice or fax message, the subscriber clicks the "Compose E-mail" button in the browser window. If the recipient is a recognized unified messaging subscriber, his or her address can be selected from the directory. If the recipient is not a recognized subscriber, the e-mail address must be entered manually. After composing the e-mail address, the subscriber clicks the "Send" button to deliver the e-mail message. If the original message was an e-mail message instead of a voice or fax message, the return address is automatically entered into the reply whether or not the addressee is a unified messaging subscriber.

Messages also can be forwarded from the browser interface. A "Forward" button enables subscribers to send an incoming e-mail message to another subscriber. If the subscriber wants to forward a voice or fax message, the subscriber saves the voice message in an audio file (or the fax in a graphic file) format and sends it as an attachment to an e-mail message. Subscribers can use the print capabilities of their Web browsers to output hard copies of received fax and e-mail messages.

There are now options for translating voice messages to text for output to a printer or word processing application. There are also systems that translate text messages for audio output. Avaya, Inc., for example, offers speech recognition for its Intuity Audix Multimedia Messaging System. The system is used primarily for voice messaging but is designed to handle and combine various communication forms. It lets users attach e-mail or faxes to voice messages and listen to speech translations of e-mail messages, and the speech-recognition capability can be used to forward attachments to other users.

While certain levels of access are not yet available—such as a viewing a fax or e-mail message over a standard telephone handset—subscribers can use the handset to manipulate even these types of messages. With a unified messaging system between the PBX and LAN, the system can report who sent the fax or e-mail message and what time it arrived.

Via the handset, subscribers also can redirect fax messages for output on whatever fax machine they designate. The user simply dials the phone number of the desired fax machine and presses the "Enter" key. The fax machine could be in a hotel lobby, at a remote office, at a customer site, or at the subscriber's own home. Incoming e-mail messages can be output in a similar manner, except that an image of the e-mail message arrives at the specified fax machine.

By pressing appropriate buttons in response to system prompts, a subscriber can record and send a voice-mail response to a voice- or e-mail message sent from another subscriber within the unified messaging environment. The unified messaging system addresses the reply automatically, so the subscriber does not have to remember (or even know) the address of the person to whom he or she is sending the reply. To reply to an externally originated message, the subscriber must dial the external number directly and rely on the recording capabilities attached to the phone service at the other end of the line.

Although there is usually no support for sending a voice message from the handset in direct response to an incoming fax message, a subscriber can forward voice and fax messages to another recognized unified messaging subscriber by entering that subscriber's mailbox number at the prompt. Subscribers also can attach voice annotations to messages they forward from the handset.

Subscribers also can forward fax and e-mail messages to systems outside the unified messaging environment by using the print capabilities of the unified messaging system. The subscriber forwards the fax by designating a remote recipient's fax machine or fax mailbox as the target output device. This feature also enables handset users to forward e-mail messages to recipients outside the system. As noted, the unified messaging system actually faxes an image of the e-mail message to a remote recipient's fax machine. Beyond providing a message to be faxed, the e-mail component of the unified messaging system does not play a role in forwarding messages outside the environment from the handset.

The unified messaging system ensures that actions initiated via the browser and handset interfaces are kept closely synchronized. If a subscriber listens to a new voice message through the browser interface, it is flagged as "read" and is not announced as a "new" message when the subscriber later accesses the in-box via the handset interface. And, as noted earlier, if a subscriber deletes a message via one interface, it is deleted from the list of messages accessible via any other interface.

Summary

While unified messaging solutions are available as a service or dedicated system, they are also available over the Internet. Numerous companies offer services that consolidate voice mail, e-mail, and faxes in one mailbox. Ranging in price from $10 to $20 a month, these services provide individuals and companies with a telephone number for incoming voice messages and faxes and an e-mail address. Despite the benefits of unified messaging, a strong market has failed to materialize. Part of the problem may be generational—younger people seem more inclined to accept unified messaging because they have grown up with the technology and seem more adept at using it, whereas older people are less inclined to accept unified messaging because they have grown up using separate communications tools and are less trusting of "all in one" solutions. Further study in this area would be required to draw definitive conclusions.

See also

Electronic mail

W

WIRELESS FIDELITY

Wireless Fidelity (Wi-Fi) refers to a type of Ethernet specified under the IEEE 802.11a and 802.11b Standards for local area networks (LANs) operating in the 5- and 2.4-GHz unlicensed frequency bands, respectively. Wi-Fi is equally suited to residential users and businesses, and equipment is available that allows both bands to be used to support separate networks simultaneously.

The IEEE 802.11 Standard makes the wireless network a straightforward extension of the wired network. This has allowed for a very uncomplicated implementation of wireless communication with obvious benefits—they can be installed using the existing network infrastructure with minimal retraining or system changes. Notebook users can roam throughout their sites while remaining in contact with the network via strategically placed access points that are plugged into the wired network.

Wireless users can run the same network applications they use on an Ethernet LAN. Wireless adapter cards used on laptop and desktop systems support the same protocols as Ethernet adapter cards. For most users, there is no noticeable functional difference between a wired Ethernet desktop computer and a wireless computer equipped with a wireless adapter other than the added benefit of the ability to roam within the wireless cell. Under many circumstances, it may be desirable for mobile network devices to link to a conventional Ethernet LAN in order to use servers, printers, or an Internet connection supplied through the wired LAN. A wireless access point (AP) is a device used to provide this link.

The IEEE 802.11b Standard designates devices that operate in the 2.4-GHz band to provide a data rate of up to 11 Mbps at a range of up to 300 feet (100 meters) using direct-sequence spread-spectrum technology. Some vendors have implemented proprietary extensions to the IEEE 802.11b Standard, allowing applications to burst beyond 11 Mbps to reach as much as 22 Mbps. Users can share files and applications, exchange e-mail, access printers, share access to the Internet, and perform any other task as if they were directly cabled to the network.

The IEEE 802.11a Standard designates devices that operate in the 5-GHz band to provide a data rate of up to 54 Mbps at a range of up to 900 feet (300 meters) using direct sequence spread spectrum technology. Sometimes called "Wi-Fi5," this amount of bandwidth allows users to transfer large files quickly or even watch a movie in MPEG format over the network without noticeable delays. This technology works by transmitting high-speed digital data over a radio wave utilizing Orthogonal Frequency Division Multiplexing (OFDM) technology.

OFDM works by splitting the radio signal into multiple smaller subsignals that are then transmitted simultaneously at different frequencies to the receiver. OFDM reduces the amount of interference in signal transmissions, which results in a high-quality connection. Wi-Fi5 products automatically sense the best possible connection speed to ensure the greatest speed and range possible with the technology. Some vendors have implemented proprietary extensions to the IEEE 802.11a Standard allowing applications to burst beyond 54 Mbps to reach as much as 72 Mbps.

IEEE 802.11 wireless networks can be implemented in infrastructure mode or ad-hoc mode. In infrastructure mode—referred to in the IEEE specification as the "basic service set"—each wireless client computer associates with an AP via a radio link. The AP connects to the 10/100-Mbps Ethernet enterprise network using a standard Ethernet cable and provides the wireless client computer with access to the wired Ethernet network. Ad hoc mode is the peer-to-peer network mode, which is suitable for very small installations. Ad-hoc mode is referred to in the IEEE 802.11b specification as the independent basic service set.

Security for Wi-Fi networks is handled by the IEEE standard called Wired Equivalent Privacy (WEP), which is available in 64- and 128-bit versions. The more bits in the encryption key, the more difficult it is for hackers to decode the data. It was believed originally that 128-bit encryption would be virtually impossible to break due to the large number of possible encryption keys. However, hackers have since developed methods to break 128-bit WEP without having to try each key combination, proving that this system is not totally secure. These methods are based on the ability to gather enough packets off the network using special eavesdropping equipment to then determine the encryption key. Although WEP can be broken, it does take considerable effort and expertise to do so. To help thwart hackers, WEP should be enabled and the keys rotated on a frequent basis.

The wireless LAN (WLAN industry has recognized that WEP is not as secure as once thought and is responding by developing another standard, known as IEEE 802.11i, that will allow WEP to use the Advanced Encryption Algorithm (AES) to make the encryption key even more difficult to determine. AES replaces the older 56-bit Digital Encryption Standard (DES), which had been in use since the 1970s. AES can be implemented in 128-, 192-, and 256-bit versions. Assuming a computer with enough processing power to test 255 keys per second, it would take 149 trillion years to crack AES.

Summary

Wi-Fi is a certification of interoperability for IEEE 802.11b systems, awarded by the Wireless Ethernet Compatibility Alliance (WECA), now known as the Wi-Fi Alliance. The Wi-Fi seal indicates that a device has passed independent tests and will reliably interoperate with all other Wi-Fi certified equipment. Customers benefit from this standard by avoiding becoming locked into one vendor's solution—they can purchase Wi-Fi–certified APs and client devices from different vendors and still expect them to work together.

See also

Access Points
Wireless LANs

WIRELESS LANS

A WLAN is a data communications system implemented as an extension—or as an alternative—to a wired LAN. Using a variety of technologies including narrowband radio, spread spectrum, and infrared, WLANs transmit and receive data through the air, minimizing the need for wired connections.

Applications

WLANs have become popular in a number of vertical markets, including health care, retail, manufacturing, and warehousing. These industries have profited from the productivity gains of using handheld terminals and notebook computers to transmit real-time information to centralized hosts for processing. WLANs allow users to go where wires cannot always go. Specific uses of WLANs include

- Hospital staff members can become more productive when using handheld or notebook computers with a wireless LAN capability to deliver patient information, regardless of their location.

- Consulting or accounting audit teams, small workgroups, or temporary office staff can use WLANs to quickly set up for ad hoc projects and become immediately productive.
- Network managers in dynamic enterprise environments can minimize the overhead cost of moves, adds, and changes with WLANs, since the need to install or extend wiring is eliminated.
- Warehouse workers can use WLANs to exchange information with central databases, thereby increasing productivity.
- Branch office workers can minimize setup requirements by installing preconfigured WLANs.
- WLANs are an alternative to cabling multiple computers in the home.

While the initial investment required for WLAN hardware can be higher than the cost of conventional LAN hardware, overall, installation expenses and life-cycle costs can be significantly lower. Long-term cost savings are greatest in dynamic environments requiring frequent moves, adds, and changes. WLANs can be configured in a variety of topologies to meet the needs of specific applications and installations. They can grow by adding access points and extension points to accommodate virtually any number of users.

Technologies

There are several technologies to choose from when selecting a WLAN solution, each with advantages and limitations. Most WLANs use spread spectrum, a wideband radio frequency technique developed by the military for use in reliable, secure, mission-critical communications systems. To achieve these advantages, the signal is spread out over the available bandwidth and resembles background noise that is virtually immune from interception.

There are two types of spread-spectrum radio-frequency hopping and direct sequence. Frequency-hopping spread spectrum (FHSS) uses a narrowband carrier that changes frequency in a pattern known only to the transmitter and receiver. Properly synchronized, the net effect is to maintain a single logical channel. To an unintended receiver, FHSS appears to be short-duration impulse noise.

Direct-sequence spread spectrum (DSSS) generates a redundant bit pattern for each bit to be transmitted and requires more bandwidth for implementation. This bit pattern, called a "chip" (or "chipping code"), is used by the receiver to recover the original signal. Even if one or more bits in the chip are damaged during transmission, statistical techniques embedded in the radio can recover the original data without the need for retransmission. To an unintended receiver, DSSS appears as low-power wideband noise.

Another technology used for WLANs is infrared (IR), which uses very high frequencies that are just below visible light in the electromagnetic spectrum. Like light, IR cannot penetrate opaque objects—to reach the target system, the waves carrying data are sent in either directed (line-of-sight) or diffuse (reflected) fashion. Inexpensive directed systems provide very limited range of not more than 3 feet and typically are used for personal area networks (PANs) but occasionally are used in specific WLAN applications. High-performance directed IR is impractical for mobile users and is therefore used only to implement fixed subnetworks. Diffuse infrared WLAN systems do not require line-of-sight transmission, but cells are limited to individual rooms. As with spread-spectrum LANs, infrared LANs can be extended by connecting the wireless access points to a conventional wired LAN.

Operation

As noted, wireless LANs use electromagnetic waves (radio or infrared) to communicate information from one point to another without relying on a wired connection. Radio waves are often referred to as "radio carriers" because they simply perform the function of delivering energy to a remote receiver. The data being transmitted are superimposed on the radio carrier so that they can be extracted accurately at the receiving end. This process is generally referred to as "carrier modulation." Once data are modulated onto the radio carrier, the radio signal occupies more than a single frequency, since the frequency or bit rate of the modulating information adds to the carrier.

Multiple radio carriers can exist in the same space at the same time without interfering with each other if the radio waves are transmitted on different frequencies. To extract data, a radio receiver tunes into one radio frequency while rejecting all other frequencies.

In a typical WLAN configuration, a transmitter/receiver (transceiver) device, called an "access point," connects to the wired network from a fixed location using standard cabling. At a minimum, the access point receives, buffers, and transmits data between the WLAN and the wired network infrastructure. A single access point can support a small group of users and can function within a range of less than 100 to several hundred feet. The access point (or the antenna attached to the access point) is usually mounted high but may be mounted essentially anywhere that is practical as long as the desired radio coverage is obtained.

Users access the WLAN through WLAN adapters. These adapters provide an interface between the client network operating system (NOS) and the airwaves via an antenna. The nature of the wireless connection is transparent to the NOS.

Configurations

WLANs can be simple or complex. The simplest configuration consists of two PCs equipped with wireless adapter cards that form a network whenever they are within range of one another (Figure W-1). This peer-to-peer network requires no administration. In this case, each client would

Figure W-1 A wireless peer-to-peer network created between two notebook computers equipped with external wireless adapters.

only have access to the resources of the other client and not to a central server.

Installing an access point can extend the operating range of the wireless network, effectively doubling the range at which the devices can communicate. Since the access point is connected to the wired network, each client would have access to the server's resources as well as to other clients (Figure W-2). Each access point can support many clients—the specific number depends on the nature of the transmissions involved. In some cases, a single access point can support up to 50 clients.

Depending on the manufacturer and the frequency band used in its products, access points have an operating range of about 500 feet indoors and 1000 feet outdoors. In a very large facility such as a warehouse or on a college campus, it probably will be necessary to install more than one access point (Figure W-3). Access point positioning is determined by a site survey. The goal is to blanket the coverage area with overlapping coverage cells so that clients can roam throughout the area without ever losing network contact. Access points hand the client off from one to another in a way that is invisible to the client, ensuring uninterrupted connectivity.

To solve particular problems of topology, the network designer might choose to use extension points (EPs) to augment the network of access points (Figure W-4). These devices look and function like access points (APs), but they

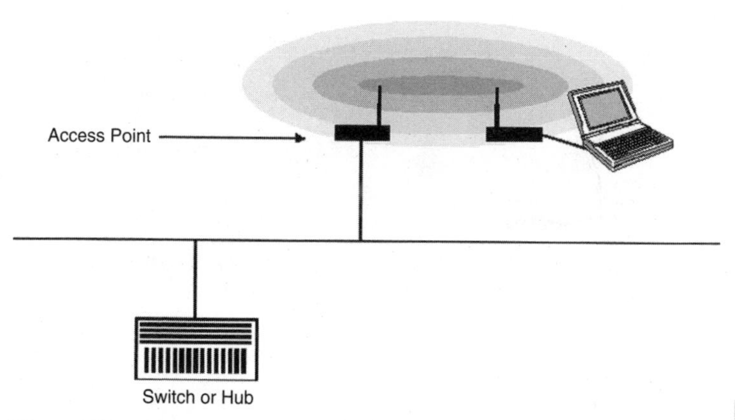

Figure W-2 A wireless client connected to the wired LAN via an access point.

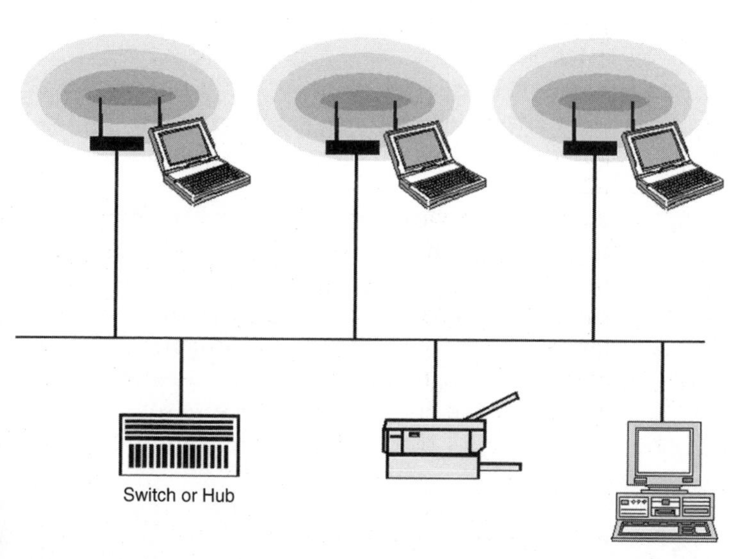

Figure W-3 Multiple access points extend wireless coverage and enable roaming.

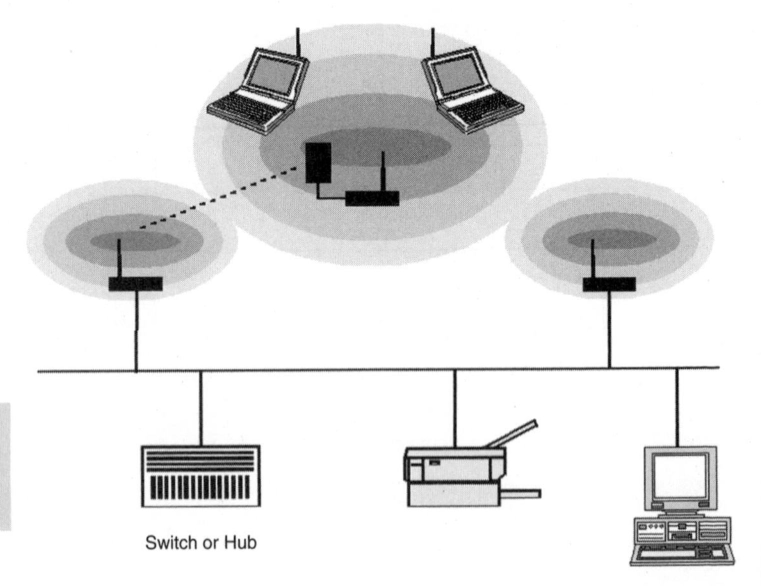

Figure W-4 An extension point can be used to extend the range of a wireless network.

are not tethered to the wired network, as are APs. EPs function as repeaters by boosting signal strength to extend the range of the network by relaying signals from a client to an AP or another EP.

Another component of WLANs is the directional antenna. If a WLAN in one building must be connected to a WLAN in another building a mile away, one solution might be to install a directional antenna on the two buildings—each antenna targeting the other and connected to its own wired network via an access point (Figure W-5).

WLAN Standards

There are several WLAN standards, each suited for a particular environment: IEEE 802.11a and 802.11b, HomeRF, and Bluetooth.

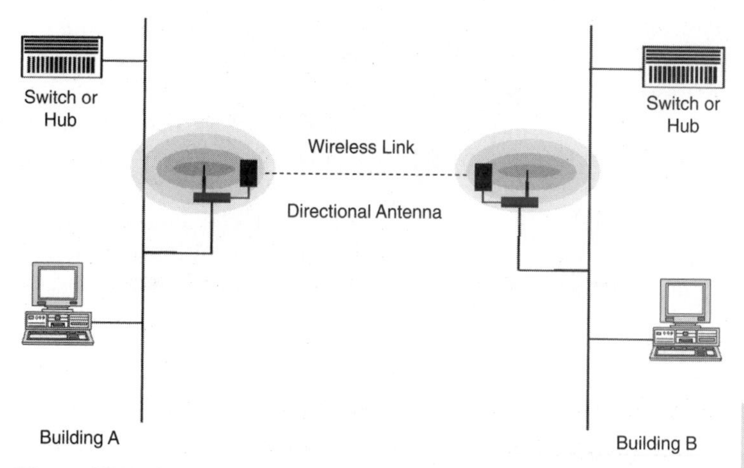

Figure W-5 A directional antenna can be used to interconnect WLANs in different buildings.

For the residential and office environments, the IEEE 802.11b offers a data transfer rate of up to 11 Mbps at a range of up to 300 feet from the base station. It operates in the 2.4-GHz band and transmits via the DSSS method. Multiple base stations can be linked to increase that distance as needed, with support for multiple clients per access point.

The HomeRF 2.0 Standard draws from IEEE 802.11b and Digital Enhanced Cordless Telecommunication (DECT), a popular standard for portable phones worldwide. Operating in the 2.4-GHz band, HomeRF was designed from the ground up for the home market for both voice and data. It offers throughput rates comparable to IEEE 802.11b and supports the same kinds of terminal devices in both point-to-point and multipoint configurations. HomeRF transmits at up to 10 Mbps over a range of about 150 feet from the base station, which makes it suitable for the average home. HomeRF transmits using spread-spectrum frequency hopping; i.e., it hops around constantly within its prescribed bandwidth. When it encounters interference, like a microwave oven or an adjacent WLAN, it adapts by moving to another frequency.

The key advantage that HomeRF has over IEEE 802.11b in the home environment is its superior ability to adapt to interference from devices such as portable phones and microwaves. As a frequency hopper, it coexists well with other frequency-hopping devices that proliferate in the home. Another advantage of HomeRF is that it continuously reserves a chunk of bandwidth via “isochronous channels” for voice services. Speech quality is high; there is no clipping while the protocol deals with interference.

The IEEE 802.11b Standard does not include frequency hopping. In response to interference, IEEE 802.11b simply retransmits or waits for the higher-level TCP/IP protocol to sort out signal from noise. This works well for data but can result in voice transmissions sounding choppy. Voice and data are treated the same way, converting voice into data packets but offering no priority to voice. This results in unacceptable voice quality. Another problem with IEEE 802.11b is that its Wired Equivalent Protocol (WEP) encryption, designed to safeguard privacy, has had problems living up to its claim.

Bluetooth also operates in the 2.4-GHz band but was not created originally to support WLANs; it was intended as a replacement for cable between desktop computers, peripherals, and handheld devices. Operating at the comparatively slow rate of 30 to 400 kbps across a range of only 30 feet, Bluetooth supports “piconets” that link laptops, personal digital assistants (PDAs), mobile phones, and other portable devices on an as-needed basis. It improves on infrared in that it does not require a line of sight between the devices and has greater range than infrared’s 3 to 10 feet. Bluetooth also supports voice channels.

While Bluetooth does not have the power and range of a full-fledged LAN, its master/slave architecture does permit the devices to face different piconets, in effect extending the range of the signals beyond 30 feet. Like HomeRF, Bluetooth is a frequency hopper, so devices that use these two standards can coexist by hopping out of each other's way. Bluetooth has the faster hop rate, so it will be the first to sense problems and act to steer clear of interference from HomeRF devices.

The three standards each have particular strengths that make them ideal for certain situations, as well as specific shortcomings that render them inadequate for use beyond their intended purpose:

- While suited for the office environment, IEEE 802.11b is not designed to provide adequate interference adaptation and voice quality for the home. Data collisions force packet retransmissions, which is fine for file transfers and print jobs but not for voice or multimedia that cannot tolerate the resulting delay.
- HomeRF delivers an adequate range for the home market but not for many small businesses. It is better suited than IEEE 802.11b for streaming multimedia and telephony, applications that may become more important for home users as convergence devices become popular.
- Bluetooth does not provide the bandwidth and range required for WLAN applications but instead is suited for desktop cable replacement and ad-hoc networking for both voice and data within the narrow 30-foot range of a piconet.

WLAN technology is continually improving. The IEEE 802.11b Standard developers seek to improve encryption (IEEE 802.11i) and to make the standard more multimedia-friendly (IEEE 802.11e). Dozens of vendors are shipping IEEE 802.11b products, and the standard's proliferation in corporate and public environments is a distinct advantage. An office worker who already has an IEEE 802.11b–equipped notebook will not likely want to invest in a different network for the home.

Furthermore, the multimedia and telephony applications HomeRF advocates tout have not yet arrived to make the technology a compelling choice. Although HomeRF currently beats IEEE 802.11b in terms of security, this is not a big issue in the home. For these and other reasons, industry analysts predict that IEEE 802.11b will soon overtake HomeRF in the consumer marketplace, especially since the price difference between the two has just about reached parity.

Summary

Once expensive, slow, and proprietary, WLAN products are now reasonably fast, standardized, and priced for mainstream business and consumer use. WLAN configurations range from simple peer-to-peer topologies to complex networks offering distributed data connectivity and roaming. To solve problems of vendor interoperability, the Wi-Fi Alliance offers a certification program that tests vendor-submitted products. Those that pass a battery of tests receive the right to bear the Wireless Fidelity (Wi-Fi) logo of interoperability.

See also

Bluetooth

Home RF

Infrared Networking

Wireless Fidelity

ACRONYMS

A

AAL	ATM Adaptation Layer
ABR	Available Bit Rate
AC	Access Control
AC	Address Copied
ac	Alternating Current
AC	Authentication Center
ACD	Automatic Call Distributor
ACELP	Algebraic Code Excited Linear Predictive
ACP	Access Control Point
ADM	Add-Drop Multiplexer
ADPCM	Adaptive Differential Pulse Code Modulation
ADSL	Asymmetric Digital Subscriber Line
AIN	Advanced Intelligent Network
ANR	Automatic Network Routing (IBM Corp.)

ANSI	American National Standards Institute
ANT	ADSL Network Terminator
AOL	America Online
APC	Access Protection Capability (AT&T)
API	Application Programming Interface
APPC	Advanced Program-to-Program Communications (IBM Corp.)
APPN	Advanced Peer-to-Peer Network (IBM Corp.)
APC	Automatic Protection Switching
ARCnet	Attached Resource Computer Network (Datapoint Corp.)
ARB	Adaptive Rate Based (IBM Corp.)
ARPA	Advanced Research Projects Agency
ARQ	Automatic Repeat Request
ARS	Action Request System (Remedy Systems, Inc.)
AS	Autonomous System
ASCII	American Standard Code for Information Interchange
ASIC	Application-Specific Integrated Circuit
ASN.1	Abstract Syntax Notation 1
ATM	Asynchronous Transfer Mode
AUI	Attachment Unit Interface
AWG	American Wire Gauge

B

B8ZS	Binary Eight Zero Substitution
BBS	Bulletin Board System
BECN	Backward Explicit Congestion Notification
Bellcore	Bell Communications Research, Inc.
BER	Bit Error Rate

BERT	Bit Error Rate Tester
BIB	Backward Indicator Bit
BIOS	Basic Input-Output System
BLEC	Building Local Exchange Carrier
BOC	Bell Operating Company
BONDING	Bandwidth on Demand Interoperability Group
BootP	Boot Protocol
BPDU	Bridge Protocol Data Unit
bps	Bits per Second
BPV	Biploar Violation
BRI	Basic Rate Interface (ISDN)
BSC	Binary Synchronous Communication
BSE	Basic Service Element
BSN	Backward Sequence Number

C

CAD	Computer-Aided Design
CAM	Computer-Aided Manufacturing
CAN	Campus Area Network
CAP	Carrierless Amplitude/Phase (Modulation)
CBR	Constant Bit Rate
CCC	Clear Channel Capability
CCS	Common Channel Signaling
CD	Compact Disc
CDCS	Continuous Dynamic Channel Selection
CD-R	Compact Disc–Recordable
CD-ROM	Compact Disc–Read-Only Memory
CDMA	Code Division Multiple Access

CDPD	Cellular Digital Packet Data
CHAP	Challenge Handshake Authentication Protocol
CIF	Common Intermediate Format
CIR	Committed Information Rate
CLEC	Competitive Local Exchange Carrier
CLP	Cell Loss Priority
CMI	Cable Microcell Integrator
CMIS	Common Management Information Services
CO	Central Office
CON	Concentrator
COS	Class of Service
CP	Coordination Processor
CPE	Customer Premises Equipment
cps	Cycles per Second (Hertz)
CPU	Central Processing Unit
CRC	Cyclic Redundancy Check
CRM	Customer Relationship Management
CSA	Carrier Serving Area
CSM	Communications Services Management
CSMA/CD	Carrier Sense Multiple Access with Collision Detection
CSU	Channel Service Unit
CTI	Computer-Telephony Integration
CVSD	Continuously Variable Slope Delta (modulation)

D

DA	Destination Address
DACS	Digital Access and Cross-Connect System (AT&T)

DAP	Demand Access Protocol
DAS	Dual Attached Station
DASD	Direct Access Storage Device (IBM Corp.)
DAT	Digital Audio Tape
dB	Decibel
DBMS	Database Management System
DCE	Data Communications Equipment
DCE	Distributed Computing Environment
DCF	Data Communication Function
DCS	Digital Cross-Connect System
DES	Data Encryption Standard
DFSMS	Data Facility Storage Management Subsystem (IBM Corp.)
DHCP	Dynamic Host Control Protocol
DIF	Digital Interface Frame
DLCS	Digital Loop Carrier System
DLEC	Data Local Exchange Carrier
DLL	Data Link Layer
DLL	Dynamic Link Library
DLSw	Data Link Switching (IBM Corp.)
DLU	Digital Line Unit
DM	Distributed Management
DME	Distributed Management Environment
DMI	Desktop Management Interface
DMT	Discrete Multitone
DMTF	Desktop Management Task Force
DNS	Domain Name Service
DOCSIS	Data Over Cable Service Interface Specification
DoD	Department of Defense
DOS	Disk Operating System

DOV	Data Over Voice
DQPSK	Differential Quadrature Phase-Shift Keying
DS0	Digital Signal Level 0 (64 kbps)
DS1	Digital Signal Level 1 (1.544 Mbps)
DS1C	Digital Signal Level 1C (3.152 Mbps)
DS2	Digital Signal Level 2 (6.312 Mbps)
DS3	Digital Signal Level 3 (44.736 Mbps)
DS4	Digital Signal Level 4 (274.176 Mbps)
DSI	Digital Speech Interpolation
DSL	Digital Subscriber Line
DSLAM	DSL Access Multiplexer
DSML	Directory Services Markup Language
DSN	Defense Switched Network
DSP	Digital Signal Processor
DSS	Decision Support System
DSU	Data Service Unit
DSX1	Digital Systems Cross-Connect 1
DTE	Data Terminal Equipment
DTR	Dedicated Token Ring
DTU	Data Transfer Unit
DWDM	Dense Wavelength Division Multiplexing
DWMT	Discrete Wavelet Multi Tone
DXI	Data Exchange Interface

E

E-mail	Electronic Mail
EBCDIC	Extended Binary Coded Decimal Interexchange Code (IBM Corp.)
ED	Ending Delimiter
EDI	Electronic Data Interchange

EEROM	Electronically Erasable Read-Only Memory
EFRC	Enhanced Full-Rate Codec
EFT	Electronic Funds Transfer
EIA	Electronic Industries Alliance
EISA	Extended Industry Standard Architecture
EMI	Electromechanical Inteference
EMS	Element Management System
EOC	Embedded Overhead Channel
EOT	End of Transmission
ESCON	Enterprise System Connection (IBM Corp.)
ESD	Electronic Software Distribution
ESF	Extended Superframe Format

F

FASTAR	Fast Automatic Restoral (AT&T)
FAT	File Allocation Table
FC	Frame Control
FC	Fibre Channel
FC-0	Fibre Channel Layer 0
FC-1	Fibre Channel Layer 1
FC-2	Fibre Channel Layer 2
FC-3	Fibre Channel Layer 3
FC-4	Fibre Channel Layer 4
FCC	Federal Communications Commission
FCIA	Fibre Channel Industry Association
FCS	Frame Check Sequence
FDDI	Fiber Distributed Data Interface
FDL	Facilities Data Link
FECN	Forward Explicit Congestion Notification
FEP	Front-End Processor

FFDT	FDDI Full-Duplex Technology
FIB	Forward Indicator Bit
FIB	Forwarding Information Base
FIFO	First In, First Out
FITL	Fiber in the Loop
FRAD	Frame Relay Access Device
FS	Frame Status
FSN	Forward Sequence Number
FTAM	File Transfer, Access, and Management
FT1	Fractional T1
FTP	File Transfer Protocol
FTS	Federal Telecommunications System

G

GBIC	Gigabit Interface Converter
GFC	Generic Flow Control
GHz	Gigahertz (Billions of Cycles per Second)
GIS	Geographical Information System
GloBanD	Global Bandwidth on Demand
GSM	Global System for Mobile (GSM) Telecommunications (Formerly Groupe Spéciale Mobile)
GUI	Graphical User Interface

H

H0	High-Capacity ISDN Channel Operating at 384 kbps
H11	High-Capacity ISDN Channel Operating at 1.536 Mbps
HDSL	High-Bit-Rate Digital Subscriber Line

HEC	Header Error Check
HFC	Hybrid Fiber/Coax
HIC	Head-End Interface Converter
HPR	High-Performance Routing (IBM Corp.)
HSM	Hierarchical Storage Management
HST	Helical Scan Tape
HTML	HyperText Markup Language
HTTP	HyperText Transfer Protocol
HVAC	Heating, Ventilation, and Air Conditioning
Hz	Hertz (Cycles per Second)

I

I/O	Input-Output
ID	Identification
IDLC	Integrated Digital Loop Carrier
IDPR	Interdomain Policy Routing
IDSL	ISDN Digital Subscriber Line
IEEE	Institute of Electrical and Electronic Engineers
ILEC	Incumbent Local Exchange Carrier
IMA	Inverse Multiplexing over ATM
IMAP	Internet Mail Access Protocol
IMS/VS	Information Management System/Virtual Storage (IBM Corp.)
INMS	Integrated Network Management System
IOC	Interoffice Channel
IP	Internet Protocol
IPH	Integrated Packet Handler
IPI	Intelligent Peripheral Interface
IPN	Intelligent Peripheral Node

IPX	Internet Packet Exchange
IrDA	Infrared Data Association
IrLAN	Infrared LAN
IrLAP	Infrared Link Access Protocol
IrLMP	Infrared Link Management Protocol
IrPL	Infrared Physical Layer
IRQ	Interrupt Request
IrTTP	Infrared Transport Protocol
IS	Information System
IS	Industry Standard
ISA	Industry Standard Architecture
ISDN	Integrated Services Digital Network
ISM	Industrial, Scientific, and Medical (Frequency Bands)
ISO	International Organization for Standardization
ISP	Internet Service Provider
IT	Information Technology
IXC	Interexchange Carrier

J

JPEG	Joint Photographic Experts Group
JTAPI	Java Telephony Application Programming Interface

K

k (kilo)	One Thousand (e.g., kbps)
kB	Kilobyte
kHz	Kilohertz (Thousands of Cycles per Second)

L

L2F	Layer 2 Forwarding
L2TP	Layer 2 Tunneling Protocol
LAN	Local Area Network
LANCES	LAN Resource Extension and Services (IBM Corp.)
LAPB	Link Access Procedure–Balanced
LAT	Local Area Transport (Digital Equipment Corp.)
LATA	Local Access and Transport Area
LCD	Liquid-Crystal Display
LCN	Local Channel Number
LCP	Link Control Protocol
LD	Laser Diode
LDAP	Lightweight Directory Access Protocol
LEC	Local Exchange Carrier
LED	Light-Emitting Diode
LI	Length Indicator
LIB	Label Information Base
LIFO	Last In, First Out
LIPS	Lightweight Internet Person Schema
LLC	Logical Link Control
LMDS	Local Multipoint Distribution Service
LSI	Large-Scale Integration
LSMS	Local Service Management System
LSO	Local Serving Office
LSP	Label-Switched Path
LSR	Label-Switched Router
LU	Logical Unit (IBM Corp.)

M

M (Mega)	One Million (e.g., Mbps)
MAC	Media Access Control
MAC	Moves, Adds, Changes
MAN	Metropolitan Area Network
MAPI	Messaging Applications Programming Interface (Microsoft Corp.)
MAU	Multistation Access Unit
MB	Megabyte
MCA	Microchannel Architecture (IBM Corp.)
MCU	Multipoint Control Unit
MD	Mediation Device
MDF	Main Distribution Frame
MDS	Multipoint Distribution Service
MF	Mediation Function
MFJ	Modified Final Judgement
MHz	Megahertz (Millions of Cycles per Second)
MIB	Management Information Base
MIF	Management Information Format
MIME	Multipurpose Internet Mail Extensions
MIPS	Millions of Instructions per Second
MIS	Management Information Services
MISR	Multiprotocol Integrated Switch-Routing
MJU	Multipoint Junction Unit
MMDS	Multichannel Multipoint Distribution Service
Modem	Modulation/Demodulation
MPEG	Moving Pictures Experts Group
MPLS	Multiprotocol Label Switching
ms	Millisecond (Thousandths of a Second)

MSN	Microsoft Network
MTBF	Mean Time Between Failure

N

NAP	Network Access Point
NAT	Network Address Translation
NAU	Network Addressable Unit (IBM Corp.)
NAUN	Nearest Active Upstream Neighbor
NC	Network Computer
NCP	Network Control Program (IBM Corp.)
NCP	Network Control Point
NE	Network Element
NEBS	New Equipment Building Specifications
NECA	National Exchange Carrier Association
NetBIOS	Network Basic Input-Output System
NEF	Network Element Function
NFS	Network File System (or Server)
NIC	Network Interface Card
NID	Network Interface Device
NIST	National Institute of Standards and Technology
NLM	NetWare Loadable Module (Novell, Inc.)
nm	Nanometer
NM	Network Manager
NMS	NetWare Management System (Novell, Inc.)
NMS	Network Management System
NNM	Network Node Manager (Hewlett-Packard Co.)
NOC	Network Operations Center
NOS	Network Operating System

NPC	Network Protection Capability (AT&T)
NSA	National Security Agency
NSF	National Science Foundation

O

OAM	Operations, Administration, Management
OAM&P	Operations, Administration, Maintenance, and Provisioning
OC	Optical Carrier
OC-1	Optical Carrier Signal Level 1 (51.84 Mbps)
OC-3	Optical Carrier Signal Level 3 (155.52 Mbps)
OC-9	Optical Carrier Signal Level 9 (466.56 Mbps)
OC-12	Optical Carrier Signal Level 12 (622.08 Mbps)
OC-18	Optical Carrier Signal Level 18 (933.12 Mbps)
OC-24	Optical Carrier Signal Level 24 (1.244 Gbps)
OC-36	Optical Carrier Signal Level 36 (1.866 Gbps)
OC-48	Optical Carrier Signal Level 48 (2.488 Gbps)
OC-96	Optical Carrier Signal Level 96 (4.976 Gbps)
OC-192	Optical Carrier Signal Level 192 (9.952 Gbps)
OC-256	Optical Carrier Signal Level 256 (13.271 Gbps)
OC-768	Optical Carrier Signal Level 768 (40 Gbps)
OC-1536	Optical Carrier Signal Level 1536 (80 Gbps)
ODBC Corp.)	Open Database Connectivity (Microsoft
OEM	Original Equipment Manufacturer
OLAP	Online Analytical Processing
OLE	Object Linking and Embedding

OMAP	Operations, Maintenance, Administration, and Provisioning
ORB	Object Request Broker
OS	Operating System
OS/2	Operating System/2 (IBM Corp.)
OSI	Open Systems Interconnection
OSS	Operations Support Systems
OTDR	Optical Time Domain Reflectometry
OTN	Optical Transport Network

P

PA	Preamble
PAP	Password Authentication Protocol
PBX	Private Branch Exchange
PC	Personal Computer
PCB	Printed Circuit Board
PCM	Pulse Code Modulation
PDA	Personal Digital Assistant
PDN	Packet Data Network
PDU	Payload Data Unit
PEM	Privacy Enhanced Mail
PGP	Pretty Good Privacy
PHY	Physical Layer
PIM	Personal Information Manager
PIM	Protocol-Independent Multicast
PIN	Personal Identification Number
PIN	Positive-Intrinsic-Negative
PMD	Physical Media Dependent
PnP	Plug and Play
PON	Passive Optical Network

POP	Point of Presence
POP	Post Office Protocol
POS	Point of Sale
POTS	Plain Old Telephone Service
PPP	Point-to-Point Protocol
PPTP	Point-to-Point Tunneling Protocol
PRI	Primary Rate Interface (ISDN)
PSN	Packet-Switched Network
PSTN	Public Switched Telephone Network
PT	Payload Type
PU	Physical Unit (IBM Corp.)
PVC	Permanent Virtual Circuit

Q

QAM	Quadrature Amplitude Modulation
QoS	Quality of Service
QPSK	Quadrature Phase Shift Keying

R

RADSL	Rate-Adaptive Digital Subscriber Line
RAID	Redundant Array of Inexpensive (or Redundant) Disks
RAM	Random Access Memory
RAS	Remote Access Server
RBES	Rule-Based Expert Systems
RCU	Remote Control Unit
RDBMS	Relational Database Management System
RF	Radio frequency
RF	Routing Field

RFI	Radiofrequency Interference
RIP	Routing Information Protocol
RISC	Reduced-Instruction-Set Computing
RMON	Remote Monitoring
ROI	Return on Investment
ROM	Read-Only Memory
RPC	Remote Procedure Call
RSVP	Resource ReSerVation Protocol
RT	Remote Terminal
RTNR	Real-Time Network Routing (AT&T)
RTP	Rapid Transfer Protocol (IBM Corp.)
RX	Receive

S

SA	Source Address
SAFER	Split Access Flexible Egress Routing (AT&T)
SAR	Segmentation and Reassembly
SAS	Single Attached Station
SBCCS	Single-Byte Command Code Set (IBM Corp.)
SCC	Standards Coordinating Committees (IEEE)
SCSI	Small Computer Systems Interface
SD	Starting Delimiter
SDCCH	Standalone Dedicated Control Channel
SDH	Synchronous Digital Hierarchy
SDLC	Synchronous Data Link Control (IBM Corp.)
SDM	Subrate Data Multiplexing
SDP	Service Delivery Point
SDSL	Symmetric Digital Subscriber Line
SFD	Start Frame Delimiter

SHTTP	Secure HyperText Transfer Protocol
SIF	Signaling Information Field
SIG	Special Interest Group
SLIC	Serial Line Interface Coupler (IBM Corp.)
SLIP	Serial Line Internet Protocol
SMT	Station Management
SMTP	Simple Mail Transfer Protocol
SN	Switching Network
SNA	Systems Network Architecture (IBM Corp.)
SNAGAS	SNA Gateway Access Server
SNMP	Simple Network Management Protocol
SONET	Synchronous Optical Network
SPI	Service Provider Interface
SPX	Synchronous Packet Exchange (Novell, Inc.)
SQL	Structured Query Language
SS	Switching System
SS7	Signaling System No. 7
SSCP	System Services Control Point (IBM Corp.)
SSCP/PU	System Services Control Point/Physical Unit (IBM Corp.)
SSL	Secure Sockets Layer
SSP	Service Switching Point
STDM	Statistical Time Division Multiplexing
STP	Shielded Twisted Pair
STP	Spanning Tree Protocol
STS	Synchronous Transport Signal
STX	Start of Transmission
SVC	Switched Virtual Circuit
SWC	Serving Wire Center

T

T1	T-Carrier Service at the DS1 Rate of 1.544 Mbps
T3	T-Carrier Service at the DS3 Rate of 44.736 Mbps
TAPI	Telephony Application Programming Interface (Microsoft Corp.)
TASI	Time-Assigned Speech Interpolation
TB	Terabyte (Trillion Bytes)
TBOS	Telemetry Byte-Oriented Serial
Tbps	Terabit per Second
TCP	Transmission Control Protocol
TDD	Time Division Duplexing
TDM	Time Division Multiplexer
TDMA	Time Division Multiple Access
TDR	Time Domain Reflectometry
TFTP	Trivial File Transfer Protocol
TIA	Telecommunications Industry Association
TIB	Tag Information Base
TIMS	Transmission Impairment Measurement Set
TSAPI	Telephony Services Application Programming Interface (Novell, Inc.)
TSI	Time Slot Interchange
TSR	Terminal Stay Resident
TTRT	Target Token Rotation Time
TV	Television
TX	Transmit

U

UBR	Unspecified Bit Rate

UDP	User Datagram Protocol
UDWDM	Ultra Dense Wavelength Division Multiplexing
UMS	Universal Messaging System
UPS	Uninterruptible Power Supply
UTP	Unshielded Twisted Pair
UDWDM	Ultra Dense Wavelength Division Multiplexing

V

VAR	Value-Added Reseller
VBR	Variable Bit Rate
VC	Virtual Circuit
VCI	Virtual Channel Identifier
VDSL	Very High-Speed Digital Subscriber Line
VF	Voice Frequency
VFIR	Very Fast Infrared (16 Mbps)
VLAN	Virtual Local Area Network
VLSI	Very Large-Scale Integration
VM	Virtual Machine
VMS	Virtual Machine System (Digital Equipment Corp.)
VoFR	Voice over Frame Relay
VPI	Virtual Path Identifier
VP	Virtual Path
VPN	Virtual Private Network
VT	Virtual Terminal
VT	Virtual Tributary
VTAM	Virtual Telecommunications Access Method (IBM Corp.)

W

W3C	World Wide Web Consortium
WAN	Wide Area Network
WDCS	Wideband Digital Cross-Connect System
WDM	Wave Division Multiplexing
WECA	Wireless Ethernet Compatibility Alliance
Wi-Fi	Wireless Fidelity
WLAN	Wireless Local Area Network
WWW	World Wide Web

X

XML	eXtensible Markup Language

INDEX

Boldface page range indicates a main entry.

D

E

F

O

P

About the Author

Nathan J. Muller is an independent consultant specializing in telecommunications technology marketing, research, and education. A resident of Sterling, VA, he serves on the Editorial Board of the *International Journal of Network Management* and the Advisory Panel of Faulkner Information Services. Among the 24 books he has authored are *The Desktop Encyclopedia of Telecommunications* and *Network Manager's Handbook*.